PLINY THE ELDER

Natural History

A SELECTION

Translated with an Introduction and notes by
JOHN F. HEALY

PENGUIN BOOKS

PENGUIN BOOKS

Published by the Penguin Group
Penguin Books Ltd, 80 Strand, London WC2R ORL, England
Penguin Putnam Inc., 375 Hudson Street, New York, New York 10014, USA
Penguin Books Australia Ltd, 250 Camberwell Road, Camberwell, Victoria 3124, Australia
Penguin Books Canada Ltd, 10 Alcorn Avenue, Toronto, Ontario, Canada M4V 3B2
Penguin Books India (P) Ltd, 11 Community Centre, Panchsheel Park, New Delhi – 110 017, India
Penguin Books (NZ) Ltd, Cnr Rosedale and Airborne Roads, Albany, Auckland, New Zealand
Penguin Books (South Africa) (Pty) Ltd, 24 Sturdee Avenue, Rosebank 2196, South Africa

Penguin Books Ltd, Registered Offices: 80 Strand, London WC2R ORL, England

www.penguin.com

First published 1991
Reprinted with revisions in Penguin Classics 2004

032

Copyright © John F. Healy, 1991, 2004
All rights reserved

The moral right of the translator has been asserted

Printed and bound in Great Britain by Clays Ltd, Elcograf S.p.A.
Filmset in 10/11pt Monophoto Bembo

ISBN-13: 978-0-14-044413-1

www.greenpenguin.co.uk

Penguin Books is committed to a sustainable
future for our business, our readers and our planet.
This book is made from Forest Stewardship
Council™ certified paper.

PENGUIN  CLASSICS

NATURAL HISTORY: A SELECTION

PLINY THE ELDER, Gaius Plinius Secundus (AD 23–79), holds a place of exceptional importance in the tradition and diffusion of Western culture. Born at Novum Comum, in northern Italy, he was an outstanding example of a successful member of the Equestrian Order. Pliny began his career in the army in Germany and later held a number of procuratorships in Gaul, Africa and Spain (70–75). On his return to Rome he devoted his talents to writing. Seven works are known but only one, the *Natural History*, survives. This encyclopedic account of the state of Roman scientific knowledge in the first century AD contains material from works no longer extant and is of unique value for our assessment of early imperial science and technology. The *Natural History* also reveals its author's personality and character, thus complementing what we learn from the *Letters* of his nephew, Pliny the Younger. Pliny describes his theme as 'unpromising' and clearly underestimates his own achievement. While he was in charge of a Roman fleet at Misenum his scientific curiosity, ironically, led to his death. He lingered too long observing the eruption of Vesuvius and died on 24 August.

JOHN HEALY was an Exhibitioner and Senior Scholar of Trinity College, Cambridge. During war service, in the Intelligence Corps in the Far East, he visited Mohenjodaro and Harappa, in the Indus valley, (Pakistan), sites which stimulated his early interest in archaeology. He subsequently became Lecturer in Classics and Classical Archaeology at Manchester University (1953–61), Reader in Greek at Bedford College, London (1961–6), Professor of Classics at Royal Holloway College, London (1963–85), and, after the merger of the two London colleges in 1985, the first Chairman of the combined Departments of Classics. His main interests are in Greek numismatics and the history of science and technology – especially economic geology, mining and metallurgy. In 1983 Professor Healy was joint organizer of a colloquium on Pliny the Elder and Roman science, held at the Royal Institution. This encouraged a renewed interest in the *Natural History* and an ongoing reassessment of Pliny's achievements. Much of his 'science' has been validated as a result of modern laboratory experiments. His most recent books are *Pliny the Elder on Science and Technology* (Oxford, 1999) and *Miniere e Metallurgia nel Mondo Greco e Romano*, a revised and enlarged edition of his earlier work, *Mining and Metallurgy*. He has contributed numerous articles to classical, art and scientific journals.

Studiorum otiosis

CONTENTS

CONTENTS

INTRODUCTION

The Life and Character
of Pliny the Elder

Pliny the Elder, Gaius Plinius Secundus, was born at Novum Comum, a town of mixed population in Transpadane Gaul (northern Italy), in AD 23 or 24 during the principate of Tiberius. This was a period of great political unrest, mutiny within the legions and rivalries in the struggle for imperial power. Pliny came from a wealthy family which belonged to the municipal governing class, and was of equestrian rank. This order was open to all Roman citizens who were at least eighteen years of age, of free birth, good character and with a property rating of 400,000 sesterces. Pliny's preoccupation with his status and public image is very evident from his digression in the *Natural History* on the history of the order (XXXIII, 32 ff.) Tacitus, writing a generation later than Pliny, observes that men of municipal origin tended to favour stricter codes of behaviour than those characteristic of members of the imperial court and the Roman aristocracy of the middle of the first century AD. Pliny's criticism, therefore, of extravagant life styles, luxury, avarice and greed is not surprising.

Pliny was introduced to Rome at an early age and writes as an eyewitness of events in the 30s AD. At Rome he found much to interest him in literary movements, oratory, philosophy and science. His formal education began there under the well-known soldier and tragic poet Publius Pomponius Secundus, whose biography he subsequently wrote – as a debt to friend and, possibly, to ensure his patronage. Following the fashion of his time, an important part of Pliny's training was in rhetoric, which greatly influenced his literary style.

The Emperor Augustus had encouraged the idea that equestrian status carried a certain obligation to seek at least one army posting. Not everyone, however, took this 'duty' seriously and some found safe niches in administration, or in ceremonial and non-combatant roles. Some continued to live extravagantly, as an officer of Pliny's acquaintance who carried an expensive dinner service with him on active service (XXXIII, 143). Pliny, by contrast, took his obligations seriously. When twenty-three or twenty-four years of age, he began his career in the army in the province of Germany – a potentially dangerous posting – taking part in a campaign against the Chauci, under Domitius Corbulo, governor of Upper Germany (47–59). Later he served in Lower Germany, probably during the governor-ship of Pomponius Secundus (50/51). On yet a further tour of duty he was a fellow officer of Titus, son of the future Emperor Vespasian, to whom he dedicated his *Natural History*. In spite of the campaign against the Chauci, the frontier was relatively settled and Pliny had the opportunity of writing. His military experiences led to his first work, a treatise on the use of the javelin as a cavalry weapon (*De iaculatione equestri*), and his substantial history of Rome's wars against the Germans (*Bella Germaniae*]. These monographs confirm his serious attitude to his military duties and his general interest in the province of Germany and its history.

In 59 Pliny returned to Rome with the intention of practising law. Several reasons have been advanced to explain why no pro-curatorship fell to him at this point in his career. Among those suggested by Syme, namely 'the loss of a patron, some minor indiscretion in society at Rome, a sudden distaste for affairs, or a prescience of the dangers which lurked in the path of industry and integrity', the last explanation would seem to have been the most likely.[1] At any rate Pliny retired temporarily from public service, and, keeping a low profile, devoted his talents to the safer field of writing books on grammar.

The Neronian tyranny and catastrophes such as the Pisonian con-spiracy against Nero (65), which involved a large number of knights as well as senators, confirmed Pliny's luck or, more probably, his sound judgement. The upheavals of the decade ended in civil war: Pliny, however, opportunely emerged in 69 as a partisan of the Emperor Vespasian and recommenced his public career, benefiting, no doubt, from the patronage of the emperor's eldest son Titus. He obtained a series of procuratorships in the next six years, namely Gallia Narbonensis (70), Africa (72), Hispania Tarraconensis (73), where he was responsible for Vespasian's financial affairs, and Gallia Belgica (75). These gave Pliny an opportunity to visit different regions of the Roman world, which, together with his keen powers of observation, accounts for the accuracy of his descriptions of

---

1. Syme, *Tacitus*, pp. 60–61.

mining and associated metallurgical processes, his knowledge of agriculture and viticulture, and his original contribution to the study of palaearctic fauna. Pliny's treatment, however, of astronomy and geography, which is almost exclusively based on secondary sources (often out of date), is in marked contrast. On his return from Gallia Belgica, Pliny was put in charge of the Roman fleet stationed at Misenum on the Bay of Naples. This was essentially an administrative appointment which carried responsibility for ship-building, repairs, victualling and the provision of chandlery.

Throughout his career Pliny had been interested in natural science, especially in occurrences of the more impressive and unusual phenomena of nature. At the time of the eruption of Vesuvius in 79, which destroyed Pompeii, Herculaneum, Stabiae and Oplontis, while going to the aid of survivors and spurred on, no doubt, by his scientific curiosity, Pliny lingered to observe the volcano at close quarters. Landing at Stabiae from a small boat he was overcome by the sulphurous fumes and died on 24 August 79.[1]

Unfortunately no portrait head, statue, or other representation of Pliny the Elder, survives from Roman times: nor is there any literary description of his physical appearance. Pliny the Younger, however, in his *Letters* (III, V, 7 ff.) gives a vivid and revealing picture of his uncle's idiosyncratic life style. He writes:

You may wonder how such a busy man was able to complete so many volumes, so many of them involving detailed study; and wonder still more when you learn that up to a certain age he practised at the Bar, that he died at the age of fifty-five and throughout the intervening years his time was much taken up with the important offices he held and his membership of the emperor's advisory council. But he combined a penetrating intellect with amazing powers of concentration and the capacity to manage with the minimum amount of sleep ... Admittedly he fell asleep very easily and would often doze and wake up again during his work ... In summer, when he was not too busy, he would often lie in the sun and a book was read aloud while he made notes and extracts. He made extracts of everything he read, and always said that there was no book so bad that some good could not be got out of it ... When travelling he felt free from other responsibilities to give every minute to work; he kept a secretary at his side with book and notebook, and in winter saw that his hands were protected by long sleeves, so that even bitter weather should not rob him of a working hour. For the same reason, too, he used to be carried about Rome in a sedan-chair. I can remember how he scolded me for walking; according to him I need not have wasted those hours, for he thought any time wasted which was not devoted to work.

1. The eruption is dramatically described by Pliny the Younger in *Letters*, VI, 16.

Pliny the Younger continues (ibid., 18):

When you consider the extent of his reading and writing I wonder if you feel that he could never have been a public official or a member of the emperor's council, but, on the other hand, now that you know about his application, that he should have achieved more? In fact his official duties put every possible obstacle in his path; and yet there was nothing that his energy could not surmount.

The *Natural History* also provides a substantial volume of evidence about Pliny's character, temperament and attitude to life. The deterioration of the political situation in the second half of the first century AD, especially under Nero (54–68), had far-reaching consequences on the general content and character of the literature of the period. Nero's jealous fear of all eminence of birth or success in the military field drove him to a policy of persecution. Moreover, his philhellenic outlook was resented by Romans generally. Pliny the Younger (*Letters*, III, V, 5) summarizes the situation: 'The slavery of the last years of Nero had rendered dangerous every study of a free and elevated character.' High-level opposition to the Flavians was mostly seen in a small but closely knit group of Stoics. Indeed Stoicism provided a philosophical foundation for the aristocratic opposition to those who tried to rule without, or against, the Senate. The basis for the opposition apparently lay in a sentimentalized view of the Republic and in the belief, less philosophical than political, that government should be in the hands of the best men, or best man, which of course to the aristocrats meant the old oligarchy of the Republic: their Republican heroes were Cato the Younger,[1] Marcus Brutus and Gaius Cassius. Such dissension, however, was ruthlessly suppressed, and a number of opponents of the Empire paid with their lives, among them Paetus Thrasea,[2] Helvidius Priscus,[3] Rubellius Plautus[4] and Junius Rusticus.[5]

1. Cato the Younger (95–46 BC) was tribune-designate at the time of the Catilinarian Conspiracy (63 BC) and helped to secure the execution of the conspirators. On the side of Pompey in the Civil War, he committed suicide after the Battle of Thapsus (46 BC).

2. A high-minded senator and Stoic who fearlessly opposed Nero. He was condemned to death on a charge of treason (AD 66).

3. Helvidius continued the opposition of his father-in-law Thrasea to imperial rule. He was denounced in AD 74 by Mucianus and later put to death by Vespasian.

4. The son of Gaius Rubellius Blandus and a descendant of the Emperor Augustus. This imperial connection made some regard him as a possible rival to Nero. In AD 60, on Nero's advice, he withdrew to Asia. Two years later he was forced to commit suicide by Nero's henchman Tigellinus.

5. Junius Rusticus wrote a *Panegyric* on his friend Paetus Thrasea, for which he

But by no means all thinking men and women turned against the emperors; Agricola, the father-in-law of Tacitus the historian, successfully pursued a military career and became governor of the province of Britain in 78. Likewise Pliny found it possible to serve the state and follow the path of scholarship. Pliny confirms his nephew's vivid sketch, assuring his readers that his literary work did not entail any withdrawal from public life: far from it! His research was done at night, as he explains, (*Preface*, 18). 'The days I devote to you, and I reckon up the sleep I need consistent with keeping well. I am content with this reward alone, in that while – in Varro's words – I occupy myself with these trifles, I prolong my life by many hours. For assuredly to live is to be awake.'

Pliny's attitude to imperial rule and his political views are plain to see in the *Natural History*. Unusually for someone with his Stoic beliefs, he was close to the Flavians, as the tone of the *Preface* addressed to the emperor makes clear. He had easy access to Vespasian, and was also a member of the emperor's advisory council. He accepted the imperial system as indispensable to Rome, being at the same time grateful for the security afforded by the *Pax Romana*. 'Italy is the ruler and second mother of the world,' he writes in the *Natural History* (XXXVII, 201). 'Italy is the most beautiful of all lands, endowed with all that wins Nature's crown.'

The role of Nature clearly had an important place in Pliny's thoughts, and was central to his attitude towards religion. Typically Roman, it was a blend of credulity and scepticism. In the *Natural History* he abhors the popular clamour over 'the gods' and includes as one of those popular quirks the ubiquitous Fortune. Pliny writes (VII, 130): 'There is always a fear that Fortune may grow tired, and, once this is entertained, happiness has no sure foundation.' The examples that follow (VII, 133–46) are reminiscent of the *loci de Fortuna* so popular in Seneca's *Suasoriae* and *Controversiae*. Nature, Pliny concludes, is what mortals call God. He stresses the gifts and tricks of Fortune throughout Book VII, which is dedicated to a study of struggling humanity; humans are constantly bested in the struggle with external forces. In his description of the situation he uses a figure that would occur naturally to any educated Roman of his day: the vicissitudes of Fortune were a commonplace beloved by the rhetoricians and worked almost to death in school exercises.

---

was condemned to death by the Emperor Domitian (AD 93?). He was a friend of Pliny the Younger and approved of Helvidius' opposition to Vespasian.

Pliny appears as a man torn between respect for knowledge and his background of rhetorical training, which he could not totally forget in spite of his dedication to science. Like Lucretius, Cicero, Seneca and Juvenal, he ridicules the terrors found in Greek mythology and uses all his powers to destroy the widespread Roman belief of life after death. He disposes unequivocally of man's claim to immortality (VII, 188):

All men are in the same state from their last day forward as they were before their first day, and neither body nor mind has any more sensation after death than it had before birth. But wishful thinking prolongs itself into the future and falsely invents for itself a life that continues beyond death, sometimes by giving the soul immortality or a change of shape, sometimes by according feeling to those below, worshipping spirits and deifying one who has already ceased to be even a man.

Pliny's attitude was further moulded by his Stoic beliefs, a philosophy that not only stressed the virtues of duty and the attainment of virtue through wise conduct, but encouraged the study of Nature. The philosophy, which accorded well with Pliny's simple tastes and industry, is a major element in the *Natural History*, and its influence surfaces again and again in Pliny's digressions on social life and attitudes under the early Empire.

The *Natural History* and Other Works

In the ancient world the dividing line between the arts and sciences was not so rigidly drawn as in more recent times. Poets and prose authors alike frequently crossed the boundaries. Thus Aristotle could write the *Poetics* on the nature of drama, as well as such scientific works as the *Physica* or the *Meteorologica*. Similarly, Virgil composed not only the *Aeneid*, a national epic glorifying Rome and the Emperor Augustus, but also the *Georgics*, a didactic poem devoted to agriculture.

Pliny the Elder's works (listed by his nephew Pliny the Younger, *Letters* III, V, 5) cover a wide range of topics: (i) a treatise on the use of the javelin as a cavalry weapon (*De iaculatione equestri* – one book); (ii) a biography of Pomponius Secundus, his friend, literary preceptor and possibly patron (*De vita Pomponi Secundi* – two books); (iii) an account of Rome's wars in Germany (*Bella Germaniae* – twenty books); (iv) a treatise on rhetorical training, with examples of declamations (*Studiosus* – three books); (v) a work on grammar

(*Dubius sermo* – eight books); (vi) a history of Rome, commencing with the close of Claudius' principate (*A fine Aufidi Bassi* – thirty-one books); and (vii) the *Natural History* (thirty-seven books). Material for the treatise on the javelin and the account of Rome's wars in Germany was probably collected during his military service in that province, and the rhetorical handbook was the forerunner of Quintilian's *Institutio oratoria*. Of all these works only the *Natural History*, Pliny's encyclopedic account of the state of science, art and technology in the first century AD, survives.

'Science' did not burst upon the ancient world fully developed, like Athena from the head of Zeus, but was the product of a long struggle. Greek and Roman mythology had defined five ages of mankind,[1] and it is significant that *four* were named after metals. The absolute rule, however, of Olympian religion and its morality, reflected in Homer's description of the fourth – the Heroic Age – initially hampered the development of scientific thought until trade and colonization brought Greece into contact with Ionia. In the seventh century BC Ionia was the crossroads between East and West, and contact with new ideas, not least in the field of comparative religion, caused the Greeks to re-examine their long-accepted beliefs. This led, in time, to the beginnings of rationalism and the emergence of a number of philosophical explanations of the origin of the universe. Whereas formerly it had been the poet's role to act both as educator and guardian of morals, this was gradually taken over by the philosopher.

The first attempts to explain the universe and man's role in the natural order of things began with the research of the Ionian philosophers known as 'natural scientists'. They believed in the existence of four basic 'elements' – fire, air, earth and water – and each philosopher chose *one* as the original material (*he arche*) or underlying substance (*to hypokeimenon*) of the universe. Water was the most logical choice in that it can readily be observed in the three states of matter, namely liquid, solid and vapour. Henceforth, the paths of natural science and religion diverged. In the fifth century BC the Sophists took over higher education and claimed, like Hippias, to impart all subjects. Socrates, by contrast, apparently soon became disenchanted with science, as Xenophon informs us (*Memorabilia*, I, I, 11): 'He did not even discuss that topic favoured by other speakers, that is, the nature of the universe, and avoided speculation on how the so-called cosmos of the professors works and on the laws that

---

1. These were the ages of Gold, Silver, Bronze, Heroes and Iron.

govern the phenomena of the heavens: indeed he would argue that to trouble one's mind with such problems is sheer folly.' In the fourth century BC Aristotle refers to 'branches of knowledge essential for a free man' (*eleutherai epistemai*) – astronomy, geometry, arithmetic, music and grammar. The *Meteorologica* treats some of these subjects, namely astronomy, geology, seismology and meteorology. The work is typical of a stage when the natural sciences had not become fully differentiated from philosophy, but none the less embraced a wide range of scientific knowledge. No fully comprehensive work of reference was produced in classical Greece. The appreciation, however, of the problems to be solved and the ideas of matter, change, elements and atoms, mostly despised by Plato, are the abiding positive contributions of Greek philosophy to science.

During the Hellenistic period there was a widening of learning: Alexandrian literature reflects a growing interest in aetiology, antiquarianism and, above all, the collection of exotic and unusual information. These trends are reflected in Aratus' *Phaenomena*,[1] a versification in hexameters of Eudoxus' old *Star Catalogue*, and Nicander's *Theriaca*, a scientific treatise on poisons and antidotes. Other works covered a wide range of topics including astronomy, geography, fishing, agriculture and bee-keeping. Research was conceived of as an imitative rather than a creative activity, and so scholarly works were little more than the results of wide reading and note-taking without critical or scientific evaluation.

In Rome there was an early interest in the compilation of facts, as Pliny records (*Preface*, 17): 'We need works of reference not books.' Cato the Censor (234–149 BC) and Varro (116–27 BC) established the genre of the encyclopedia at Rome. In the first century BC Lucretius (c. 94–c. 55) set out to expound the system of the Greek philosopher Epicurus, and achieved a harmonious interplay of poetry and instruction never subsequently surpassed.[2] Lucretius, following

---

1. The success of the *Phaenomena*, a dry astronomical work, is puzzling, but is probably due to its illustration of the Stoic doctrine of Providence, taking for its example the utility of the stars to sailors and farmers alike.

2. Epicurus of Samos (341–270 BC) travelled in Lesbos and Troas and settled at Athens (306/7 BC). The Epicureans believed that philosophy had the practical purpose of securing a happy life: the ideal state was *ataraxia*, 'freedom from disturbance'. In order to attain this state they recommended, among other things, the avoidance of deep emotional attachments and the abstention from the competitive world of politics. They also believed that the soul, composed of atoms like the body, died with it, and that the gods did not interfere in the physical world, which owed its origin to natural causes.

Epicurus, offered a prescription for happiness, in particular asserting that the mental fears and forebodings that 'religion' instils (*religio* is equated with superstition) can be dispelled by a knowledge of the laws of 'physics', which leave no place for divine intervention in life or for human survival after death. Lucretius exhibits the recognizable marks of a scientific spirit: he observes the natural world closely and critically, and insists on the universality of natural causes; he has an over-mastering faith in reasoned arguments and a concern for clear and systematic exposition, matched by a capacity to achieve this. His work *On the Nature of the Universe* thus paradoxically combines the passionate intensity of a proselytizing tract with the qualities that place it firmly among the forerunners of modern scientific publications.

Pliny and Lucretius share a number of themes. Both wish to explain the universe and its phenomena in rational terms and to free the minds of men from fear through a greater understanding of the world. But their treatments of the subject differ. Lucretius writes imaginatively, as a poet and Epicurean, while Pliny reveals himself as a scientist with Stoic inclinations. He equates God with Nature and, an ecologist at heart, develops the theme of mankind's abuse of earth's resources in its pursuit of an ever more extravagant lifestyle. Perhaps the most significant difference, however, is the wider scope of Pliny's treatment, in which he attempts to provide an exhaustive catalogue of nature. His encyclopedic account of the natural sciences, interspersed with essays and digressions on the achievements of man, contains, according to Pliny himself, some 20,000 facts, from 2,000 works by 100 chosen authors. These figures, however, represent a rather conservative estimate since no fewer than 146 Roman and 327 foreign authors are quoted.

The title of the work, *Historia Naturalis*, is of particular interest. *Historia* is clearly derived from the Greek word *historia*, which can mean not only 'history' but also 'inquiry' or 'research'. It is with this sense that Herodotus uses the word at the beginning of the work known to us now as the *Histories*. Indeed Pliny's treatment and presentation of material are both reminiscent of the early Greek historian. Both are avid cataloguers whose approach is often ingenuous and uncritical. This assessment, however, is not intended to detract from their original contributions to our knowledge of the ancient world. The adjective *naturalis* means 'belonging to nature' or 'to the nature of things', and although we may adhere to the long-standing tradition of translating the title as *Natural History*, Beyet's version *Recherches sur le monde* more accurately reflects the scope of the work and accords with Pliny's own definition of his subject as 'the natural world, or life' (*Preface*, 13).

Pliny shows a commendable attitude towards his sources (*Preface*, 21 ff.):

You will count as proof of my professionalism the fact that I have prefaced these books with the names of my authorities. In my opinion such acknowledgement of those who have contributed to one's success – unlike the practice of most of the authors I have mentioned – is a not ungracious gesture and abounds with honourable modesty.

For you ought to know that when I compared authorities, I found that writers of bygone times had been copied by the most reliable and modern authors, word for word, without acknowledgement. Nor was this done in that well-known Virgilian spirit of rivalry, or in that artless manner of Cicero, who declares himself a companion of Plato, and adds, in his *Consolation* to his daughter, 'I follow Crantor.'

Surely it is characteristic of a mean spirit and of an unfortunate attitude to prefer to be caught committing a theft rather than to repay a loan, especially as capital accumulates from interest.

In Book I, where Pliny lists his sources, he includes not only major authors but many lesser authorities and yet others – for example, King Juba – who appear, as it were, *honoris causa*. As well as named authorities, Pliny, like all authors from Herodotus onwards, cites additional anonymous sources – 'someone', 'some people', 'Greeks' and 'local people' say – and includes statements prefaced by 'so the story goes', 'it is reported', and 'the traditional account is'.

Research and scholarship are clearly matters close to Pliny's heart. In Book XIV he digresses from his discussion of trees to complain that few people are acquainted with the works of writers of former times, whose research was more productive and successful. And he enlarges upon what he considers to be the moral degeneration of the times with an undeniable zeal (XIV, 4–5):

In earlier times peoples had their power limited to their own boundaries, and for that reason their talents were circumscribed; there was no scope for amassing a fortune, so they had to exercise the positive quality of respect for the arts. Accordingly, they put these arts first, when displaying their resources, in the belief that the arts could bestow immortality. This was the reason why life's rewards and achievements were so plentiful.

The expansion of the world and the growing extent of our resources proved harmful to subsequent generations. Senators and judges began to be chosen by wealth, and wealth was the only embellishment of magistrates and commanders ... In such a climate the only pleasure consisted in possession, whereas the true prizes of life went to rack and ruin and all the arts that were called 'liberal' . . . became quite the opposite.

In traditional Greek education, astronomy, arithmetic, geometry and the theory of music were classed as part of the liberal arts. The structure, however, and content of the *Natural History* reveal that Pliny consciously rejected the liberal arts as an adequate framework for human knowledge. These subjects had become specialized and carried a high degree of abstraction. Although accessible to scholars, they were difficult to grasp for ordinary people with practical rather than theoretical needs. Moreover, they contained only a small part of all available knowledge, ignoring, as they did, most of the manifold phenomena of the natural world. Accordingly, Pliny decided to concentrate on topics of more immediate importance for human life in general. One consequence of this choice was that disciplines such as arithmetic, geometry and harmonics were discarded in favour of subjects more directly related to everyday life, such as the study of places, where man lives and works, and of the animals and plants that surround him and provide him with the most urgent necessities of his existence – shelter, food, drink and medicines. All these fields of knowledge had their own literature and authorities. But, as Pliny rightly claims, no one had tried to assimilate and survey these topics within the compass of a single work.

The *Natural History* is not, as has been suggested, a work intended as a vade-mecum for a provincial governor. Far from this, as Pliny explains categorically (*Preface*, 6), it is 'written for the masses, for the horde of farmers and artisans and, finally, for those who have time to devote to these pursuits'.

ASTRONOMY

The first subject to be considered by Pliny in the *Natural History* is astronomy. Pliny follows accepted Hellenistic cosmology, namely that the world is spherical and thus perfectly adapted to its daily rotation which produces the rising and setting of the sun and stars. His views on the nature of constellations are difficult to follow and seem to be based on mythic or animistic conceptions of nature long since abandoned by Greek astronomy.

The rest of the universe is the realm of the four elements – fire, air, water and earth. Pliny's account shows that he did not understand the principles by which Aristotle had connected the physics of the elements with the structure of the universe. He follows Aristotle in holding that the universe is eternal, and he is aware of the seven 'classical' planets.[1] Pliny discusses periods of planetary revolution but there is no evidence to suggest that he knew that Hipparchus[2] had already discovered the phenomenon of precession, nor does he show any real understanding of 'motion in longitude'. Again, he is aware of the variation of the length of the longest day with the noon altitude of the sun at the equinoxes, but is unable to explain this relationship in precise terms. With regard to the seasons, Pliny, unlike Hipparchus, adheres to antiquated concepts, as do Manilius, Geminus and Columella. His account of the moon is more detailed – a medley of facts, physical speculations and astrological asides. He describes how it completes its orbit in twenty-seven and a third days, is invisible for two days and sets out on its new course on the thirtieth day – this is the nearest Pliny comes to realizing the connection between the sidereal and synodic month. The effect of the moon on tides and the role of the sun and moon in eclipses are explained. It is interesting that his record of the eclipse that occurred before Alexander the Great's victory at Arbela (20 September 331 BC) is more precise than the account in Ptolemy's *Geography*. But, although he records detailed information about eclipses, there is little evidence that he understands the details of their geometry. Nor does he understand the relationship between latitude, gnomon shadow and the longest day, unlike Hipparchus who possessed this knowledge some two centuries earlier. Indeed, in Book II Pliny bestows lavish praise on Hipparchus' achievements, among them the discovery of a nova (95).

It would be unfair to judge Pliny's astronomy from the theoretical sections in Book II. For him, as for many others in the ancient world, there were other more practical reasons for studying astronomy. Civilized life depended on agriculture and commerce. Both farmer and sailor needed to observe celestial phenomena as part of their daily routine. So Pliny attaches more weight to the practical

1. Saturn, Jupiter, Mars, Sun, Venus, Mercury, Moon.
2. The famous astronomer from Nicaea, Bithynia (*c.* 190 BC–after 126 BC), whose only extant work is his commentary on the *Phaenomena* of Eudoxus and Aratus, containing criticisms of the descriptions and placings of constellations and stars by these two authorities and a list of simultaneous risings and settings. Hipparchus was the first to construct a theory of the motion of the sun and moon that was properly based on observation. He discovered the precession of the equinoxes by comparing his observation of the distance of the star Spica from the autumnal equinox with that of Timocharis made about 160 years before. He was also the first person known to have made systematic use of trigonometry.

utility of the subject and is mindful of the difficulties involved in its application to everyday pursuits: 'it is an arduous and vast aspiration to succeed in making the divine science of the heavens known to the ignorant rustic, but it must be attempted, owing to the great benefit that it confers on life' (XVIII, 206). He mentions the calendars (*parapegmata*) used by farmers and, since farming, like navigation, depends on the weather, includes a long section on meteorology to show how the weather is influenced by the sun, moon and stars.

GEOGRAPHY

Although Pliny has a positive contribution to make in a number of scientific fields, the quality of his information, where he relies on secondary sources, is inferior to that acquired through personal experience. This is particularly noticeable in his books on geography (III–VI). Pliny is not a geographer in the modern sense and his account is written with a literary rather than a scientific approach. He accepts, like Claudius Ptolemaeus, the sphericity of the earth and the zones into which it was customarily divided, but, in doing so, merely follows views accepted since the time of the Pythagoreans and Plato. Pliny makes no attempt to evaluate the evidence he derives from his sources: in estimating the circumference of the earth he is content merely to quote Eratosthenes' calculations;[1] in describing south India, although it had become well known by the first century AD through commerce, he continues to follow Megasthenes, a writer of the third century BC, with the result that his information is often 200 years out of date! The geographical books are basically a gazetteer. The distances recorded are, more often than not, inaccurate, but the descriptions of peoples and their way of life – for example, in countries bordering the Ganges – are extremely colourful and reminiscent of Herodotus' account of Egypt. This interest in faraway peoples and places is reflected also in Pompeian and other wall-painting.[2]

The very limited contribution made by Pliny to the science of geography is all the more surprising in the light of the knowledge

1. Eratosthenes (*c.* 275–194 BC) was a scholar of many talents and the first systematic geographer. He wrote the *Geographica* in which he dealt with physical, mathematical and ethnographical geography, and sketched the history of the subject.

2. See generally A. Maiuri, *Roman Painting* (London, 1957); A. Stenico, *Roman and Etruscan Painting* (London, 1963); D. Strong, *Roman Art* (Harmondsworth, 1976).

that must have accumulated from the spread of the Roman Empire and from Rome's extensive trading contacts by the early imperial period.

ZOOLOGY

Pliny devotes nine books to animals; he treats them from a zoological point of view (VIII–XI) and explains their role in pharmacology (XXVIII–XXXII). His data, observations and comments on both wild and domesticated animals, embrace aspects as widely separated as anatomy, magic and popular beliefs. In addition he discusses man at some length (VII).

Throughout these books, as elsewhere, Pliny makes no attempt to conceal his limitations. He is indebted to many authors, in particular Aristotle, but he differs from the latter in the presentation of his material. Pliny wishes to offer for the first time in Latin a compendium of Aristotle's famous works on zoology with additional information unknown to Aristotle. He does not, however, search for any comprehensive classification of the animal kingdom, nor does he try to put forward any systematic description of the characteristics by which animals may be differentiated. Such organization as there is in Books VIII–XI is disarmingly simple: the account begins with the largest of the groups considered – namely, elephants, whales and ostriches – and ends with the smallest – bees, 'the chief insect species'. Characteristically, zoological facts are interspersed with digressions – for example, on the building of aviaries in Rome, Roman taste for rare and exotic birds and the role of birds in divination.

In the second group of books (XXVIII–XXXII) Pliny turns to parts and products of animals, the properties of which are supposed to be able to cure or relieve illnesses, diseases and physical accidents. Some further facts about well-known animals are included, as well as data about new animals of special interest due to their particular use in medicine – for example, the spitting cobra, fresh-water turtles and corals. As Bodson writes, 'Pliny enlarges Aristotle's recording of animals (either species, genus, or family, according to modern terminology) by more than 150 units. The extension of the world known at the time and the exploration of the parts of the Roman empire, explain these figures. They should not be minimized despite the errors which mar the text more than once and prevent it being correctly interpreted.'[1] Indeed Pliny's evidence is a unique guide to

1. 'Aspects of Pliny's Zoology', in R. French, *Science in the Early Roman Empire* (London, 1986), pp. 98–110.

those animals with which the Romans had become familiar. Apart from the Magi, anonymous and popular hearsay, Pliny relies on standard authors whose works are known or were subsequently lost, on material from unnamed sources, and on personal observation. He sometimes includes the first and often only record in antiquity of species unknown to Aristotle and his successors.

Pliny's description of the fauna of the Alps is of particular interest as an example of his original contribution to the description of the palaearctic fauna and for the light it throws on his intention and method.[1] The tribes of the Alpine region were pacified by Augustus in 7 BC and ruled by a *praefectus* until Nero replaced this office by a procuratorship in AD 63. The region, however, was known to the Romans from much earlier times, due to trade links established with Transalpine Gaul, the unrest of the inhabitants of Cisalpine Gaul and, of course, Hannibal's expedition in the Second Punic War. Strabo mentions wild horses and cattle in the Alps and quotes Polybius' description of a wild, deer-like mammal, most probably the European elk (*Alces alces*).

All the other ancient evidence is to be found in the *Natural History*. Four wild mammals: chamois (*Rupicapra rupicapra*), ibex (*Capra ibex*), mountain hare (*Lepus timidus*), and marmot (*Marmota marmata*), besides goats and cattle; four native birds: ptarmigan (*Lagopus mutus*), black grouse (*Lyrurus tetrix*) and capercaillie (*Tetrao urogallus*), Alpine chough (*Pyrrhocorax graculus*), and a migratory species from Egypt, either *Plegadis falcinellus* (glossy ibis), or perhaps the famous *Gerontius eremita* (bald ibis); one fish, burbot (*Lota lota*), and one species of snail are listed. Most of them are so exactly described, at least in regard to their most meaningful characteristics, that the identification of the species can be made beyond all doubt.[2]

These descriptions are so accurate that they are likely to have been the result of eyewitness reports, supported by personal and careful observation in the field, perhaps during his military service from AD 46 onwards. Pliny may also have supplemented his research with further information collected in Rome, where several of those species were imported either for the circus parades and games, or for the aviaries, or, in the case of game birds, as delicacies for imperial banquets. In the mention of one bird by the general term ibis, Pliny, whatever his sources may have been – he refers to only

1. ibid., pp. 102–3.
2. ibid.

Egnatius Calvinus – proves that he was well-informed in his account, which has become of greater historical importance since the animals included have, over the centuries, become endangered species and even threatened with extinction.

Pliny's list of Alpine fauna is limited. There is no place given to insects, reptiles and fish in general, and he omits a great number of birds and mammals: his list is, in fact, virtually confined to animals and game birds. His account of German fauna is similarly limited. He says that Germany 'produces few animals'. In spite of this assertion, however, he records information on 'the most remarkable breeds of wild oxen' – namely, the European bison and aurochs. The omission of any account of wolves, bears, boars and deer, which Pliny must have seen in abundance in Germany, shows a conscious discrimination: clearly these animals were too common to warrant special mention. Pliny is content in his selection to concentrate on species that were remarkable or best suited to complete Aristotle's descriptions.

In Book VII the subject is man, the most important member of the animal kingdom. His account of man's reproduction, physiognomy, mental powers and character follows a predictable course, with little serious consideration of physiology. Pliny is clearly obsessed with the unusual and freakish. As well as cannibals and pygmies, he describes one tribe with 'umbrella feet', another with dogs' heads, another with only one leg, and so on. He gives examples of exceptional physical attributes and achievements. He lists discoveries in the arts and sciences that made a significant contribution to civilization, and those responsible for them. Even such oddities as a parchment copy of the *Iliad* small enough to fit into a nutshell find mention in his account. His 'facts' add up to something like an ancient *Guinness Book of Records* and, although unlikely in the extreme, have a compelling interest. Characteristically he intersperses these 'facts' with digressions that have as their theme the capriciousness of Fortune – the life of the Emperor Augustus providing the chief example.

BOTANY

In his introduction to the science of botany Pliny writes (XIV, 7):

I will pursue research into topics that have been forgotten, nor will the relative insignificance of certain things deter me any more than when I was dealing with animals, although I observe that Virgil, the most outstanding poet, for this reason omits some of the resources of the garden and, in those that he recounts, picked only the flower of his material – happy and fortunate in the regard that he enjoys with others.

Pliny's interest in plants derived from their importance to mankind, not from the general principles that arise from their scientific study.

The Hippocratic Corpus[1] contains the first scientific approach to plants. At the same time as lists of medicinal plants were compiled, some attempts were made to compare the structure and physiology of plants with that of animals and man. Trade, colonization and military conquest led to a knowledge of increasing numbers of exotic plants and products obtained from them: some of these plants, indeed, were brought under cultivation. This accumulation of data, together with observations and research, led Theophrastus, following Aristotle's methodology, to create in his lectures on plants at the Lyceum, the first synoptic treatment of scientific botany. In the two books of the *De Plantis* (*History of Plants*), published *c.* 300 BC, he embraces what we may call descriptive and physiological botany. The work contains a wealth of detailed, accurate observations, and, with its critical rigour, anticipates, in embryo at least, almost every branch of modern botanical inquiry.

Pliny devoted sixteen books of the *Natural History* to plants. To judge from his list of authorities, he had access to a vast literature, although he relies, in general, on a few major writers. He consulted the *De Plantis* and is heavily dependent on Theophrastus for his general description of plants (XII–XIX). In the case of medicinal plants (XX–XXVIII) he draws on Sextius Niger and the evidence of Greek herbals. Occasionally, he includes new material. For example, he mentions the agricultural use of marl in Gaul and Britain, and names the Celtic terms for the different types – an account that is probably derived from the works of Pompeius Trogus, a native of Gaul. Indeed, Pliny uses many sources other than Theophrastus, and this, together with the way in which such information is used, imparts a lasting value to the botanical sections of the *Natural History*. It is clear that he also placed considerable reliance on personal observation. He records that he had inspected almost all the plants in the garden of Antonius Castor (an expert horticulturist in Rome), and, doubtless, found other opportunities to have plants

1. A collection of writings that deals with all medical subjects, including prognostics, dietetics, surgery and pharmacology. The views expressed within the collection differ widely and so cannot be attributed to Hippocrates in their entirety.

shown to him, even picked specimens. His evidence supplements the knowledge we have derived from Pompeian and other wall-paintings that depict gardens.

Morton cites two cases where Pliny certainly contributed new botanical facts derived from his own observations during official postings. He refers, for the first time in ancient literature, to the exclusively maritime pea, growing on islands in the North Sea (XVII, 121). Secondly, while in Hispania Tarraconensis, he saw the genista, used for rope-making (XIX, 26; XXIV, 65). He describes its cultivation, use and economic importance. In Spain, he also saw rosemary shrubs (*ulex*), used in the processing of gold-ores. He is at his best in recording these species, but does not make any meaningful, detailed botanical investigations of the plants themselves. In the words of Morton: 'Theophrastus was an original observer and scientist, who brought the judgement of experience to every aspect of botanical research. Pliny has no such basis of specialist experience, and this, even more than the differences in intention, temperament and intellect, explains his blindness to the great theoretical advances which entitle Theophrastus to be called "the father of scientific botany". The vital contributions made by Theophrastus to the methods and general principles of botany were either ignored by Pliny, or were misunderstood and robbed of their meaning and importance.'[1]

Pliny's grouping of plants is arbitrary and bizarre, and lacks the principles of identification and classification to be found in Theophrastus. Thus in his treatment of agricultural and garden plants (XVII–XIX) he not surprisingly neglects to repeat Theophrastus' remarkable theoretical ideas in relation to plant nutrition, metabolism and adaptation to the environment. There is, however, in these sections a wealth of detail. His sources include Androtion, Cato, Varro, Virgil and, above all, Columella. Pliny records the introduction of new agricultural and horticultural species into Italy, and includes much interesting information about olive-growing, oil-production, viticulture and wines: his debt to Varro is freely acknowledged. In one digression (XVIII, 296) he describes an ox-driven grain-harvester from Gaul, which was basically a large metal comb (*pecten*) mounted on wheels, an implement that had been considered a figment of Pliny's imagination until illustrations of it were discovered in southern Belgium (in 1958) on a stone relief

---

1. 'Pliny on Plants: His Place in the History of Botany', in R. French, op. cit., p. 88.

datable to the first century AD.[1] Three other examples are now known, and these machines were used on large estates in Gaul. Other fascinating digressions from Pliny's main theme are the descriptions of varieties of paper manufactured from the papyrus plant and the use of mistletoe by the Druids.

The Greek corpus of plants had included about 800 species. Pliny records 1,300, but many are clearly duplications and his actual total probably does not exceed 800. Many of his descriptions, almost all copied from his numerous authorities, are too vague and lacking in detail, so that identification of these plants is difficult, if not impossible. Morton sums up Pliny's importance to the study of botany. 'With all his weaknesses and omissions, Pliny did one very important thing: he kept alive the concept of botany as a broad, unitary science of plants (as Aristotle and Theophrastus had conceived it) not to be reduced to a knowledge of drug-plants, or the operations of cultivation.'[2]

MEDICINE

One of the most fascinating sections of the *Natural History* is Pliny's account of the history of medicine (XXIX, 1–28), from the time of the Trojan War, when 'medicine assigned to its earliest practitioners a seat among the gods and a place in heaven' (XXIX, 3), to his own day. It is Pliny, above all, who provides us with a framework of both fact and interpretation, which can be used to build up a proper description and history of medicine in the late Republic and earlier imperial period.

Hippocrates of Cos was one of the first Greeks to advance the science of medicine which became a near monopoly of the Greeks in succeeding centuries. The Roman assessment of the benefits of medicine varied. Cicero and the hypochondriac Seneca gave doctors a good press. The latter writes in glowing terms about his personal physician (*De Beneficiis*, VI, 15, 4): 'He spent more time than the average doctor on me . . . he listened to my groans with sympathy; amid a crowd of patients, my health was his first concern; he attended others only when *my* health permitted it; I was bound to him, not as to a doctor, but by ties of friendship!' Pliny, by contrast, followed the views of Cato and condemned doctors out of hand.

1. See K. D. White, *Roman Farming* (London, 1970), pp. 182–3 and figs. 36–7.
2. In R. French, op. cit., p. 95.

Some of the criticisms have a familiarly modern ring. When he deals with doctors one can hear his authentic voice; the views he expresses are undoubtedly his own even where he obtains his facts from an earlier author. As Nutton observes, it is here too that his interest in the curious workings of humanity links closest with his own sentiments and, possibly, his own experience. Who could forget his denunciation of the royal physician, grasping and adulterous, or his sturdy opposition to trendy innovations, whether they were cold baths, the prescribing of wine, or Thessalian methodism.[1] The passion of his involvement becomes all the more apparent when contrasted with the calm exposition of the development of medicine by his near contemporary Cornelius Celsus.

Medicine was not *just* un-Roman, it was Greek![2] Rome had become increasingly subjected to Greek influence from the end of the fourth century BC, and the cult of Asclepius was introduced in 292 BC. Pliny says (XXIX, 12) that in 219 BC Archagathus enjoyed the distinction of being the first doctor in Rome. Although there is evidence to suggest that there were others who preceded him, his career marks an important step, since he was given Roman citizenship and provided with a surgery at the public expense. This was certainly the first occasion on which the state employed a health officer, an indication of the degree of acceptance of Greek tradition and influence. The most widespread complaint against doctors seems to have been that they were paid too much, as, for example, Quintus Stertius who earned no less than 500,000 sesterces a year (more than the *census* rating of a knight), or Crinias who left 10 million sesterces, having contributed as much to rebuilding the walls and fortifications of his native city. But their treatments also received heavy criticism. Some doctors' excessive recourse to surgery and cautery earned them the nickname 'Executioner' or 'Butcher' (*Carnifex*). Pliny comments with feeling, 'Only doctors can kill a man with impunity' (XXIX, 18). 'Some tombs,' he writes, 'carried the inscription "A gang of doctors killed me"' (XXIX, 11).

In Pliny's opinion Romans had survived without medicine for 600 years and its introduction into Italy merely contributed to the decay in morals. His account may betray, as Nutton comments, considerable ignorance of actual medical practice, but, none the less,

1. V. Nutton, 'The Perils of Patriotism: Pliny and Roman Medicine', in R. French, op. cit., pp. 31–2.
2. ibid., p. 33.

it is a fascinating exposé of the ethos of medicine and of the relationship between doctor and patient in the first century AD.

PHARMACY

Pliny's catalogue of drugs and medicinals (XX–XXXII) appears to cover the entire scope of pharmacy in the classical world. He refers to magical remedies that are part of the folk tradition not only of Italy but also of Rome's Mediterranean dominions. Altogether the *Natural History* lists over 900 substances used as drugs (Dioscorides describes 600, Theophrastus 550, Galen 650 and Paul of Aegina 500).

There is abundant evidence that Pliny was well aware of Greek medical writers from Diocles and the Hippocratics to the Greek physicians of his own day, and many of the 'medical' sections in the *Natural History* are taken up with drugs and pharmaceuticals drawn from a huge variety of sources in both Greek and Latin. There are, however, many strange errors in the facts and observations about drugs. Pliny's unbridled curiosity led to a too rapid compilation of data. He pays scant attention to specific detail and seems to have shied away from cross-checking references and making any necessary revisions. Some of these mistakes can be corrected through parallel readings in the *Materia Medica* of Dioscorides, who used some of the same sources. It is surprising that Pliny did not know the work of Scribonius Largus, whose *Compositiones* (*Compounds of Drugs*) was written in Claudius' reign. The contrast between Scribonius' drug lore and that of Pliny is instructive. While Scribonius combined book-learning with experienced, practical knowledge of drugs, so that his work records what a professional druggist would do in the classification of medicinals, Pliny is, on the whole, content simply to follow his sources verbatim. The *Natural History* fails to provide a reasoned assimilation of pharmacological texts that antedate Augustus – with the possible exception of works of Theophrastus, Nicander and Apollodorus.

CHEMISTRY

There is nothing in the *Natural History* that satisfies any modern definition of analytical chemistry – the subject, as a clearly defined area of study, dates from the eighteenth-century Enlightenment. There is, however, evidence of well-developed techniques that would one day underlie the modern, scientific approach. Pliny records a number of tests intended to identify a material as correct for its proposed use, including tests of smell, taste, feel, colour, melting-point and the effect of heat. These tests represented the foundations of an orderly system, and some, such as the use of the touchstone in the qualitative examination of gold (XXXIII, 126) and the streak test for minerals (see p. xxxi) signified extremely important advances.

MINERALS AND METALS

In Books XXXIII–XXXVII Pliny turns to earth sciences with a survey of minerals, mining, metals and metallurgy.[1] He also includes sections on art, architecture and Roman coinage.

In the field of mineralogy the pre-Socratics produced little of consequence. Aristotle (384–322 BC) was the first authority to give a detailed account. His theory of the production of minerals and metals from vaporous exhalations (*Meteorologica*, III, 378a–b) comes very near to describing a process known as pneumatolysis, by which vaporous or heated waters discharged from igneous magmas by volcanic action are deposited on cooling in rock fissures to form minerals. The oldest extant scientific treatise that deals expressly with minerals is the *De Lapidibus* (*On Stones*), of Theophrastus (*c.* 370–*c.* 287 BC). This work lists and describes sixteen mineral species, grouping them as 'metals, stones and earths'. The minerals recorded within these categories were by way of illustration and clearly did not represent the total of all that were known at the time. This systematic method of classification, although based on superficial characteristics rather than on any concept of chemical composition, was itself a considerable advance.

Pliny's account is heavily dependent on Theophrastus, and he quotes from him extensively. However, he also mentions minerals not found in Theophrastus' catalogue – for example, rhombic sulphur (*sulphur*), fluorspar (*murra*) and tourmaline (*lychnis*). Pliny certainly relies on a number of other authorities, known to us by name but whose works are no longer extant. In particular he outlines, with some scepticism, Posidonius' explanation of how transparent and semi-transparent stones are formed. Posidonius, like other authorities, believed that quartz – known as *krystallos* because of its ice-like appearance – was formed by moisture that froze in the sky at

---

1. See generally J. F. Healy, 'Pliny on Mineralogy and Metals', ibid., pp. 111–146, and *Mining and Metallurgy in the Greek and Roman World* (London, 1978).

the very low temperatures existing there. Similar explanations may be found in Diodorus Siculus (II, 52, 1–4) and Seneca (*Naturales Quaestiones*, II, 54, 1).

In his discussion of alum Pliny shows an awareness of the process of crystallization by precipitation. This, together with the actual appearance of quartz crystals, may have led to the commonly held view that all minerals were in a constant state of growth and could regenerate themselves. It is a view that Pliny accepts without criticism. He mentions the assertion of the natural philosopher Papirius Fabianus that marble grows in quarries, and describes the experience of quarrymen as supporting this thesis.

Pliny and most ancient authorities are prone to inconsistency in the nomenclature of minerals: *schistos*, for example, can mean alum, a salt of alum or even an ore of iron. Usually the Romans adopted the Greek terms ending in *-ite*, *-itis*, and *-ites*. Some of these terms are related to the characteristics of the mineral referred to: the name haematite, for example, which refers to a reddish ore of iron, is derived from *haima*, the Greek word for blood. Some are generic: *adamas*, for example, refers to hard substances including minerals and metals. Occasionally problems arise from the fact that modern and ancient nomenclatures differ. *Chrysocolla*, loosely applied to any bright green copper mineral by the Greeks and Romans, but mainly to malachite, is currently a term restricted to hydrous copper sulphate.

In general Pliny shows a much wider understanding than his predecessors of both the physical properties and the optical characteristics of minerals. He gives examples of cleavage (the sarcophagus stone and selenite), of hardness (fluorspar and onyx marble), and, by implication, of density (when he describes tin and gold as being heavier than accompanying gangue minerals). He is familiar with the pyroelectrical properties of amber and tourmaline, and is aware that the streak (i.e., the trace of a mineral that is left on the surface of a hard stone, when the mineral is drawn across it) is characteristic of the individual mineral.

In his description of the crystal systems of quartz, diamonds, beryls and the 'rainbow stone' (*iris*), Pliny enters the field of crystallography, anticipating the work of Romé de l'Isle and abbé Häuy in the eighteenth century. 'Why rock-crystal is formed with hexagonal faces,' he writes (XXXVII, 26), 'is not easy to explain, and the difficulty is complicated by the fact that its terminal points are not symmetrical, while its faces are so perfectly smooth that not even the most skilful lapidary could achieve such a finish.' His account of the octahedral Indian diamond (XXXVII, 56) is of particular interest because of the accuracy of his description: '[It] has a certain affinity with rock-crystal, which it resembles in its transparency and its smooth faces meeting at six corners. It tapers to a point in two opposite directions and is all the more remarkable because it is like two spinning-tops joined together at their bases. It can even be as large as a hazel-nut.' Pliny even anticipates the modern industrial use of diamonds (XXXVII, 60) where he describes how diamond splinters inserted into iron tools were used by engravers. But his observations concerning the physical properties of the diamond are grossly incorrect. He accurately describes beryls – which, like diamonds, belong to the hexagonal system – but believes that the prisms in this case are the work of skilled Indian lapidaries!

Pliny's account of the main metals known in the ancient world is not only intrinsically interesting but provides valuable further evidence for our understanding of imperial Roman technology. He has a particular interest in gold, since, as procurator in Hispania Tarraconensis, he had the opportunity of studying mining operations and the technology of refining. The accuracy of his record and the inclusion of local Spanish and other non-Roman mining terms confirm that Pliny is here writing from personal observation. The main method used by the Romans in refining gold – namely, salt cementation – is directly comparable with that currently employed in Brazil, at Serra Pelada, where neither cyanide nor mercury recovery processes are allowed because of the potential risks involved. Surprisingly, after his extensive treatment of gold, Pliny dismisses silver in a few paragraphs, and appears to be more concerned with other metals, or minerals, associated with the ores of silver. The manner in which this discussion interrupts the main theme is further evidence of his lack of revision. His account of mercury closely follows that of Theophrastus.

Copper, together with tin and lead, occupies almost the whole of Book XXXIV. In Pliny, no less than in other ancient authors, the terminology for tin and lead provides a number of problems. These metals are usually described as white (*album* or *candidum*) and black (*nigrum*) lead respectively. A further term 'bright lead' (*plumbum argentarium*) had been interpreted by scholars to be either a mixture of tin and lead, or just tin. Rottländer has shown that both views are incorrect and that the term refers to lead itself.[1] When argentiferous galena is oxidized it forms litharge, leaving the silver as an unoxidized residue. The lead can be recovered by reduction with

1. See R. C. A. Rottländer, 'The Pliny Translation Group of Germany', in R. French, op. cit., pp. 11–19.

charcoal and does not so readily oxidize in air, so that it remains bright for a long time. Lead containing many traces of other elements is more easily oxidized and blackens. 'Bright lead' is a by-product, therefore, of the silver industry, hence lead ingots from Roman Britain were stamped *ex argent*.

The last substance to be dealt with in Pliny's account of minerals and metals was arguably the most important to the Romans. Iron was used in the manufacture of weapons and, increasingly under the Empire, in civil engineering. Pliny names the principal ores and distinguishes iron with a high carbon content suitable for forging cutting edges (*ferri nucleus*) from steel (*acies*). The analysis of archaeological finds, among them Roman scythes, confirms the existence of such iron and steel.

ART AND ARCHITECTURE

The sections of Books XXXIII–XXXVII devoted to art, architecture and related topics provide in many instances our earliest references to works and buildings that no longer exist, as well as a valuable insight into the techniques used by the craftsmen of the time.

In Book XXXIV Pliny describes the types of bronze used by Greek and Roman sculptors (6 ff.) and the method used in casting this alloy. The major sculptors and their works in this medium are listed. In Book XXXV he considers the controversial question of how painting originated, but his treatment is cursory and superficial. He declares the Egyptians' assertion that they invented painting 6,000 years before it passed to Greece to be untenable (XXXV, 15), but does not explain why. However, the outline that follows of Roman art from earliest times to his own day provides some useful historical information. In Book XXXVI Pliny turns to the different kinds of marble and sculptors who worked in this medium. Many statues existing today can be identified from his list, but many more are lost. The book also includes descriptions of such celebrated buildings as the Pyramids (XXXVI, 76 ff.), the Pharos at Alexandria (XXXVI, 83) and the Temple of Diana at Ephesus (XXXVI, 95 ff.). Not surprisingly buildings in Rome receive special treatment, among them theatres (XXXVI, 113 ff.) and aqueducts (XXXVI, 121 ff.).

These books contain a wealth of information, but reveal a characteristic absence of revision or logical exposition. There are numerous digressions, as, for example, concerning the origins of Roman coinage or the history of the *Ordo Equester* (XXXIII, 29 ff.).

LANGUAGE AND STYLE[1]

Pliny's language and style, which some scholars have unfairly criticized as being the worst found in any Latin author, were influenced by many factors, among them his military service, his experience as a procurator in several provinces and his rhetorical education. Stoicism also made an impact on his writing, a philosophy which, not surprisingly, appealed to a man of Pliny's character and ethos. Among Roman writers, he was particularly indebted to Lucretius, Virgil, Columella and Seneca the Younger.

Lucretius had been the first to complain of the poverty of the Latin language (*patrii sermonis egestas*) in regard to technical vocabulary, a problem also alluded to by Seneca. Pliny, however, is less specific but observes that the *Natural History* uses 'rustic or foreign terms – indeed barbarian words that have to be introduced with "if you'll pardon the expression"' (*Preface*, 13). Some Roman words adequately convey the meanings of foreign terms – *mundus* and *vita*, for example, can readily be used for *kosmos* and *bios* – but many precise equivalents are lacking. In the *Natural History* numerous expedients are adopted to overcome the problem, including the assimilation of transliterated Greek terms in the books on medicine and architecture, and the adoption of local words (Greek, Spanish and Celtic) in those on mining, metallurgy, mineralogy and agriculture. Existing words are used in a specialized, or technical sense beyond that of their everyday meaning, as *ferri nucleus* for high-carbon iron and *acies* for steel; occasionally, as in the description of lead-processing, where both Spanish and Greek words occur, two sets of terms are employed. Pliny also uses a number of interesting neologisms. Greek words are directly incorporated into the text (as in Cicero's *Letters*) in the same way that French and other foreign words appear in English. Lucretius, restricted by metrical considerations, often employs special archaic forms of words or periphrases to represent single nouns, whereas Pliny tends to use abstract nouns for concrete and often to create new substantives from the neuter plurals of adjectives. Some technical terms he shares with Varro, Columella and Seneca the Younger, but others are often exclusively his own invention. He also includes a wide range of adjectives and adverbs that he has coined himself.

*

1. See J. F. Healy, 'The Language of Pliny the Elder', *Filologia e Forme Letterarie*, vol. 4 (Urbino, 1988), pp. 1–24.

The main stylistic features of Silver Latin regularly appear in the *Natural History* and the influence of rhetoric is well-marked, sometimes giving rise to affectation and artificiality, as, for example, in the apostrophe of Cicero (VII, 116 f.): 'But what excuse could I have for not mentioning you, Marcus Tullius? Or by what mark of distinction can I proclaim your outstanding excellence? By what else than the most outstanding evidence of that whole nation's decree, selecting from your entire life only the achievements of your consulship?' But more often Pliny is to the point, cataloguing facts in a dry style with abrupt sentences that create a staccato effect. He is conscious of his own limitations, describing his talent (*Preface*, 12 f.) as 'exceedingly unremarkable' and his theme as 'somewhat light weight', an 'unpromising subject', for which an elevated style would be out of place. Brevity and compression, hallmarks of the writing of the age, are noteworthy features, as are the well-known epigrams, such as 'To live is to be awake' (*Preface*, 18) or 'God is man helping man' (VIII, 57). As a persistent critic of human pride, ambition and extravagance, freely commenting on man's frailty and his cruel impact on the world, Pliny resorts to keywords such as *avaritia* (greed), *luxuria* (extravagant life style) and *fortuna* (fortune). These are constantly personified, sometimes to great rhetorical effect. In one personification of fortune Pliny writes (II, 22):

Fortune alone is invoked, alone commended, alone accused and subjected to reproaches; deemed volatile and indeed, by most men, blind as well, wayward, capricious, fickle in her favours and favouring the unworthy. To her is debited all that is spent, and to her is credited all that is received; she alone fills both pages in the ledger of mortals' accounts; and we are so subject to chance that Chance herself takes the place of God.

As often, what Pliny lacks in elegance is compensated for by the passion and conviction of his tone.

Humour of a sardonic kind is to be found throughout the *Natural History*: wine from Surrentum is described as 'vintage vinegar', and, in reference to Hades, Pliny observes (II, 158): 'If there were any beings in the nether world, assuredly the tunnelling brought about by greed and luxury would have dug them up.' It is interesting to compare this with Lucretius' argument (III, 978 ff.): 'As for all those torments which are said to take place in the depths of Hell, they are actually present here and now, in our lives. There is no wretched Tantalus, who fears a rock suspended above him, as the myth relates ... there is no Tityos in Hell.' Sarcasm, anticipating that displayed by Tacitus, is to be found in such observations as 'The chief consolation for Nature's shortcomings in regard to man is that not even God can do all things' (II, 27); or 'Not even the fury of storms, however, closes the sea; pirates first compelled men by threat of death to rush into death and venture on the winter seas, but now avarice exercises the same compulsion' (II, 125).

Chief among Pliny's literary merits are an instinct for the startling story and a strong sense of atmosphere. In XIV, 142 he gives this telling and vividly drawn picture of the results of over-indulgence in wine-drinking:

Even in the most favourable circumstances the intoxicated never see the sunrise and so shorten their lives. This is the reason for pale faces, hanging jowls, sore eyes and trembling hands that spill the contents of full vessels; this the reason for swift retribution consisting of horrendous nightmares and for restless lust and pleasure in excess. The morning after, the breath reeks of the wine-jar and everything is forgotten — the memory is dead. This is what people call 'enjoying life'; but while other men daily lose their yesterdays, these people also lose their tomorrows.

The *Natural History* is peppered with such vivid scenes — compare, for example, Pliny's descriptions of St Elmo's fire on the lances of soldiers on guard (II, 101), the giant whale at Ostia (IX, 5), Agrippina's dress of pure gold (XXXIII, 63) — and Pliny's comments are refreshing for their bluntness. The reader is left in no doubt as to where he stands. It is perhaps appropriate that he should end the *Natural History* (XXXVII, 205) with a valediction to Nature, the force that seems so central to all that he writes: 'Greetings, Nature, mother of all creation, show me your favour in that I alone of Rome's citizens have praised you in all your aspects.'

PLINY IN THE MIDDLE AGES AND LATER[1]

As early as the third century AD Solinus assembled his *Collectanea Rerum Memorabilium*, an account of the regions of the world with the fables and marvels traditionally attributed to them. Although the material is taken from Pliny, the compilation is different and does not include any of that author's occasional scepticism. The *Collectanea* was very popular in Europe and presented a world of people with huge feet or dogs' heads, legendary animals like the

---

1. See generally G. Holmes, *The Oxford Illustrated History of Medieval Europe* (Oxford, 1988); F. A. Wright and T. A. Sinclair, *A History of Later Latin Literature from the Middle of the Fourth to the End of the Seventeenth Century* (London, 1931).

griffin and manticore, and magic stones. Books on the healing qualities of precious stones were common in Egypt and Syria and were used by Pliny, Dioscorides and Galen. The lore transmitted by Byzantine, Arab and Jewish sources came back to the West, in translation, in the twelfth and thirteenth centuries when Marbod of Rennes wrote his treatise *Liber Lapidum* (*On Precious Stones*) based on Solinus and Isidorus.

Some time after AD 370 Symmachus sent a copy of the *Natural History* to Ausonius (*Letters*, I, 24). In the fourth century learned familiarity with Plinian material was impressive and Ammianus Marcellinus relates a curious but telling incident involving Symmachus' father (XXVII, 3, 4). He was driven from Rome for an alleged remark to the effect that he would rather use his wine to make concrete than sell it at a discount to the people. This has rightly been taken as a specific reference to a passage in the *Natural History* (XXXVI, 181) where lime is slaked in wine. Ausonius' *Mosella* likewise contains various allusions that seem to have been derived from the *Natural History*. Roman authors who helped to transmit the more academic classical knowledge include Macrobius, Martianus Capella, Chalcidius and Isidorus of Seville. Macrobius' *Commentarii in Somnium Scipionis* (*Commentary on the Dream of Scipio*) and Capella's *De Nuptiis Philologii et Mercurii* (*On the Marriage of Philology and Mercury*) both include cosmography that helped to determine medieval ideas on the world. Like Pliny they were convinced that the world was a sphere and agreed on the zones (*climata*) into which it was divided – ideas originally derived from Plato and the Pythagoreans. Of ancient scholars only the Jews maintained a flat-earth theory.

Bede possessed about half the books of the *Natural History* and rated it a work of the highest merit. His *De Rerum Natura* relied heavily on Pliny's astronomical and meteorological material. Bede also adapted and corrected Pliny's work on tides, and referred to the evidence of *Natural History*, XXXVII, when discussing the various gems mentioned in the Book of Revelation. Bede helped to popularize the *Natural History* rather than to ensure its survival.

Manuscripts of the *Natural History* multiplied and in the ninth century were to be found at Corbie, St Denis, Lorches, Reichenau and Monte Cassino. Some of its books became an established part of monastic culture, being used for astronomy and medicine and as a source for illustrations for biblical commentaries – even for sermons.

During the eleventh and twelfth centuries schools increased in number, and this led to a revival of the study of little-known classical texts through the medium of Latin translations from Greek and Arabic. Pliny was read at Chartres, where John of Salisbury knew of the *Natural History*. At Oxford, Robert Crichlade produced the *Defloratio Historiae Naturalis Plinii Secundi* (1141), an abridgement of the *Natural History* in nine books, which omitted material no longer relevant. In addition, by careful deletions and the replacement of incompatible sections by ones of his own entitled *De errore ipsius Plinii*, he was able to give a Christian interpretation to Stoic beliefs.

In the Middle Ages encyclopedias were popular sources of instruction. Thomas of Cantimpré acknowledges his debt to Aristotle, Pliny and Solinus. Bartholomew of England also had an extensive first-hand knowledge of Pliny; he wrote his *De proprietatibus Rerum* for beginners and non-specialists, and it became a popular textbook. Vincent of Beauvais produced the most comprehensive medieval encyclopedia, his *Speculum Mundi* (in four parts, *c.* 1244) and drew on many sources. On natural phenomena, such as earthquakes, volcanoes, whirlwinds, thunder and lightning, Vincent drew on Aristotle's *De Caelo* and *Meteorologica*, Seneca's *Naturales Quaestiones* and Pliny's *Natural History*, as well as the encyclopedias of Isidorus and William of Conches. Botany was in its infancy and so Vincent in his account of this subject was further indebted to Pliny, as well as to Dioscorides Palladius, Rhazes and Avicenna. In spite of the spread of Arab medicine and translations of Galen, Pliny was none the less an important source to encyclopedias for pharmacological information, and was quoted at length by Vincent on the curative properties of minerals, animal products and plants.

Pliny's astronomy enjoyed its highest reputation in the early Middle Ages, since its heliodynamic theme gave it a sense of coherence. The work, however, of William of Conches (*c.* 1128), with its elemental theory eventually replaced the *Natural History*. Nine commentaries primarily or solely on Book II were published between *c.* 1480 and 1556.

An interesting light is thrown on the state of Pliny's text by Petrarch who complained about a codex of the *Natural History* purchased in Mantua on 6 July 1350. 'What', he wrote, 'would Cicero, or Livy, or the other great men of the past, Pliny above all, think if they could return to life and read their own books?' He answered his question with the surmise that they would scarcely recognize these corrupt and barbarous texts as theirs. Petrarch began work on emending and annotating the text of the *Natural History* in an attempt to sort out the disorder that had been introduced into a number of manuscripts of the ninth to the eleventh centuries.

Certainly, by the end of the fifteen century the whole of the *Natural History* had come to be greatly admired by scholars. It was one of the earliest Latin texts to be printed; the first edition appeared in Venice in 1469, offering, however, a distinctly imperfect text. The most eminent early scholars to work on the recension of the text were Hermolaos Barbarus, who wrote his *Castigationes Plinii* in 1492–3, claiming to have corrected some 5,000 textual errors, and Beatus Rhenanus, who corrected the text for Erasmus' edition published in 1525.

The great challenge to Pliny came with Leoniceno's publication of his *De Erroribus Plinii* in 1509. Leoniceno was a fervent Hellenist, and so not surprisingly, Pliny fell short of his ideal in a number of ways. Pliny was not Greek, and he did not write in Greek, although he was able to read the Greek sources on which much of the *Natural History* was based. Leoniceno attacked Pliny for having no adequate scientific method, unlike the older and more authoritative Theophrastus and Dioscorides. In particular, he asserted, Pliny had no special knowledge of philosophy and medicine – essential components of Leoniceno's Hellenist outlook. Leoniceno, a physician by profession, could hardly have been sympathetic to Pliny in the light of his scathing criticisms of Greek doctors and Greek medicine. He similarly attacked Alessandro Benedetti, Pliny's 'medical patron'. By contrast, Theophrastus and Galen were upheld as models.

In spite of his obvious shortcomings, however, Pliny was not without his defenders. Collenuccio appointed himself Pliny's 'patron' or advocate. 'Even in cases of confessed crime,' he asserted, 'the defendant escapes punishment if he is sufficiently useful to, or a great ornament of society.' All who defended Pliny saw him as such an ornament of 'Latin culture' (*Res Latina* – a term reflecting a reaction against the *Graecitas* of the Hellenists).

CONCLUSION

The *Natural History* received a new lease of life in 1601 when Philemon Holland published his translation, and it has continued to influence literature down to the present time. It is a tribute in itself that Pliny's great work has outlived the compilations of such near contemporaries as Varro and Verrius Flaccus, and is far better known today than some of the most popular medieval encyclopedias. But perhaps the most appropriate *epilogos* is that written by his nephew Pliny the Younger in a letter to the historian Tacitus (*Letters*, VI, 16, 3):

'The fortunate man ... is he to whom the gods have granted the power either to do something which is worth recording, or to write what is worth reading, and most fortunate of all is the man who can do both. Such a man was my uncle, as his own books and yours will prove.

FURTHER READING

WORKS OF GENERAL INTEREST

JOHN BOARDMAN, JASPER GRIFFIN and OSWYN MURRAY, eds., *The Oxford History of the Classical World* (Oxford, 1986)

MICHAEL GRANT and RACHEL KITZINGER, eds., *Civilization of the Ancient Mediterranean*, 3 vols. (New York, 1988)

S. HORNBLOWER and A. SPAWFORTH eds., *The Oxford Classical Dictionary* (Oxford, 2003)

G. HOLMES, ed., *The Oxford Illustrated History of Medieval Europe* (Oxford, 1988)

— ed., *Oxford Illustrated History of the Classical World* (Oxford, 1988)

WORKS ON SPECIAL TOPICS

MARY BEAGON, *Roman Nature. The Thought of Pliny the Elder* (Oxford, 1992)

A. LE BOEUFFLE, *Le Ciel des Romains* (Paris, 1989)

SORCHA CAREY, *Pliny's Catalogue of Culture; Art and Empire in the Natural History* (Oxford, 2003)

M. H. CRAWFORD, *Roman Republican Coinage* (Cambridge, 1974)

D. R. DICKS, *Early Greek Astronomy to Aristotle* (London, 1970)
The Geographical Fragments of Hipparchus (London, 1960)

I. EDELSTEIN, Ancient Medicine (Baltimore, 1967)

R. FRENCH, *Ancient Natural History* (London, 1994)

— ed., *Science in the Early Roman Empire* (London, 1986)

MOTT T. GREENE, *Natural Knowledge in Preclassical Antiquity* (Baltimore-London, 1992)

J. F. HEALY, *Mining and Metallurgy in the Greek and Roman World* (London, 1978)

J. F. HEALY, *Pliny the Elder on Science and Technology* (Oxford, 1999)

— *Miniere e Metallurgia nel Mondo Greco e Romano* (Rome, 1994) revised and enlarged edition with colour illustrations

G. ISAGER, *Pliny on Art and Society. The Elder Pliny's Chapters on the History of Art* (London, 1991)

J. G. LANDELS, *Engineering in the Ancient World* (London, 1978)

G. E. R. LLOYD, *Early Greek Science: Thales to Aristotle* (London, 1970)
Greek Science after Aristotle (London, 1973)

F. DE OLIVIERA, *Les Idées Politiques et Morales de Pline l'Ancien* (Coimbra, 1992)

E. D. PHILIPS, *Greek Medicine* (London, 1978)

A. RADL, *Der Magnetstein in der Antike* (Stuttgart, 1988)

J. SCARBOROUGH, *Roman Medicine* (London, 1969)

W. D. SMITH, *The Hippocratic Tradition* (London, 1979)

R. SYME, Tacitus (Oxford, 1957)

W. W. TARN and G. T. GRIFFITH, *Hellenistic Civilisation* (London, 1966)

J. M. C. TOYNBEE, *Animals in Roman Life and Art* (London, 1970)

K. D. WHITE, *Roman Farming* (London, 1970)

F. A. WRIGHT and T. A. SINCLAIR, *A History of Later Latin Literature from the Middle of the Fourth to the End of the Seventeenth Century* (London, 1931)

ARTICLES

H. L. BONNIEC, 'Bibliographie de l'histoire naturelle de Pline l'Ancien', *Collections d'études Latines*, vol. 1 (1946), *Série scientifique*, p. xxi

P. M. GREEN, 'Prolegomena to the Study of Magic and Superstition in the Natural History of Pliny the Elder' (doctoral thesis, Cambridge, 1952)

R. HALLEUX, 'Les deux métallurgies du plomb-argentifère dans l'histoire naturelle de Pline', *Revue de Philologie*, vol. 49 (1975), pp. 72–88

J. F. HEALY, 'Pliny the Elder and Ancient Mineralogy', *Inter-disciplinary Science Review*, vol. 6, no. 2 (June 1981), pp. 166–180
'Mines and Quarries', in Michael Grant and Rachel Kitzinger, eds., *Civilization of the Ancient Mediterranean: Greece and Rome* (New York, 1988), vol. 2, pp. 779–93.

A. LOCHER, 'Plinius der Ältere uber das Eisen', *Archäologisches Eisenhuttenwes*, vol. 51, pp. 487–92

E. PAPARAZZO, 'Organic substances at Metal Surfaces: Archaeological Evidences and the Elder Pliny's account' (*Archaeometry*, 2003)

R. C. A. ROTTLÄNDER, 'Glasherstellung bei Plinius dem Älteren', *Glastechnische Berichte*, vol. 52 (1979), pp. 265–70

J. SCARBOROUGH, 'The Preface of Dioscorides' Materia Medica: Introduction, Translation, Commentary', *Transactions and Studies of the College of Physicians of Philadelphia*, vol. 4, no. 3 (September 1982), pp. 187–227

J. STANNARD, 'Roman Botany', *Isis*, vol. 56 (1965), pp. 420–25

O. VITTORI, 'Pliny the Elder on Gilding', Endeavour, vol. 3, no. 3 (1979), pp. 128–31

A. WALLACE-HADRILL, 'Pliny the Elder and Man's Unnatural History' (*Greece and Rome*, NS 37, 1990), pp. 80–96

C. ZIRKLE, 'The Death of Gaius Plinius Secundus', *Isis*, vol. 58 (1967), pp. 553 ff.

TRANSLATOR'S NOTE

The first duty of any translator is to preface his work with an *apologia* or, at least, statement of intent.

The following translation is based on the text of Karl Mayhoff (1905 and 1909) with emendations derived from modern linguistic research and from data obtained by scientific experiment carried out in the hope of elucidating problems that occur in Pliny's sections on minerals and metals.

In presenting a selection of major topics from the *Natural History* which, it is hoped, will be of interest to a wide readership, I have tried to maintain a balance between Pliny's record of Greek and Roman science and his observations of Roman life and mores under the Empire. This has necessitated occasional modifications, or abridgements, as in his account of astronomy, when the explanatory commentary would greatly have exceeded the volume of text translated. Similarly I have 'cut the Gordian knot' by leaving aside whole books where the subject-matter is highly specialized. In the interest, however, of providing a continuous text, omissions within books are not generally indicated.

The encyclopedic *Natural History*, more perhaps than any other Latin prose work, presents exceptional difficulties for the translator: one should, however, in fairness to Pliny, add that these are not always of his own making! Foremost is the fact that the text was unrevised at the time of his death in AD 79 and did not have the benefit of an expert reader or vigilant copy-editor. As a result the narrative is often disjointed, discontinuous and not in a logical order: it sometimes appears to be a collection of unassimilated notes and arguments that are incomplete or circular.

Pliny's technical vocabulary represents a 'language within a language', with highly specialized meanings not found in literary Latin. In addition there are borrowings from Greek, Celtic and Spanish that have added to the confusion of some commentators. In my translation I have used modern terminology, if appropriate: for example, although the science of crystallography was not established until the nineteenth century, Pliny correctly describes the *crystal-systems* of the diamond, beryl and quartz, although he does not understand the principles involved; he refers to the *precipitation* of crystals in an *evaporating solution*. Errors in transmission of the unrevised text by generations of copyists have been compounded by emendations put forward by editors with inadequate scientific expertise. But, arguably, the greatest difficulty in translating Pliny stems from his epigrammatic style, his rhetorical training and his apparent obsession with an elliptical brevity that can leave the reader confused as to the direction and even the nature of his argument.

Finally some practical points. Place-names retain their Roman form except where a modern name is more generally accepted – for example Athens, Syracuse and the Black Sea. Dates are given as Olympiads, consular years or AUC (from the foundation of Rome) in the text, and as BC or AD in the footnotes. Weights and measures are imperial rather than metric on the grounds that 1 foot, for example, seems more meaningful than 0.305 m.

I would like to record my thanks to the many colleagues mentioned in this work, in particular to Professor J. B. Hall and Mr J. M. Carter for their valuable advice on many occasions, and to my wife, Barbara – on Pliny's behalf, for her patience and, on my own, for helpful suggestions in respect of a number of problems in translation. I am also indebted to Mrs Andrée Morcom for typing much of the manuscript.

My special gratitude is due to Mr Charles Drazin and Mr Anthony Turner for their general oversight of a very difficult manuscript: I have greatly benefited from their expertise, vigilance and considered opinions relating to many aspects of this work. Any errors or infelicities that remain are solely my own.

JOHN F. HEALY
ROYAL HOLLOWAY AND
BEDFORD NEW COLLEGE, 1990

I am grateful to Helen Conford for her invaluable help in seeing this revised edition through the press. It is updated in the light of recent research on the *Natural History* and contains additional bibliographical material for Further Reading.

JOHN F. HEALY
MACCLESFIELD, 2004

Natural History

A Selection

PREFACE

The Preface to the Natural History, in the form of a covering letter addressed to the Emperor Titus, provides an introduction to Pliny's encyclopedic work which embraces all aspects of Nature and natural science

Pliny the Elder sends greetings to his friend the Emperor Titus. Most august ruler, let this title – a very true one – be yours, while your father continues to enjoy the title of 'most distinguished' to a ripe old age. I have decided to tell you, in this rather presumptuous letter, about my most recent production, my books of *Natural History*, a novel venture for the Muses who inspire your Roman citizens. 'For you used to reckon my little efforts of some worth' – to soften, *en passant*, my 'oppo' Catullus (you recognize this army slang); for, as you know, by changing round the first syllables of line 4 of his first poem, he made himself a little harsher than he wished his 'good chums the Veraniuses and Fabulluses'[1] to think him.

2–4. At the same time I am doing this so that my impertinence may achieve what another presumptuous letter of mine almost failed to do, so you complain, namely some concrete result, and so that everyone may know how the Empire lives with you on equal terms. You have a triumph to your credit, you are a censor and have been consul six times, you share tribunician power and are commander of your father's praetorian guard: all this belongs to your political life. And what a splendid comrade you were when we were on active service together! Nor has the extent of your good fortune brought about any changes in you, except to make it possible for you to distribute such benefits as you wish. And so, although all ways of paying you respect are open to the rest of mankind, for me there remains only the presumption of showing respect on a more intimate basis. You will take the blame for this and forgive yourself for my offence.

1. Both Veranius and Fabullus served in Spain in 57–56 BC. See Catullus, *Poems*, XII, XVI and IX, 1 ff.

5. The supreme power of your rhetoric, the tribune-like authority of your wit, never shone more truly in any other man. With what eloquence you praise your father, with what eloquence, your brother's renown! How you excel in the poet's art! O great and fertile genius, you have devised a way of imitating your brother also!

6. But who could review these works without a qualm, when about to submit to judgement by your talent, especially when this has been solicited? Formally dedicating the *Natural History* to you puts one in a different situation from merely publishing the work. In the latter case I could have said, 'Why do *you* read this, my Emperor? It is written for the masses, for the horde of farmers and artisans and, finally, for those who have time to devote to these pursuits. Why do you appoint yourself a juror? You were not on the roll when I undertook this work.'

12. I have compounded my presumption by dedicating these books to you – a somewhat light-weight work and one not abounding in talent – which, in my case, is exceedingly unremarkable. There is no place for digressions, speeches, discourses, miraculous happenings, or *faits divers* – although such matters might have been pleasant for me to recount, and a source of entertainment for my reader.

13. My subject, the natural world, or life (that is life in its most basic aspects), is a barren one and, in very many instances, employs rustic or foreign terms – indeed barbarian words that have to be introduced with 'if you'll pardon the expression'.

14. Moreover, my path is not well-worn by writers, nor the kind along which the mind aspires to wander. No Roman author has attempted the same project, nor has any Greek treated all these matters single-handed. Many of us seek pleasant fields for research, while others deal with matters of immense complexity where one is overwhelmed and cannot see the wood for the trees. First and foremost I must deal with subjects that are part of what the Greeks term an 'all-round education',[1] but which are unknown or have been rendered obscure by scholarship.

15. Yet, other subjects have been so over-exposed by publication as to become boring. For it is difficult to give a new look to things that are old hat, an air of authority to what is novel, lustre to what is *passé*, light to the obscure, acceptability to things that arouse aversion, credibility to matters open to question – and indeed to

1. *enkyklios paideia.*

4

give to all things Nature, and to Nature herself, all her intrinsic qualities. And so, even if I have not succeeded, willingness to make the attempt constitutes an excellent and high-minded endeavour.

16. For my part, I think that those people have their own special place in learning who prefer the useful service of providing help in overcoming difficulties to the popularity that accrues from pleasing. I have followed this practice myself in my other works and confess that I am surprised at Livy, a most renowned author, when he begins one book of his *History of Rome from the Foundation of the City* as follows: 'I have gained enough fame already and I might have settled down to retirement, were it not for my restless mind which feeds on work.' Assuredly he ought to have written his history for the glory of the Roman people, conqueror of nations, and for the Romans' reputation, not for his own glory. It would have been more meritorious to have persevered because of love of the work rather than for his own peace of mind, and better to have done this for the Roman people than merely for himself.

17. In the words of Domitius Piso, we need works of reference, not books. In the thirty-seven books of my *Natural History*, I have, therefore, included the material derived from reading 2,000 volumes, very few of which scholars ever handle because of the recondite nature of their contents – some 20,000 facts worthy of note, from 100 authors whom I have researched. To these I have added very many facts that my predecessors did not know or that I have subsequently discovered from my own personal experience.

18. I do not doubt that many facts have eluded me. For I am only human and busy with official duties. I pursue my research in odd hours, that is at night – just in case any of you think I pack up work then! The days I devote to you and I reckon up the sleep I need consistent with keeping well. I am content with this reward alone, in that while – in Varro's words – I occupy myself with these trifles, I prolong my life by many hours.

19. For assuredly to live is to be awake. For these reasons and difficulties I do not dare to make any promises: you furnish the very words I write to you. This is the guarantee of my work and an indication of its value. Many objects are regarded as very valuable because they have been dedicated to a temple.

20. I have written about all of you, about your father, your brother and yourself, in an appropriate work, *The History of Our Times*, which begins where the history of Aufidius Bassus leaves off. 'Where is this work?' you will ask. It was finished long since and is sacrosanct. I had resolved to entrust it to my heir so that no one

might think that my life was dedicated to ambition. So I am well-disposed towards those who seize this vacant position and indeed towards our descendants who, I know, will challenge us in a fight to the finish as we ourselves have similarly challenged our predecessors.

21. You will count as proof of my professionalism the fact that I have prefaced these books with the names of my authorities. In my opinion such acknowledgement of those who have contributed to one's success – unlike the practice of most of the authors I have mentioned – is a not ungracious gesture and abounds with honourable modesty.

22. For you ought to know that when I compared authorities, I found that writers of bygone times had been copied by the most reliable and modern authors, word for word, without acknowledgement. Nor was this done in that well-known Virgilian spirit of rivalry, or in that artless manner of Cicero, who declares himself a companion of Plato, and adds, in his *Consolation* to his daughter, 'I follow Crantor.'[1] It is the same with Panaetius in his *On Duties* – volumes that should be learned by heart and not merely held in one's hands every day.

23. Surely it is characteristic of a mean spirit and of an unfortunate attitude to prefer to be caught committing a theft rather than to repay a loan, especially as capital accumulates from interest. The Greeks have a wonderful flair for inventing titles for books. They called one *Honeycomb*; others called their works *Violets, Muses, Encyclopedias, Manuals, Meadow, Plague, Impromptu* – eye-catching titles that might cause a man to jump bail. But when you dip inside what emptiness awaits you!

26. For my part I am not ashamed of failing to dream up a more arresting title.

28. I openly confess that I could have added many things to my books, and not only to my *Natural History*, but to all my published works, so that, *en passant*, I might be on my guard against those 'Scourges of Homer' that you mention. For I hear that the Stoics, members of the Academy and Epicureans – I have always anticipated this from philologists – are in labour producing a reply to the books I have published on grammar. Indeed during the past decade they

---

1. A philosopher of the Old Academy from Soli, Cilicia (*c.* 335 – *c.* 275 BC), whose commentary on Plato's *Timaeus* was the first in a long line of ancient commentaries on the Platonic dialogues. His work *On Grief* was much admired by later writers, among them Cicero.

have suffered a series of miscarriages, although even elephants give birth more quickly.

29. As if I did not know that a woman had written a book against Theophrastus, who was so outstanding as an orator that he acquired the epithet 'out of this world'. This was the origin of the expression 'hoist with one's own petard'.

30. I cannot resist quoting the very words of Cato the Censor, which are relevant here, to show that his monograph *On Military Discipline* found critics ready to attack it. These were the sort of critics who try to gain prestige for themselves by disparaging another person's expertise. Yet Cato learned his military skill under Scipio Africanus,[1] or rather under him and Hannibal as well, and yielded nothing even to Africanus when the latter won a triumph as supreme commander. What is the upshot of all this? Cato says in the aforementioned work: 'I know that if certain writings are published, there will be many to quarrel and pick faults in an unseemly manner, but these are mostly persons who lack any real distinction. For my part I have allowed their eloquence to go in one ear and out of the other.'

31. When told that Asinius Pollio[2] was composing declamations against him, to be published by himself or his children after his death (so that he would not be able to reply), Plancus said, not inelegantly: 'Only ghosts fight with the dead!' This retort had such an effect on those declamations that in learned circles nothing was judged more shameful.

32. So, secure from such nit-pickers (an expressive compound invented by Cato – for what else do these people who enter into disputes do except pick quarrels over trifles?), I shall carry on with the rest of my proposed plan.

33. Since it is in the public interest to save your time, I have appended to this letter a list of contents of the books of my *Natural History* and have taken the greatest care to prevent your having to read them from cover to cover. By the same token you will ensure that others do not need to peruse the whole work, but only have to look for whatever each needs, and they will know in what place to find it.

---

1. Scipio defeated Hannibal at Zama (202 BC), thus bringing the Second Punic War to an end and also Hannibal's dream of establishing a Carthaginian empire.

2. Asinius Pollio (76 BC – AD 4) supported Caesar in 45 BC but later joined Antony. He saved Virgil's estate from confiscation (41 BC). From the booty derived from his victory over the Parthini of Illyria (39 BC), Pollio built the first public library in Rome. Thereafter he devoted himself to literature.

THE UNIVERSE
AND THE WORLD

BOOK II
ASTRONOMY

Pliny begins his Natural History with an account of the world and the heavens

1. The world and this expanse – or whatever other name men are pleased to call the sky that covers the universe with its vault – are properly held to be a deity,[1] everlasting, boundless, an entity without a beginning and one that will never end. Men are not concerned to explore what is extraterrestrial, nor can the human mind make a guess about such things.

2. The world is sacred, eternal, boundless, self-contained, or, one should say, complete in itself, finite yet resembling the infinite, of all things certain yet resembling the uncertain, embracing in its grasp all things without and within. The world is the work of Nature and, at the same time, the embodiment of Nature herself.

3. It is madness that certain men have occupied their minds with measuring the world and have dared to publish their results. Again it is madness that others,[2] seizing the opportunity, or following the lead of those just mentioned, have propounded the existence of countless other worlds so that one must believe in a matching number of systems, or if one system controls all worlds, the same number of suns, moons and other innumerable heavenly bodies as there are already in a single world.

4. It is madness, absolute madness, I repeat, to go out of our world to examine what lies beyond as if our knowledge of all within it were perfect. As if, indeed, man, who does not know the measure of himself, could measure anything else, or the mind

1. This is in accordance with Pythagorean and Stoic doctrine.
2. Namely Leucippus (second half of the fifth century BC) and Democritus of Abdera. The latter postulated an 'atomic theory' in which atoms were absolutely indivisible units. The atoms differ from each other in size and shape only, and the only change they undergo is in their relative and absolute position, through movement in space. By their changes in position the atoms produce compounds of the changing sensible world. Compounds differ in quality according to the shape and arrangement of the component atoms, their congruence or otherwise, and the amount of space between them.

of man could see things that the world itself does not possess.

5. The world appears round like a perfect sphere. This is demonstrated in the first place by the use of the word 'sphere', the term generally employed to describe it. The evidence of the facts also confirms this.

6. The rising and setting of the sun leave us in no doubt that the world is this shape and that it revolves eternally, without rest and at an indescribable speed, each revolution taking twenty-four hours.

The Kosmos, sky and signs of the Zodiac

8. The Greeks have called the world by a name that means 'ornament' while we have given it the name *mundus*, because of its perfection and consummate grace. The word for sky is undoubtedly derived from the term 'engraved',[1] as Marcus Varro explains.

9. Its orderly structure supports this view – its circle, called the Zodiac, is divided up into the likenesses of twelve animals, and the sun has made consistent progress through these signs over many centuries.

The four elements, planets and sun

10. I observe that there is no doubt about there being four elements. The uppermost is fire, source of those eyes of the great array of blazing stars. The next element is a vapour which the Greeks and ourselves call by the same name, 'air'. This is the principle of life that permeates every part of the universe and is intermingled with the whole. Suspended by its force the earth is balanced in the middle of space, together with the fourth element – its waters.

12. The seven stars, which we call 'planets'[2] because of their movement (though no stars wander less), hang suspended between heaven and earth by this same air and are separated by fixed intervals. The sun, the ruler not only of the seasons and of the lands, but even of the stars themselves and the sky, moves in the midst of the planets: the sun is greatest in size and power.

13. When we consider his functions we must believe that the sun is the soul, or, more intelligibly, the mind of the universe, the ruling principle and divinity of Nature. The sun provides the world with light and takes away darkness: he blacks out and lights up the

1. Pliny confuses the word *caelum* with *caelare*, 'to engrave'.
2. Derived from the Greek verb *planasthai*, 'to wander'.

THE UNIVERSE AND THE WORLD

rest of the stars. The sun controls the change of the seasons and the continual regeneration of the year, following Nature's practice. He dispels the gloomy aspect of heaven and lightens the clouds over men's minds. He lends his light also to the rest of the stars, is splendid, supreme and sees and hears everything; I observe that Homer, the most distinguished person in the field of literature, held this view about the sun's function.

The search for God

14. I think it a sign of human weakness to try to find out the shape and form of God. Whoever God is — provided he does exist — and in whatever region he is, God is the complete embodiment of sense, sight, hearing, soul, mind and of himself. To believe in either an infinite number of deities corresponding to men's vices, as well as their virtues — like the goddesses of Chastity, Concord, Intelligence, Hope, Honour, Mercy, Faith — or, like Democritus, in only two, namely Punishment and Reward, plumbs an even greater depth of foolishness.

15. Frail, toiling men, mindful of their own weakness, have separated these deities into groups in order to worship them piecemeal — each person worshipping the deity he most needs. Thus different races have different names for the same deities, and we find innumerable gods in the same races. Even the gods of the lower world, together with diseases and many kinds of plagues are listed in groups in our fearful anxiety to appease them.

16. For this reason there is a Temple of Fever, on the Palatine, dedicated by the state, one of Bereavement, at the Temple of the Household Gods, and an Altar of Bad Luck, on the Esquiline. One could take this to mean that there are more heavenly beings than men, since individuals also make as many gods, by adopting Junos and Genii for themselves. Some nations have animals — even repulsive creatures — as gods, and many things more disgraceful to relate; they swear by rotten food and other such things.

17. Indeed to believe in marriages between gods and goddesses and that, over so long a period of time, no children resulted from these unions; to believe that some gods are always old and greyhaired, while others are young men and boys; to believe in gods who are black, winged, lame, born from eggs, or who live and die on alternate days — such beliefs are little short of the fantasies of children. But the invention of adulterous acts between gods and goddesses themselves, as well as quarrels and hatred, and the invention of gods of theft and crime, surpasses all shamelessness.

18. God is man helping man: this is the way to everlasting glory. This is the road our Roman leaders took, the road by which Vespasian, the greatest ruler of all time, journeys with heavenly step and, accompanied by his children, comes to the rescue of a tired world.

19. The apotheosis of such men is the oldest method of rewarding them for their good deeds. Who would not admit that men call one another Jupiter or Mercury or other names according to their character, and that this practice is the origin of heaven's nomenclature?

20. Moreover, it is ridiculous to think that a supreme being – whatever it is – cares about human affairs. Don't we believe that it would be defiled by so gloomy and complex a responsibility? Can we doubt this? It is hardly a useful exercise to judge what is more profitable to mankind when some men have no respect for the gods, while the regard shown by others is shameful.

21. They wait upon gods with foreign rituals, they wear their images on their fingers; they pass sentence on the monsters they worship and invent food for them; they inflict dire tyrannies on themselves, resting only fitfully even when asleep; they do not make decisions about marriages, about having children, or any other matter, unless instructed by sacrifices. Some swear false oaths on the Capitol and perjure themselves by Jupiter Tonans and profit from their crimes, while others are punished for attending to their sacrifices.

Fortune

22. Men have devised for themselves a deity half-way between these two concepts so that our conjectures about God are even less clear. Throughout the whole world, in all places and at all times, Fortune alone is invoked, alone commended, alone accused and subjected to reproaches; deemed volatile and indeed, by most men, blind as well, wayward, capricious, fickle in her favours and favouring the unworthy. To her is debited all that is spent, and to her is credited all that is received; she alone fills both pages in the ledger of mortals' accounts; and we are so subject to chance that Chance herself takes the place of God; she proves that God is uncertain.

23. Some men banish even Fortune, and attribute happenings to their star and to the laws governing birth; these people believe that God has made his decree – once only – for all men who will be born, and has been granted a life of ease for the rest of time. This belief is beginning to become established and the learned and the uneducated masses alike make a bee-line for it.

24. Look at the warnings inferred from lightning, the prognostications of oracles, the prophecies made by soothsayers, and even minor things rated as omens – a sneeze and missing one's footing. The late Emperor Augustus told how, on the day on which he was almost removed from power by a military coup, he had put his left boot on his right foot.

25. These various happenings catch men unawares so that, among these things, this alone is certain, namely that there is no such thing as certainty, and that nothing is more wretched or more conceited than man. Indeed the remainder of living creatures have food as their only anxiety, a department in which Nature's largesse is itself sufficient. And the good thing preferable to all others is the fact that these creatures do not think about glory, money, ambition nor, above all, about death.

The power of the gods

26. The belief that in these matters the gods show concern in human affairs corresponds with life's experience: punishment for wrongdoing, although sometimes long after the event – for God is busy with such a mass of things – nevertheless comes without fail. Man was not born closely related to God to be worthless and no better than wild beasts.

27. The chief consolation for Nature's shortcomings in regard to man is that not even God can do all things. For he cannot, even if he should so wish, commit suicide, which is the greatest advantage he has given man among all the great drawbacks of life. God cannot give mortals the gift of everlasting life, or recall the dead, or cause a man who has lived not to have lived, or someone who has held office not to have held it. He has no power with respect to the past, except to forget it, and, to underline our association with God by facetious arguments, he cannot make twice ten other than twenty, or achieve many similar things. These facts show without a shadow of doubt the power of Nature and prove that this is what we mean by 'God'. This digression will not have been inappropriate, since these matters have already been widely publicized through our continuous examination of the nature of God.

The planets

28. Let us retrace our steps to other matters concerning Nature. The

stars, as I have already stated, are attached to the world and not, as the man in the street thinks, assigned to each of us, shining in proportion to our individual lot, brightly for the rich, less brightly for the poor, and dimly for those who are weak. The stars do not come into being with the birth of human 'protégés', nor when they fall do they signify someone's death.

29. There is no such close association between the sky and ourselves so that the bright light of the stars up there is also mortal, as we are fated to be.

32. The following points have been established beyond doubt: the planet called Saturn is the highest and, for this reason, appears to be the smallest, orbiting in the largest circle and returning to its starting-point in thirty years at the shortest. All the other planets – among them the sun and the moon – orbit in the opposite direction to the earth, that is, to the left, while the earth always moves to the right.

33. Although they are borne on by it and carried westwards with a continuous revolution of immeasurable velocity, the planets travel with opposite motion along their respective paths. So the air does not become rolled up into a sluggish ball through revolving in the same direction as a result of the continuous rotation of the earth, but is scattered into separate parts by the head-on impact of the planets.

34. Saturn is naturally cold and frozen. The orbit of Jupiter is much below it and so revolves with faster motion – once every twelve years. The third planet is Mars, which some people call Hercules. Since it is close to the sun it blazes with a fiery glow; it revolves once about every two years. Because of Mars' excessive heat and Saturn's cold, Jupiter, which is situated between them, is influenced by both and consequently is healthy.

35. The sun's course is divided into 360 degrees but so that an observation of its shadows may coincide with its starting-point, five days are added each year and an intercalary day to every fourth year to make our measurement of time agree with the sun's orbit.

36. Below the sun revolves a large planet called Venus, with alternate orbits, and indeed its alternative names confirm its rivalry with the sun and moon. When it comes in advance and rises before dawn, it is called Lucifer, as being another sun and bringing the dawn. When, on the other hand, it shines after sunset, it is called Vesper, as prolonging the daylight or performing the function of the moon.

37. Pythagoras of Samos[1] discovered the nature of Venus in about the 42nd Olympiad – 142 years after the foundation of Rome.[2] Venus is greater in size than all other planets and is so bright that it is the only planet to cast a shadow by its rays.

39. The planet next to Venus is Mercury, called by some Apollo; its orbit is similar, but its size and power are by no means similar.

41–3. The last planet, which surpasses everyone's wonder, the one most familiar to earth and devised to remedy darkness, is the moon. Her many faces have greatly puzzled the intelligence of those who observe her and are ashamed that the planet closest to earth is the one about which we are most ignorant. It is always waxing or waning, and now is curved into the shape of a sickle-blade, now divided in half, now fully circular, spotted, then suddenly shining clear, vast and full, then suddenly nothing. At one time the moon shines through the night, at another it rises late and, for part of the day, helps the light of the sun, eclipsed yet visible during the eclipse, hidden at the end of the month when it is not believed to be in eclipse. At one moment it is low in the sky, at another high – but not with any consistency, sometimes raised to the sky, sometimes touching the mountain-tops, now carried up to the north, now down to the south. Endymion[3] was the first man to observe these individual facts about the moon. This was the reason for the tradition about his love for it. But we show no gratitude to those whose toil and concern have thrown light on this source of illumination. Yet, thanks to a strange affliction of the human mind, it pleases us to record bloodshed and slaughter in our historical records so that those who have no knowledge of the world itself may know about the crimes of mankind.

44–5. The moon is nearest to the pole and, for this reason, has the shortest orbit, completing the same distance every twenty-seven and one-third days that Saturn, the highest planet, covers in thirty years,

1. Pythagoras (c. 531 BC), born on Samos, from which he emigrated to Croton, is best known as a mathematician and astronomer. His cosmology was a development of that of the Milesians, treating the pair 'Limit–Unlimited' as the primary member of a group of ten pairs which they regarded as first principles. They said that the universe was produced by the First Unit (Heaven) inhaling the Infinite (Void) so as to form groups of units or numbers, and that all things – for example, even opinion, opportunity and injustice – were numbers and had cosmic position. Pythagoras also propounded the hypothesis that the earth is spherical. He established the basis of musical intervals.

2. 612–609 BC.

3. Endymion was a beautiful young man loved by Moon. In the Elean version of the story she bore him fifty daughters, that is, the fifty months of the Olympiad.

as has been stated. Then the new moon delays two days in conjunction with the sun and, after the thirtieth day at the latest, sets out on the same course. The moon is arguably our teacher of all that can be discerned in the sky, for example, that the year should be split into twelve monthly intervals, because she herself follows the sun this number of times as he returns to his starting-point. She, like the rest of the planets, is ruled by the sun's brilliance, in that she shines with a light completely borrowed from the sun, like the light we see flickering when it strikes water; in consequence she causes water to evaporate only with a rather gentle and incomplete force and even increases the volume of water, while the rays of the sun diminish it. The moon is seen to shine with variable intensity because she is only full when facing the sun, and on other days shows to the earth only as much light from herself as she receives from the sun.

46. When in conjunction with the sun, the moon cannot be seen, because, turned towards him, she gives back the whole 'draught' of light to its source. The planets are, undoubtedly, nourished by water from the earth. The moon's eclipses and the sun's, the most breath-taking and wondrous phenomena in the whole of our contemplation of Nature, are indications of their magnitude and shadow. It is clear that the sun is hidden by the transit of the moon, and the moon by the earth coming between it and the sun.

51. The eclipse of the moon is unquestionable proof of the size of the sun, just as the sun itself, in eclipse, proves the small size of the earth.

Eclipses of the moon and sun: the researches of Sulpicius Gallus, Thales and Hipparchus

53. The first Roman to put forward an explanation for eclipses of the moon and sun was Sulpicius Gallus,[1] who was a military tribune at the time, but subsequently consul with Marcus Marcellus. On the day before King Perseus was defeated by Aemilius Paulus, Gallus was brought before the assembled troops by their commander to give prior warning of an eclipse and thereby dispelled the army's anxiety. Subsequently Gallus wrote a book on eclipses. Among the Greeks, however, Thales[2] of Miletus was the first to examine

---

1. Gallus predicted a lunar eclipse on the eve of the Battle of Pydna (168 BC), when Lucius Aemilius Paulus defeated King Perseus. The Roman army successfully countered the tactics of the Macedonian phalanx.

2. Thales, one of the Seven Sages, who came from Miletus, was expert in economics, engineering, geography, astronomy and mathematics. According to tradition he foretold, to within a year, the solar eclipse that occurred during the Battle of the

eclipses, and in the fourth year of the 48th Olympiad,[1] predicted an eclipse of the sun which occurred two years later in the reign of Alyattes.[2] After him Hipparchus recorded the orbits of both planets for 600 years prior to his own day. He included an almanac of nations, gazetteer of places and descriptions of peoples, according to contemporary evidence, in such a way as to suggest that he shared in the grand designs of Nature.

In praise of natural scientists who have freed men's minds from fear

54. O great men, way above mortal estate, who, by your discovery of the laws that govern such great divinities, have freed the miserable mind of man from fear — a mind that feared eclipses of the planets signified some sort of crime or death (those exalted poets Steisichorus and Pindar clearly experienced this fear because of an eclipse of the sun), or blamed poison when the moon 'died' and consequently went to her help by making discordant noises. Through such fear and ignorance of the underlying cause, the Athenian general Nicias was afraid to lead his fleet from harbour and so destroyed the Athenians' greatness.[3] Praise be to your intellect, you interpreters of the heavens, you who comprehend the universe, discoverers of a theory by which you have bound gods and men!

56. It is established that eclipses recur within 223 months. Eclipses of the sun occur only when the moon is in her last or first phase — which people call their 'conjunction'; eclipses of the moon happen only at full moon and always within the period when they last occurred. Every year at fixed days and hours eclipses of both planets take place below the earth and, even when above the earth, they are not seen everywhere, sometimes because of clouds.

Halys (28 May 585 BC). He asserted that 'all things are full of gods' and that water is the basis of the universe. This was a logical deduction in so far as water may readily be observed in the three states of matter, namely, liquid, solid and gas.

1. 585 BC.

2. King of Lydia (c. 610–560 BC), of the house of Gyges. Alyattes drove out the Cimmerians and extended his kingdom to the Halys where he joined battle with Cyaxares (585 BC).

3. At the time of the ill-fated Athenian expedition to Syracuse an eclipse of the moon (27 August 413 BC) caused Nicias to delay abandoning the siege of that city, with disastrous results. See Plutarch, *Nicias*, 23.

Some estimates of the distances between planets

83. Many men have tried to calculate the distances of the planets from the earth, and have gone on record claiming that the distance of the sun from the moon is nineteen times the distance of the moon from the earth. Pythagoras, a man of penetrating intellect, deduced that the distance of the moon from the earth was 14,500 miles; of the sun from the moon, 29,000; and of the sun from the Zodiac, 43,500. Sulpicius Gallus, a countryman of ours, concurred with this opinion.

85. Posidonius states that mists, winds and clouds reach a height of just less than 5 miles above the earth and that from this point the air is clear, liquid and light without disturbance. He also says that the distance from the cloudy air to the moon is 230,000 miles and from the moon to the sun, 575,000 miles. This distance stops the vast mass of the sun from burning up the earth. The majority of authorities, however, have said that clouds reach a height of 103 miles.

Comets and their significance

89. There are also stars of several kinds that suddenly come into being in the sky itself. The Greeks call them 'comets'; we call them 'long-haired' stars, because they bristle with a blood-red tail that is shaggy on top, like hair. The Greeks call those that have a mane spreading out from their lower part like a long beard 'bearded stars'. 'Javelin stars' quiver like a spear and portend something extremely terrible. This was the type about which Titus Caesar wrote in his fifth consulship,[1] in his very well-known poem. That was its most recent appearance to date.

90–91. The shortest time a comet was visible is recorded as seven days, the longest, eighty. Some comets move, like planets, while others remain fixed; almost all the latter are in the North – not in any definite part of it, but mainly in that bright area that bears the name the Milky Way. Aristotle[2] states that several comets may be seen at the same time, a fact not observed by anyone else, at any rate as far as I know. This phenomenon means strong winds, or heat. Comets occur also in winter months and at the South Pole, but in that region have no rays.

---

1. AD 76. Vespasian, Titus' colleague, was consul for the seventh time.
2. *Meteorologica*, 345[a]29.

Comets as portents

The Ethiopians and Egyptians saw a terrifying comet to which Typhon, the king at that time, gave his name. The comet was fiery in appearance and twisted after the manner of a coil; it looked wild and was not really a star so much as a kind of ball of fire.

92. Planets and other stars now and then have tails that fan out like hair. Sometimes there is a comet in the western sky, generally a star, that inspires terror and is not easily propitiated. This comet appeared during the civil disturbance when Octavian was consul,[1] as earlier during the war between Pompey and Caesar;[2] in our own era it appeared about the time of the poisoning as a result of which Claudius Caesar left the Empire to Domitius Nero,[3] and then, during Nero's principate, when it was almost continuously visible and fierce in its appearance. Men think that the direction in which a comet begins to move, the star from which it draws its strength, the things it resembles and the places in which it shines, are all important factors.

93. They believe that if it looks like a flute, it is a portent for music; if it is in the private parts of a constellation, it is a portent for permissive behaviour; if it forms a triangle, or a square, in relation to certain altitudes of the fixed stars, it portends men of genius and learning; in the head of the Northern or the Southern Serpent, it brings poisonings. The only place in the whole world where a comet is worshipped is in a temple at Rome. The late Emperor Augustus[4] himself judged this comet highly propitious in that it appeared at the beginning of his principate, when, as a member of the college he had founded, he was holding games in honour of Venus Genetrix, not long after his father Julius Caesar had died.

94. He expressed his joy publicly in these words:

On the very days of my games a comet was visible for seven days in the northern region of the sky. It used to rise about an hour before dark and was bright and visible from all lands. The general populace believed that this signified Caesar's soul received among the spirits of the immortal gods,

---

1. 43 BC.

2. 49 BC.

3. AD 54. Having succeeded his stepfather, he secured his position by murdering Britannicus.

4. Augustus (63 BC–AD 14) was the son of a plebeian family from the Volscian region who was adopted into the *Gens Iulia* by his maternal uncle Julius Caesar as 'Gaius Iulius Caesar Octavianus'. The Senate conferred on him the religious title 'Augustus' (27 BC), and he became the first Roman emperor.

and this gave rise to the addition of a star to the bust of Caesar that we dedicated soon after in the Forum.

This was what he said in public; inwardly he rejoiced because he interpreted that comet as having come into being for his benefit and as embracing his own birth. Indeed, if I confess the truth, it did bring health to the world. Some believe that even comets go on for ever and in an orbit of their own, but are not visible except when deserted by the sun. Others, however, believe that they come into existence from water that happens to be about and from the force of fire, and for the same reason are dissolved.

Hipparchus' astronomy

95. Hipparchus can never receive sufficient praise. No one has done more to prove that the stars are linked to man and that our souls are a part of heaven. He discovered a nova that came into existence in his lifetime. He was led to wonder with regard to this star, because of the change in the brightness with which it flared up, whether this happened quite frequently and whether the stars we think are fixed actually move. Therefore, he dared a thing that would be astonishing even for God, namely to list the number of stars for future genera- tions and to check the constellations by name, having devised tables to indicate their positions and brightness so that it might easily be discerned not only whether stars perish and are born, but whether they transit or move, and, similarly, whether they grow larger or smaller. Thus Hipparchus left the sky as a legacy to all men − if anyone should be found to claim this inheritance.

Meteors and St Elmo's fire

96. Meteors are also visible but only when falling, just like the one that sped across the sky at midday before the very eyes of the people when Germanicus Caesar was putting on a gladiatorial show. There are two [main] kinds: one type the Greeks call *lampades*, that is 'torches'; the other *bolides* or 'missiles', like those visible at the Battle of Mutina.[1] Other meteors occur in a similar way; there is, for example, a type that the Greeks call *dokoi*, 'beams', and this resembles the one that appeared when the Spartan fleet was defeated and lost control of Greece.[2]

---

1. Decimus Brutus was besieged at Mutina by Antony in 44 BC.
2. At the battle of Cnidus in 394 BC.

97. An opening-up of the sky also occurs, which people call a chasm, as well as something with the appearance of blood and fire that tumbles down from this opening to earth; there is no more alarming cause of fear for men. This phenomenon occurred in the third year of the 107th Olympiad,[1] when King Philip was causing upheaval in Greece. I am of the opinion that these things happen, as indeed is the case with other phenomena, at fixed times and as a result of the power of Nature and not, as the majority of people think, from other causes invented by the ingenuity of men's intellect. To be sure, these meteors did foretell great misfortunes, yet I think they happened because of natural phenomena and because they were going to happen anyway. Their cause was hidden by the rarity of their occurrence, and for this reason they are not understood. Ominous and long-drawn-out eclipses of the sun occur, such as after the murder of Julius Caesar[2] and during the Antonine War when there was continuous twilight for almost a whole year.

99. Furthermore, several suns can be seen at the same time: they are visible neither above nor below the actual sun, but at an angle; they are never close by or opposite the earth, and cannot be seen at night but only at sunrise or sunset. On one particular occasion more than one sun was reported to have been seen at midday on the Bosporus, and these suns have often been seen – for example, they appeared during the consulships of Spurius Postumius and Quintus Mucius,[3] Quintus Marcius and Marcus Porcius,[4] Marcus Antonius and Publius Dolabella,[5] and Marcus Lepidus and Lucius Plancus.[6] This phenomenon occurred in our own day when Claudius was emperor, during his consulship with Cornelius Orfitus.[7] No more than three suns have ever been reported to date. Three moons appeared at the same time during the consulship of Gnaeus Domitius and Gaius Fannius.[8]

100. The phenomenon most people call 'nocturnal suns' was seen at night in the consulship of Gaius Caecilius and Gnaeus Papirius[9] and often at other times, so that it appeared as bright as day although it was night. When Lucius Valerius and Gaius Marius were consuls,[10] a blazing shield traversed the sky from west to east, at sunset, giving

1. 349 BC.
2. 44 BC.
3. 174 BC.
4. 118 BC.
5. 44 BC.

6. 42 BC
7. AD 51
8. 222 BC.
9. 113 BC.
10. 86 BC.

off sparks as it went. In the consulship of Gnaeus Octavius and Gaius Scribonius[1] a spark was seen to fall from a star and to increase in size as it drew near the earth. When this became the size of the moon, it shone as on a cloudy day and, when it returned to the sky, became a torch. This was an isolated phenomenon. The proconsul Silanus and his entourage saw it. Stars also appear to rush about, and never without good reason; for this heralds the onset of fierce winds from the same region.

101. Stars become visible at sea and on land; I have seen them with the appearance of lightning clinging to the spears of soldiers on guard-duty at night in front of the rampart. At sea I have observed St Elmo's fire[2] on the yard-arms and other parts of a ship, jumping about with a sound like a voice, just as birds hop from perch to perch. When these stars come singly they are heavy and sink ships, and, if they fall down into the bottom of the hold, they destroy them by fire. When they come in pairs, they signify safety and herald a successful voyage. Men say that the terrifying star called Helena is put to flight by their approach. This is the reason the two stars are named Castor and Pollux and men at sea call on them as gods. Stars also shine round men's heads in the evening; this constitutes an omen of great significance. All these phenomena have no certain explanation but are concealed within the majesty of Nature.

The weather and the effect of heavenly bodies on natural phenomena

102. This completes my account of the world itself and the constellations. I pass on to the remaining facts about the sky worth mentioning. Our ancestors called the 'sky' what is otherwise known as 'air' – all that resembles empty space, but which pours forth the breath of life. From this come clouds, thunderclaps, thunderbolts, hail, frost, rain, storms, whirlwinds and most human misfortunes, and the struggle between the elements of Nature.

103. The power of the constellations presses down earthly things that strive to reach for the sky, and draws to itself things that cannot reach up of their own accord. Rain falls, clouds go upwards, rivers

---

1. 66 BC.

2. St Elmo's fire is the glow accompanying the brushlike discharges of atmospheric electricity which appears as a tip of light on the masts of ships during stormy weather. St Elmo is the patron saint of sailors.

dry up, hailstones beat down: the sun's rays scorch and strike everywhere on earth in the middle of the universe and, broken, bounce back and take with them all that they have drunk. Steam falls from on high and again returns on high. Empty winds violently swoop down and go back with their plunder. So many things draw their breath from the upper air, but the air strives in the opposite direction. The earth pours breath back to the sky as if it were a vacuum.

104. So as Nature goes to and fro like a sort of ballista, discord is kindled by the speed of the world's movement. Nor is the battle allowed to stand still, but it is continually snatched up and whirled round, revealing in the huge globe that encircles the world the causes of things, covering another and another heaven with clouds. This is the kingdom of the winds. Here their nature is all-important and embraces almost all the phenomena attributable to the air. Most men attribute the hurling of thunderbolts and lightning to the violence of the winds – indeed it rains stones sometimes because these are carried aloft by the wind. So it is with many things.

The decay of research in Pliny's day

117. More than twenty Greek writers of old have published their observations on these topics. And this surprises me, that although the world was in a state of upheaval and divided piecemeal into kingdoms, so many men were concerned with facts that were so difficult to research – especially when living in the midst of wars, when hosts were untrustworthy and pirates, the universal scourge, held up the free passage of information. Consequently, today a person may learn some facts about his own region, from the notebooks of those who have never been there, more accurately than from the knowledge of the local inhabitants. Yet, nowadays, in this happy time of peace under an emperor who takes such pleasure in promoting literature and science, absolutely nothing is being added to the sum of knowledge as a result of original research; indeed not even the discoveries made by people long ago are thoroughly assimilated.

118. The rewards were not greater when the successes were divided between a large number of researchers, and indeed the majority unearthed things with no other reward than the thought of helping posterity. Men's moral fibre has diminished with time, but not their revenues, and, since every sea has been opened up and every coast provides a hospitable landing, a great many people

travel by sea not for knowledge but for profit. Their minds, so blindly obsessed with greed, do not perceive that profit can be more surely made as a result of knowledge. Therefore, I shall give a more careful account of the winds than is perhaps appropriate to the task I have undertaken.

The eight main winds

119. The earliest authorities recognized four winds in all, one for each quarter of the world (for this reason even Homer does not name any more) – a superficial system, as it was judged not long afterwards. A subsequent age added eight, but this proved too subtle and too precise. The next age decided on a middle course, adding four winds from the long list to the short one. So there are two winds from each quarter of the sky:[1] Subsolanus and Vulturnus (in Greek, Eurus and Apeliotes);[2] Auster and Africus (Notus and Libs);[3] Favonius and Corus (Zephyr and Argestes);[4] Septentrio and Aquilo (Aparctias and Boreas).[5]

Whirlwinds

131. Sudden blasts, which, as has been stated, arise as a result of the earth breathing out and fall back again to earth, obscuring it with a covering of cloud, occur in many forms. Indeed these blasts rush on unrestrained, like torrents, and cause thunder and lightning. If they blow with a heavier impact and burst a wide hole in a dry cloud, they cause a storm that the Greeks call a 'cloudburst'; if they break out from a downward curve with more limited rotation, they cause a whirlwind.

132. Cyclones are particularly disastrous for sailors, as they twist round and shatter not only the yard-arms but also the ships. The only remedy, and this a slight one, is to pour out vinegar before its approach, this substance being naturally cold. The whirlwind, when driven back by its own impact, snatches things up, carries them to the sky and sucks them aloft.

1. Listed by Vitruvius, *On Architecture*, 1, 6, 4. Cf. also the well-known representation on the frieze around the top of the octagonal Tower of the Winds in the Agora, Athens.
2. E and SE.
3. S and SW.
4. W and NW.
5. N and NE.

133. If the blast bursts out of a larger cave of down-pressing wind, but a cave less wide than in the case of a storm, and is accompanied by a crashing sound, this is what people call a 'whirlwind', which lays low everything in its path. When a whirlwind is hotter and rages with a fiery blaze, it is called a 'pillar of fire'; it burns and destroys everything with which it comes into contact. There is, however, no cyclone when the wind is northerly, nor a cloudburst when snow is falling or lying on the ground.

Thunder and lightning

135. Thunderbolts are infrequent in winter and summer, but for opposite reasons. In winter, owing to the thicker layer of cloud, the air is made dense and all the exhalation of the earth, being stiff and cold, extinguishes whatever fiery vapour it receives. This is the reason why Scythia and the surrounding frozen regions are free from the fall of thunderbolts. On the other hand, the very great heat keeps Egypt similarly free, since the hot and dry exhalations from the earth very rarely condense and only form thin, insubstantial clouds.

136. Thunderbolts are more frequent in spring and autumn since their summer and winter causes are both present in each of those seasons. This explains why they are frequent in Italy since the air moves more freely as a result of milder winters and stormy summers, making it always somewhat like spring or autumn. In some parts of Italy that slope down from the north towards the warmth, such as the regions of Rome and Campania, there is lightning in winter and summer, a phenomenon that does not occur in any other place.

138. Etruscan writers consider that nine gods employ eleven thunderbolts – because Jupiter has three kinds. The Romans have retained only two of these gods. They attribute daytime thunderbolts to Jupiter and ones at night to Summanus, the latter being less frequent, of course, because of the coldness of the sky.

140. Historical records state that thunderbolts were occasioned by certain rites and prayers, or sent in answer to them. There is an old story from Etruria of such a prayer being answered: the portent they called Olta came to the city of Volsinii when its territory had been devastated; it was sent in answer to a prayer from the city's king, Porsenna.[1] Before his time, Numa[2] frequently made such a

---

1. Tradition says that Porsenna came from the city of Clusium in central Etruria.
2. Numa succeeded Romulus as king of Rome.

prayer, as Lucius Piso, a weighty authority, tells us in Book I of his *Annals*; he adds that when Tullus Hostilius[1] followed Numa, but with too little regard for the ritual, he was struck by lightning. We also have groves, altars, and sacred rites; and Jupiter's titles include the 'Stayer', the 'Thunderer', the 'Subduer' and the 'Called-down'.

141. Mankind's opinion on this subject varies in accordance with individual character. It is the mark of a bold man to believe that Nature obeys sacred ritual. Equally, it is the mark of a dull-witted person to deny that ritual has beneficial powers, since knowledge has made such strides in the interpretation even of thunderbolts that it can foretell that others will come on a fixed day. This progress has resulted from countless public and private experiments.

Lightning and its effects

142. It is certain that in a thunderstorm the flash of lightning is seen before the peal of thunder is heard; this is not surprising as light travels faster than sound. Nature makes the stroke of a thunderbolt coincide with the sound, but the sound comes from the bolt setting out, not from its having struck. Nor does anyone who is hit ever see the lightning, or hear the thunderclap beforehand. Lightning on the left is considered lucky because the sun rises on the left side of the world.[2]

145. Lightning without thunder occurs more often at night than during the day. Man is the one living being lightning does not invariably kill, the rest die instantly. Nature evidently accords man this honour because so many creatures outstrip him in strength. All things fall in the opposite direction to the flash. Unless a man is turned round by the force of the blow when struck, he does not die; men struck from above collapse in a heap. A man struck when awake is found with his eyes closed, or, when asleep, with his eyes open. It is not lawful to cremate a man who dies in this way; religious custom requires burial. No living thing can be burnt by lightning and survive. The immediate area of the wound is colder than the rest of the body.

146. Of things growing in the ground, lightning does not strike the laurel bush, and it does not penetrate more than about four feet below ground. For this reason, when men are afraid, they think that

---

1. Hostilius, the third king of Rome, is credited with the destruction of Alba Longa.

2. i.e., looking south.

deep caves are the safest place, or tents made of the skin of creatures called sea-calves because of all sea animals they alone are not struck by lightning.

The eagle is the only bird that is not struck and is represented as armed with a thunderbolt for this reason. In Italy, during the time of the Civil War,[1] people stopped building towers between Tarracina and the Temple of Feronia because none escaped destruction by lightning.

Miraculous happenings in the sky

147. Besides these happenings in the lower sky, it is on record that it rained milk and blood during the consulship of Manius Acilius and Gaius Porcius, and often at other times it rained flesh, as when Publius Volumnius and Servius Sulpicius were consuls, and no flesh left unplundered by birds went bad. Likewise it is recorded that it rained iron in Lucania the year before Marcus Crassus was killed by the Parthians,[2] and, together with him, the Lucanian soldiers of whom there was a large number serving in the army. The appearance of the iron that rained down was like sponges; the augurs prophesied wounds from above. When Lucius Paullus and Gaius Marcellus were consuls it rained wool round about the stronghold of Compsa, near which Titus Annius Milo was killed a year later, and when Milo was pleading a case it rained fired bricks – according to the official records for that year.

148. I am informed that, during the war with the Cimbri, the din of weapons and the sound of a trumpet were heard in the sky, and that this phenomenon occurred quite a few times before and after. In the third consulship of Marius,[3] the people of Ameria and Tuder saw heavenly armies marching from the east and west to join battle; the army from the west was put to flight. The sky itself is quite often seen ablaze when the clouds have been set on fire by a rather large flame and this phenomenon is not in the least surprising.

149. The Greeks declare that Anaxagoras,[4] in the second year of

1. 49–44 BC.
2. At the Battle of Carrhae in 53 BC.
3. 103 BC.
4. Anaxagoras, from Clazomenae (*c.* 500–*c.* 428 BC), was the first philosopher to live in Athens. He claimed that in the beginning the world was a 'mixture' containing 'seeds' of every qualitatively distinct natural substance, organic and inorganic: these are infinitely divisible into parts like each other and the whole and represent Anaxagoras' elements. *Nous* starts a rotatory motion which gradually spreads. Thus

the 78th Olympiad,[1] through his knowledge of literature relating to astronomy, prophesied that within a certain number of days a stone would fall from the sun, and this happened during the day at Aegospotami in Thrace. The stone is on show even now and is as big as a cart and brown in colour. There was also a blazing comet at night at that place. If anyone believes in this prophecy, he must at the same time admit that Anaxagoras' power of divination was a greater cause for wonder, and that our understanding of the universe is overthrown and everything rendered chaotic if it is believed that the sun itself is a stone, or even that it had a stone inside it. It will not be doubted, however, that stones often fall from the sky.

150. A stone of moderate size indeed is worshipped for this reason today in the gymnasium at Abydus, which the same Anaxagoras is said to have prophesied would fall in the middle of the country. One is worshipped at Cassandria, renamed Potidaea; a colony was founded there because of this phenomenon. I myself saw one that had fallen a little while before in the territory of the Vocontii.

Rainbows

151–3. Rainbows, which occur often, are neither miraculous nor significant. They are not even a reliable indication of rain or fine weather. Clearly, a ray of the sun striking a hollow cloud has its point repelled and refracted back to the sun and the different colours result from a mixture of clouds, fires and air. Rainbows occur only opposite the sun and always have a semi-circular form. They do not occur at night, although Aristotle says that a rainbow is sometimes seen then, but admits that this only happens on the fourteenth day of the lunar month.[2]

Mother Earth

154. Earth comes next, the one part of the world of nature on which we have bestowed the title of 'Mother' out of the highest respect because she deserves this. Earth is the province of men, as

---

seeds are separated out, the dense, moist, cold and dark (*aer*), going to the centre, their opposites, (*aither*), to the circumference. The heavenly bodies are stones, torn from the earth, which motion renders red-hot. Anaxagoras follows the Ionian tradition of a flat earth, but knows the causes of eclipses.

1. 467 BC.
2. *Meteorologica*, III, 372^{a}27.

the sky is the province of God. Earth receives us when we are born and feeds us after birth and always supports us, and, at the very end embraces us in her bosom, sheltering us like a mother especially then, when we have been disowned by the rest of Nature. Sacred for no greater service than that by which she makes us also sacred – carrying our monuments and epitaphs, prolonging our name and extending our memory against the brevity of time. Her divinity is the last to which, in anger, we pray 'to lie heavily' on the deceased, as though we did not know that she is the only element which is never angry with man.

155. Water turns to rain, and freezes as hail, swells up in waves, and falls headlong as torrents. The air grows thick with clouds and unleashes its fury in storms.

Mother Earth's gifts and man's abuse of them

Earth, however, is kind, gentle, indulgent, always a servant to man's needs, productive when compelled to be, or lavish of her own accord. What scents and tastes, what juices, what things to touch, what colours! With what good faith she repays the interest on what has been loaned! What food she grows for our benefit! Living things that are noxious – creatures for which the breath of life is to blame – she has to receive when their seed is sown and to maintain them when they burst into life; but their harm lies in the evils of those who generate them. Earth does not receive back a snake when it has bitten a man and she exacts punishment even in the name of those who are lifeless.

156. Indeed Earth may be credited with having invented even poisons out of pity for us, lest, when we are tired of life, starvation, a death most out of keeping with earth's benefits, should destroy us by a slow wasting-away; lest precipices should scatter our lacerated body; lest the perverted punishment of the noose that imprisons the breath of life it seeks to extinguish should torture us; lest, if we seek death in the depths, our burial should serve as fodder; lest the torture of iron should rend our body. Thus Earth, out of pity for us, has produced a substance – a very easy draught – by drinking which, like thirsty men, we may effortlessly deprive ourselves of life without physical injury or loss of blood, in such a way that no birds or beasts should touch us when dead and that those who have killed themselves should be preserved for the earth.

157. Earth has produced a remedy for our ills, but we have made that a poison to destroy our life. Why do we not use iron, which is

indispensable, in a similar way? Nor should we be justified in complaining even if she bore poison for the purpose of wrongdoing, since we are not grateful to one of Nature's elements. For what delights and affronts does she not afford mankind? She is dumped into the sea, or excavated to provide channels. She is tortured at all hours by water, iron, wood, fire, stone and crops, and by far more besides to serve our pleasures rather than our needs.

158. Yet so that what she suffers on her surface, her outermost skin, may seem bearable by comparison, we penetrate her inmost parts, digging into her veins of gold and silver and deposits of copper and lead. We search for gems and certain very small stones by sinking shafts into the depths. We drag out Earth's entrails; we seek a jewel to wear on a finger. How many hands are worn by toil so that one knuckle may shine! If there were any beings in the nether world, assuredly the tunnelling brought about by greed and luxury would have dug them up. Are we surprised if Earth has brought forth creatures to harm us?

159. Wild animals, I believe, guard her and ward off impious hands. Do we not mine among snakes and handle veins of gold among poisonous roots? Yet this is the work of a kind goddess, because all these outlets, from which wealth is derived, lead to crime, bloodshed and wars. We sprinkle our blood on Earth and cover her with our unburied bones. Yet when our madness has, as it were, been purified, she herself covers these bones and also hides the crimes of mortals. This too I would consider among the crimes of our ungrateful minds, that we are ignorant of her nature.

The earth and its sphericity

160. The shape of the earth is the first fact about which there is general agreement. At any rate we call the earth a sphere and admit that it is included within poles. The form, however, is not that of a perfect sphere, for there are high mountains and widely spreading plains . . . But the continuous revolution of the universe around the earth forces her huge globe into the shape of a sphere.

161. There is a great conflict between the learned and the man in the street at this juncture. Scholars assert that men are spread out all round the earth and stand with their feet pointing towards each other and that the top of the sky is alike for all of them and that their feet point down towards the centre of the earth from wherever they are. An ordinary person, however, inquires why men on the opposite side do not fall off – as if there is not an equally good reason for them wondering why *we* do not fall off.

Oceans

163. But the greatest question for the masses is whether they must believe that the seas are also spherical. Yet nothing else in the world of Nature is more obvious. For droplets of liquid when hanging down take the shape of small globules. When dropped on dust or on the woolly surface of leaves, they are seen to be absolutely spherical. In filled cups there is a meniscus, although, because of the transparency of the liquid and its fluidity, it finds its own level. These facts are more easily understood from theory than by observation. A still more amazing occurrence is that when a little water is added to a cup that is already full, it overflows. Yet the opposite happens when heavy objects are added – often even as many as twenty silver coins: presumably they pass into the water and merely make its surface convex; water poured on to this surface runs off.

164. The same reason explains why land, not visible from the deck of a ship, can nevertheless be seen from the mast and why, as a ship recedes into the distance, if some shining object is tied to the top of the mast, it appears to sink slowly and finally disappears from view. Lastly, with what other configuration could Ocean, which we admit is on the outer edge, cohere and not fall away in the absence of a boundary enclosing it? Why its edge does not fall away, although the sea is globular in shape, amounts to something amazing.

The circumnavigation of the known world

167. Today the whole of the West has been circumnavigated from Gades and the Pillars of Hercules round Spain and Gaul. The larger part of the Northern Ocean was explored under the auspices of the late Emperor Augustus, when a fleet sailed round Germany to the promontory of the Cimbri and then, observing a vast sea before them, or learning about it by report, reached Scythia and places frozen with excessive moisture ... On the eastern side, the whole region, under the same star, from the Indian Ocean to the Caspian Sea was sailed throughout by Macedonian forces in the reigns of Seleucus and Antiochus, who wished that it should be called both Seleucus and Antiochus after themselves.

168. Many coasts of Ocean[1] round the Caspian Sea have been explored and nearly the whole of the North has been traversed by

---

1. The ancients thought that Ocean was an expanse of sea surrounding the perimeter of the earth's land mass.

galleys, so that there is now no room left for conjecture about the existence of the Maeotic Lake, whether it is a gulf of Ocean, as I observe many have believed, or an overflow from it, from which it is separated by a narrow strait. On the other side of Gades, from the same western point, a great part of the southern gulf is sailed over today in the circuit of Mauretania. Indeed Alexander the Great's eastern conquests explored the larger part of it as far as the Arabian Gulf. When Gaius Caesar, Augustus' son, was campaigning there, it is said that figureheads from Spanish wrecks were recognized.

169. When the power of Carthage flourished, Hanno[1] sailed from Gades to the extreme part of Arabia and published a note of his voyage, as did Himilco,[2] when sent at the same time to explore the outer coasts of Europe. Caelius Antipater states he had seen someone who had sailed from Spain to Ethiopia to trade.

Man occupies a small fraction of the earth, itself a mere dot in the universe

174. Let all these portions[3] be taken away from the earth, or rather this dot in the universe, as the majority of savants have taught – for this is all the earth is in the context of the whole universe – and this then is the substance of our glory, this is its home, here we fill positions of power and here we covet wealth and put mankind into a turmoil repeatedly and fight wars – even civil wars – and empty the land by killing one another.

175. To sum up the outward madness of nations, this is the land in which we drive out our neighbours and dig up and steal their turf to add to our own, so that he who has marked his acres most widely and driven off his neighbours may rejoice in possessing an infinitesimal part of the earth.

Sunrise and sunset vary with longitude

181. Although night and day are the same throughout the world their times vary. This is known as a result of many experiments –

1. The Carthaginian explorer (before 480 BC), whose account of West Africa – the only surviving specimen of Punic literature – is preserved in a Greek translation. He was probably influenced by Herodotus' style. See C. Müller, *Geographici Graeci Minores* (Paris, 1882), vol. 1, p. 1 ff.

2. The Carthaginian navigator (before 480 BC) who explored the western coastline of Europe from Cadiz to Brittany.

3. Pliny includes seas, channels, lakes, deserts and other uninhabited regions.

for example, with Hannibal's towers in Africa and Spain, and in Asia, where, because of the fear of pirates, similar watch-towers were built for protection. Warning fires lit on these at midday were often found to have been seen by the people furthest to the rear, at 9 p.m. Philonides, one of Alexander's runners, completed the 138 miles from Sicyon to Elis in nine hours from sunrise,[1] but, although the return journey was downhill, this took until 9 p.m. This happened often. The reason was that on the way there his route was with the sun, but on his return he passed the sun as it met him coming in the opposite direction. This is the reason that ships sailing westwards even on the shortest day beat the distance they sail at night, because they travel with the sun.

The sun's altitude varies according to latitude

182. Sundials do not register the same everywhere because the shadows cast by the sun, as they change, alter the time shown every 34 miles, or at the furthest every 57 miles. So in Egypt, at noon, at the time of the equinox, the shadow of the gnomon measures a little more than half its length, whereas at Rome the shadow is a ninth shorter than the gnomon, at Ancona a thirty-fifth longer and, in the region of Italy called Venetia, the shadow is equal to the length of the gnomon at the same hour.

Hours of daylight vary with latitude

186. So it happens that because of the varied lengthening of daylight, the longest day lasts 12 and eight-ninths equinoctial hours at Meroe, 14 hours at Alexandria, 15 in Italy and 17 in Britain, where the light summer nights confirm what theory urges us to believe, namely that because the sun approaches nearer to the top of the world on summer days, owing to a narrow circuit of light the parts of the earth that lie at the poles have continuous daylight for six months at a time and continuous night for six months when the sun has withdrawn in the opposite direction towards midwinter.

187. Pytheas[2] of Massilia writes that this happens in the island of Thule, six days north of Britain, and some say it also happens in

---

1. The actual distance from Sicyon to Elis, as the crow flies, is about 80 miles.

2. A Greek explorer who circumnavigated Britain and sailed beyond to Thule. He calculated the latitude of Massilia and laid bases for cartographic parallels through northern Gaul and Britain.

Mona, which is about 200 miles from the British town of Camulodunum.

Sundials and the measurement of a day

Anaximenes[1] of Miletus, the pupil of Anaximander (whom I have already mentioned), discovered the theory of shadows and what men call 'gnomonics'. He first put a sundial on show at Sparta, a device they call 'Hunt-the-Shadow' (*sciothericon*).

188. Different peoples measure the actual unit called 'a day' in different ways. The Babylonians reckon this as the interval between two sunrises; the Athenians, that between two sunsets; the Umbrians that from midday to midday; ordinary people everywhere, from dawn to dark. Roman priests and those who fix the 'civil day', likewise the Egyptians and Hipparchus, reckon the day from midnight to midnight. But it is clear that the interruption of daylight between sunset and sunrise is shorter near the summer solstice than at the equinoxes, because the position of the Zodiac is more oblique about its middle points, but straighter near the solstice.

Climate and its effect on racial characteristics

189. There is no doubt that the Ethiopians are burnt by the heat of the sun and are born with a burnt appearance and with curly beards and hair. In the opposite region of the world the races have white, frosty skins with flaxen-coloured hair that hangs straight.

190. In the middle of the earth, because of a healthy mixture of fire and water, there are tracts that are fertile for all things; the men are of medium stature, with a very definite blending noticeable in their complexion; even their behaviour and manners are gentle, their senses supple and intellects fertile; they are able to comprehend the whole of Nature. These people have governments, which the outermost races have never possessed, just as they have never obeyed the central people, for they are detached and solitary, in keeping with the savagery of Nature that oppresses them.

Earthquakes

191-2. The Babylonian theory is that earthquakes and fissures in the

---

1. A natural philosopher (*fl. c.* 546 BC). He believed that air was the essence of all things and that the soul was composed of air.

earth, as indeed all other natural phenomena, happen as a result of the force of the planets – the three, that is, to which they assign thunderbolts.[1] These occur when the planets are travelling with the sun. A remarkable and immortal act of divination is attributed, if we believe this, to Anaximander of Miletus, the natural scientist who, they say, warned the Spartans to protect their city and houses because an earthquake was imminent. Then the whole city fell in ruins and a large part of Mount Taygetus, which stuck out like a ship's stern, broke off and crashed on to the ruins. A further forecast is attributed to Pherecydes,[2] the teacher of Pythagoras, and this was also accurate, in that he warned his fellow citizens of an earthquake about which he had a premonition when drawing water from a well. If these assertions are true, how far can such men be thought different from a god, even during their lifetime? Although such matters ought to be left to individual judgement, I think there is no doubt that winds cause earthquakes. For earth tremors never occur unless the sea is calm and the sky so motionless that birds cannot hover, because all the air which bears them up has been taken away.

Warning signs of impending earthquakes

193. Earthquakes occur in different ways and with remarkable after-effects, in some places overturning walls, in others sucking them down into a deep fissure, in others throwing up masses of earth, in others sending forth rivers and, sometimes, even fires or hot springs, and in yet others changing the course of rivers. A terrible sound precedes or accompanies earthquakes, sometimes like a rumbling, sometimes like cattle lowing, or men shouting, or the clash of weapons struck together, according to the nature of the material that receives the shock wave and the form of the caverns or burrows through which it passes.

194. Sometimes the earth is not shaken in a simple way, but trembles and vibrates. Indeed the fissure sometimes remains open, showing the objects it has sucked in; at other times it hides them by closing its mouth and bringing soil over the gap in such a way that no traces of it remain. For the most part cities and tracts of farmland are swallowed up, although areas by the sea are especially prone to

1. Saturn, Jupiter and Mars.
2. A prose writer from Syros (c. 550 BC). Pherecydes' cosmogonic myth *Eptamychos*, described the creation of the world by one or more of a triad of eternal deities: Zas (Zeus), Chronos (Kronos) and Cthonie (Ge).

earthquakes, and mountainous parts are not free from such disasters. I have discovered that tremors have quite often been registered in the Alps and Apennines.

196. Sailors can also predict with a high degree of certainty that an earthquake will occur when a wave suddenly swells up without a wind or a shock wave shakes their ship. Even on board ships posts begin to tremble, as they do in buildings, and foretell the onset of an earthquake by rattling; timid birds perch. There is also a sign in the sky: a cloud like linen thread stretches in fine weather over a wide area in advance of an impending earthquake, either in the daytime or a little after sunset.

197. Water in wells is muddier and smells foul. In wells there is also a protection against earthquakes such as is often provided by caves for they let out the confined breath. This is observed in whole towns. Foundations of buildings pierced by a large number of channels for drainage are less shaken. Structures above sewers are much safer, as is seen in Italy, at Neapolis, whereas the solid parts of cities are more liable to suffer from disasters of this kind.

The safest parts of buildings are arches, also angles of walls, and posts, which spring back into place with each alternate thrust; and walls built of clay bricks are less damaged by tremors.

Historical earthquakes and their consequences

199. A vast earthquake of ominous significance happened during the consulship of Lucius Marcius and Sextus Julius, in the district of Mutina, as I learned in the books of Etruscan religious lore. Two mountains crashed together with a mighty sound, leaping forward then retreating, while between them flame and smoke billowed to the sky. This happened in daylight and a great crowd of Roman knights, with their friends as well as travellers, observed this from the Via Aemilia. All the farm buildings were dashed to the ground by the force of this crash and very many animals that were inside were killed. This happened in the year before the Social War,[1] which was arguably more disastrous to the land of Italy itself than were the civil wars. Our times have observed a no less amazing

---

1. Also known as the Marsic, or Italic, War. This was waged against Rome's Italian Allies (91–87 BC), among whom the Marsi were prominent. The main fighting took place in 90–89 BC. Rome gained her victory largely through the political concession of granting Roman citizenship to the Allies. This led to the consolidation of Italy south of the River Po.

sight in the last year of the Emperor Nero,[1] as I have set forth in my history of Nero's principate. Meadows and olive-trees, with a public road between, crossed to opposite sides. This happened in territory belonging to the Marrucini, on the estate of Vettius Marcellus, a Roman knight who looked after Nero's business interests.

200. Flooding accompanies earthquakes. The largest earthquake happened in the principate of Tiberius Caesar when twelve cities in Asia Minor were razed to the ground in one night. The earthquake with the record number of tremors was during the Second Punic War, when reports of fifty-seven tremors reached Rome in the space of one year.[2] This was the year in which a violent earthquake happened while Carthaginians and ourselves were locked in combat; neither side felt this. An earthquake does not represent a simple disaster, nor is the danger confined to the earthquake itself, but equally, or more so, it is dangerous as an omen for the future. The city of Rome never experienced an earthquake without this being forewarning that something was about to happen.

Products obtained from the earth

207–8. Let me speak of the marvels of the earth rather than the crimes perpetrated by Nature. And, I swear, even the heavenly phenomena could not have proved more difficult to recount. I will tell of the wealth of mines, so varied, so rich, so prolific, brought to the surface in so many ages, although every day and across the whole world fires, the collapse of buildings, shipwrecks, wars and fraud cause such great havoc. So much do extravagance and the hordes of mankind squander. I will mention the great variety of gems, the many-coloured markings in stones and, among these, the brilliance of a certain stone that only allows daylight to pass through it. Then there are the great number of medicinal springs, the flames of fire that in so many places have burnt continuously for centuries, and the lethal exhalations either emitted from chasms or present because of the mere configuration of the ground. These fumes are fatal in some places only to birds, as in the vicinity of Soracte near Rome, in others to all living creatures except man, and sometimes to man also, as in the regions of Sinuessa and Puteoli. These places some men call 'breathing holes',[3] others the 'jaws of hell' – ditches

1. AD 68.
2. 217 BC.
3. Located in the Phlegraean Fields, near Naples, where slight volcanic activity can still be observed in a number of fumaroles.

that put forth deadly fumes. There is also a place near the Temple of Mephitis at Ampsanctum in the region of the Hirpini; people die when they enter it. Similarly the hole at Hierapolis, in Asia Minor, is harmless only to the priest of the Great Mother goddess. Elsewhere there are prophetic caves in which those who are intoxicated by the fumes foretell the future, as at the celebrated oracle at Delphi. What other explanation could mortal man give except that these things are caused by the divine power of Nature, which is spread throughout the universe and bursts out in different ways.

Local marvels

210. Paphos on Cyprus[1] has a famous shrine of Venus in a certain courtyard on which no rain falls, and there is a similar courtyard around a statue of Minerva at the town of Nea in the Troad. In the same town sacrifices that have been left do not decay.

211. Near the town of Harpasa stands a rough boulder which can be moved by one finger, but resists a push made with the whole body. On the Tauric peninsula, in the state of Parasinum, there is some earth that heals all wounds. In the neighbourhood of Assos, in the Troad, there is a stone called 'Sarcophagus', by which all bodies are eaten away. Near the River Indus there are two mountains: the nature of one is to attach itself to iron, while the other rejects it. Thus if a man has nails in his shoes, on one of the mountains at each step he is unable to tear his foot away from the ground, and on the other he cannot put his foot down. It is recorded that at Locri and Croton there has never been a plague or earthquake, and that in Lycia forty days of fine weather follow an earthquake. Corn sown in the district of Arpi does not grow, and at the Altars of Mucius, in the vicinity of Veii, at Tusculum and in the Ciminian Forest there are places where stakes driven into the ground cannot be pulled out. Hay grown in the district of Crustumium is harmful there, but healthy when taken away.

Fire: petroleum, naphtha and volcanoes

235. In Samosata, the capital of Commagene, there is a marsh that exudes an inflammable mud called fossil pitch.[2] When it touches

---

1. Cyprus, according to tradition, was the birthplace of Aphrodite.
2. *maltha*.

anything solid, it sticks to it; also, when touched, it adheres to people trying to escape from it. So the people of Samosata defended their city walls with this substance when Lucullus attacked them.[1] The army was repeatedly burnt by its own weapons. Water feeds the flames. Experiments have shown that the flames can be extinguished only by earth. Naphtha has similar properties – this is the name of the substance that flows out like liquid bitumen in the vicinity of Babylon and in Parthia near Astacus. It has an affinity with fire, which leaps to it immediately it sees it, wherever it is. This is how Medea, according to the story, burnt her rival when her wreath caught fire after she had gone to the altar to make a sacrifice.

236. Among the spectacular occurrences connected with mountains, Etna always glows at night and provides enough fuel for its fires to last a very long time, although the mountain is covered by snow in winter and covers with hoar-frost the ash it has thrown up. Nor is it Etna alone with which Nature in her anger threatens the earth with conflagration. Mount Chimaera, in Phaselis, blazes day and night with a continuous flame; Ctesias[2] of Cnidus says that water increases the fire, but earth or dung puts it out. The mountains of Hephaestus in Lycia flare up so violently when touched by a blazing torch that even the stones and sand in rivers glow; the fire is fed by rain. Locals say that if someone lights a stick and draws a furrow, streams of fire follow.

237. If the pleasant Bowl of Nymphaeus (which does not burn the leaves of the thick wood above it and, although next to a cold stream, is always red hot) ceases to flow, it is an omen of terrible happenings for its neighbours in Apollonia, as Theopompus states. It is increased by rain and sends out asphalt to mix with that stream which is unfit for drinking and which, in any case, is more liquid than asphalt.

Finally, Pliny records distances between places and gives estimates of the earth's circumference

247. Eratosthenes, whom I observe is highly regarded by everyone, is an expert in every refinement of learning and in this particular

---

1. 74 BC.

2. A Greek doctor (late fifth century BC). He was a physician at the Persian court. He assisted Artaxerxes at the Battle of Cunaxa and was sent as an envoy to Evagoras and Conon (398 BC). Ctesias was the author of a history of Persia (*Persika*), in twenty-three books, written in Ionic. He also produced a geographical treatise, (*Periodos*) in three books, and a work on India (*Indika*).

matter indeed the foremost authority. He gives the earth's circumference as approximately 29,000 miles. This is a bold estimate but the result of such clever reasoning that one is ashamed not to believe it. Hipparchus, who in his refutation of Eratosthenes and in all the rest of his assiduous researches is an extraordinary man, adds a little less than 3,000 miles.

248. Dionysodorus (for I shall not leave out this outstanding example of Greek fantasy) believes differently. He came from Melos and died an old man in his native land. His female relatives who were his heirs attended to his funeral. While they were carrying out the rites on the days following the funeral they are said to have found a letter signed by Dionysodorus written to those on earth. He wrote that he had gone from his tomb to the lowest part of the earth, which was about 4,830 miles away.[1] Experts in geometry interpreted this to mean that the letter had been sent from the centre of the earth, which was the longest distance down from the surface and was also the centre of its sphere. Their calculation from this led them to state that the circumference of the earth was 29,000 miles.

1. i.e., the radius of the earth. Assuming, as the experts did, that $\pi = 3$, the circumference was $\pi \times 2r$, that is, approximately 29,000 miles.

BOOK III
SPAIN AND ITALY

Pliny begins his account of the physical, political and historical geography of the ancient world

1. Now I will describe the geography of the world, although this is reckoned to be a never-ending labour and one not to be undertaken lightly, or without some risk of criticism. Yet in no other discipline could allowances more fairly be made, if only because it is the least cause for surprise that one born a human being should not be aware of all human knowledge. I shall not, therefore, follow any single authority, but shall use the one I consider most reliable in each region, since just about all writers have this in common, that each describes most carefully the area in which he is writing.

2. For this reason I shall not blame or find fault with any one. I shall only list the names of places, as briefly as possible, deferring the reasons for their fame to the appropriate chapter. My account now includes the whole world.

The Pillars of Hercules

3. The whole earth is split into three continents: Europe, Asia and Africa. We start our journey in the west, at the Pillars of Hercules, where the Atlantic bursts through and spreads into the Mediterranean.

5. I begin with Europe, nurse of the people who have conquered all nations, and by far the most beautiful region of the earth. Most authorities consider, and rightly so, that it occupies a half, not a third, of the world (dividing the whole circle by a line from the River Tanais to the Pillars of Hercules).

In discussing Spain, Pliny explains the difficulties encountered in the exact measurement of distances

16. Marcus Agrippa gave the total length of Baetica as 475 miles and its breadth as 258 miles, but this was when the boundaries went as far as New Carthage. Such changes in the boundaries of provinces

42

often give rise to great errors in the calculation of distances, the mileage being both over- and under-estimated. For a long period of time the sea has encroached upon the shoreline or the shores have moved forward, rivers have curved or, once meandering, become straight. Furthermore, different people measure from different starting-points and the line they follow differs. The result is that no two authorities agree.

17. Who could believe that Marcus Agrippa, who was very diligent and careful in this field, and the Emperor Augustus were both in error when presenting a map of the world for the people of Rome to see? For Augustus completed the portico that contained the map of the world. This building had been begun by his sister following the design and rough notes of Marcus Agrippa.

30. Nearly all Spain is rich in lead, iron, copper, silver and gold mines: selenite is mined in Hither Spain, and cinnabar in Baetica. In addition there are marble quarries.

Pliny describes Italy in some detail

38. Next is Italy, the first part of which is inhabited by Ligurians. Then come Etruria, Umbria and Latium, in which are located the mouths of the Tiber and Rome – the capital of the world – some 16 miles inland from the sea. After these come the coastlines of the Volsci, Campania, Picenum, Lucania and Bruttium. Bruttium, the southernmost point of Italy, juts out into the sea from the almost crescent-shaped ridges of the Apennines. After Bruttium is the coast of Magna Graecia.

39. I know full well that it would be considered characteristic of an ungrateful and lazy mentality (and deservedly so), if my observations on Italy were casual and superficial – Italy, a land that is the nursling and mother of all other lands. Italy was chosen by the divine inspiration of the gods to enhance the renown of heaven itself, to unite scattered empires, to make customs and manners more gentle and, by the sharing of a common language, to bring together the disparate, wild tongues of many nations, that is to give mankind civilization. To put it succinctly, Italy was to become the sole parent of all races throughout the world.

40. But what am I to do? Who could even hint at the great distinction of all its places? The great fame of individual items and of its inhabitants holds me spellbound. How am I to describe the coast of Campania, a fertile region so blessed with pleasant scenery that it was manifestly the work of Nature in a happy mood?

41. Then indeed there is that wonderful life-sustaining and healthy atmosphere that lasts all the year through, embracing a climate so mild, plains so fertile, hills so sunny, woodlands so secure and groves so shady. Campania has a wealth of different kinds of forest, breezes from many mountains, an abundance of corn, vines and olives, splendid fleeces produced by its sheep, fine-necked bulls, numerous lakes, rich sources of rivers and springs that flow over the whole region. Its many seas and harbours and the bosom of its lands are open to commerce, while even the land eagerly runs out into the sea as if to assist mankind.

42. I shall not mention the character and customs of Campania, its men or the nations subjugated by its language and military might. The Greeks themselves have given their judgement on Italy by naming a very small part of it Magna Graecia.

The physical geography of Italy and its most important river, the Tiber

53–5. The River Tiber, formerly called Thybris and, before that, Albula, rises in Arretium, about the middle of the Apennines. At first it is a tiny stream navigable only where its waters are collected together in sluices and then discharged, just like its tributaries the Tinia and the Glanis, which have to be held back by dams for nine days unless rain comes to the rescue. And even when a river, because of its rough and uneven bed for much of its distance, the Tiber can be navigated only by rafts, or more accurately logs. It flows 150 miles, not far from Tifernum, Perusia and Ocriculum, separating Etruria from the Umbrians and the Sabines. But, below the Glanis from Arretium, the Tiber is swollen by forty-two tributaries, in particular the Nar and the Anio. The latter is navigable and encloses Latium to the rear and is much increased by the waters and springs carried into Rome by aqueducts. For this reason it can take large ships, of whatever size you care to name, from the Mediterranean. The Tiber is the most placid entrepreneur of products from all over the world, and has, perched on its banks and overlooking it, more villas than any other rivers anywhere: no river is more hemmed in on either side. Yet the Tiber does not fight, although it suffers frequent, sudden inundations, nowhere greater than in Rome itself. In fact, it is looked upon rather as a warning sign, its rising being an instrument of religion rather than of destruction.

56. Old Latium extends from the Tiber to Circeii and this is still a distance of 50 miles – so slight were the roots of the Empire at its beginning.

Campania and the Bay of Naples

60. Next comes the well-known fertile region of Campania. In its
hollows begin the vine-bearing hills and the celebrated effects of the
juice of the vine, famous the world over, and, as writers of old have
said, the venue of the greatest competition between Bacchus and
Ceres. From there stretch the regions of Setinum and Caecubum,
joined by those of Falernum and Calenum. After this rise Mount
Massicus, Mount Barbarus and the hills of Surrentum. At that point
the plains of Leborium spread out, and there the harvest of spelt is
polished to make tasty grits. These shores are watered by hot springs,
and in no seas can the repute of their famous fish and shellfish be
equalled. Nowhere is the olive-oil superior, another object of mankind's pleasure. Oscans, Greeks, Umbrians, Etruscans and Cam-
panians have lived here.

61. On the coast is the River Savus, the town of Volturnus, with
its river of the same name, Liternum, Cumae, originally founded by
the Chalcidians, Misenum, the harbour of Baiae, Bauli, the Lucrine
Lake and Lake Avernus, next to which formerly was the town of
Cimmerium, and finally Puteoli, once called the colony of Dicae-
archus. After this are the Phlegraean Fields and the Acherusian
Marsh, near Cumae.

62. On this coast is Neapolis itself, also founded by Chalcidians,
and called Parthenope after the tomb of one of the Sirens, Herculaneum, Pompeii with Mount Vesuvius nearby and watered by
the River Sarnus, the territory of Nuceria and, 9 miles from the sea,
Nuceria itself, and Surrentum with the promontory of Minerva,
formerly the abode of the Sirens.

Rome, the largest city in the world

66-7. Romulus left Rome with three, or, if we accept those who
put forward the highest number, four gates. The area enclosed by
its walls at the time of the principate and censorship of the Ves-
pasians,[1] measured 13 miles in circumference, taking in the seven
hills. Rome itself is divided into fourteen regions with 265 crossroads
complete with their lares. If a straight line is drawn from the
milestone at the head of the Roman Forum to each of the gates,
which today number thirty-seven (provided that the Twelve Gates

---

1. AD 73. Vespasian and Titus, who was associated with his father, shared the
duties of the censorship.

are counted as one each and the seven old gates, which no longer exist, are left out), the total length is 20 miles in a continuous line. If, in addition, one were to take into account the height of the buildings, one could scarcely avoid the conclusion that there has been no city in the whole world comparable to Rome in size. In the east, it is bounded by the mound of Tarquinius Superbus, a construction among the foremost wonders of the world; for, where the approach was flat and Rome was most vulnerable to attack, he made it as high as the walls. In other directions the city is protected by high ramparts or else by steep hills, although the spread of buildings has incorporated a number of communities into its suburbs.[1]

Sicily

86. Sicily outshines all islands in fame. It was called Sicania by Thucydides, and by several authors Trinacria because of its triangular appearance. Its coastline is 528 miles long, according to Agrippa.[2] Formerly it was joined to Bruttium, but subsequently the sea encroached upon the land and it became separated from the mainland by a strait next to the Royal Pillar, 14 miles long and $1\frac{1}{2}$ miles wide. Because of Sicily's 'breaking away', the Greeks named the town situated on the Italian side of the strait Rhegium.[3]

87. In those straits are the rock of Scylla[4] and the whirlpool of Charybdis,[5] both well known for their treacherousness. Sicily is, as I have mentioned, triangular in shape: the promontory of Pelorum points towards Italy, opposite Scylla, Pachynum towards Greece – the Peloponnese is 440 miles distant – and Lilybaeum towards Africa, some 180 miles from the promontory of Mercury and 190 miles from the promontory of Caralis in Sardinia.

88–9. Sicily has five colonies and sixty-three cities and states. From Pelorum, on the coast facing the Ionian Sea, is the town of

1. e.g., Tibur and Aricia.
2. As a general, under Augustus, Agrippa (64/3 BC–AD 12) organized a survey of the Roman Empire. He wrote a geographical commentary – the basis for the map of the Roman Empire, displayed in the Porticus Vipsania (built after his death).
3. The name Rhegium would have suggested the Greek verb *rhegnunai*, 'to break'.
4. Scylla, a legendary monster with six heads, living in a cave opposite Charybdis. She represented a rock, or other natural hazard of the sea.
5. According to legend, a whirlpool in a narrow channel of the sea – the Straits of Messina. Small whirlpools, known locally as *Garofali* or *Vortici*, can be seen in most parts of the Straits. These are caused by the difference in the speed of the currents.

Messana; its inhabitants, the Mamertines are Roman citizens. Next is the promontory of Trapani, the colony of Tauromenium, formerly Naxos, the River Asines and Mount Etna, a source of wonder at night because of its fires. The circumference of its crater is 2½ miles; its hot ash reaches as far as Tauromenium and Catana, and the noise is heard as far as Maroneum and the Gemelli Mountains. Then come the three rocks of the Cyclops and the rivers Symaethum and Terias.

92. The islands towards Africa include Gauli, Melita, Pandateria, Lipara, Therasia and Holy Island, so named because it is sacred to Vulcan. On Therasia is a hill that belches forth flames at night. The next island is Strongyle, 6 miles east of Lipara, over which Aeolus ruled. It differs from Lipara only in the flame of its volcano being brighter. The local inhabitants are said to be able to predict from its smoke how the winds are likely to blow; this is the origin of the belief that the winds obeyed Aeolus.

Italy from Calabria to the River Padus

117. The River Padus, called Eridanus by the Greeks, is second to none in renown, being famous through the punishment of Phaethon.[1] It increases in volume when the snows melt at the rising of the dog-star and, although its flooding has more effect on the fields than on navigation, yet it claims none of its spoil for itself and where it leaves its silt it bestows great fertility on the land. The Padus is 300 miles long as the crow flies from its source to the Adriatic and adds 88 miles by its winding course. It receives not only navigable rivers from the Apennines and the Alps but also huge lakes which discharge into it, and it carries as many as thirty streams with it to the sea.

119. The Padus flows to Ravenna by way of the canal of Augustus, and at that point is called the Padusa (formerly Messanicus). The mouth nearest to Ravenna forms the harbour of Vatrenus, and it was from there that Claudius sailed into the Adriatic, in what was really a huge floating palace rather than a ship, when he celebrated his triumph over Britain.

121. The Padus unites with these streams and flows through them into the sea, forming a triangle between the Alps and the coast, like

---

1. See Ovid, *Metamorphoses*, II, 47 ff. Jupiter hurled a thunderbolt at Phaethon who had mishandled the chariot of his father, the Sun, and caused him to fall into the River Padus (the Po) to prevent him from setting the earth ablaze.

that made by the Nile in Egypt (which people call the Delta). Its three sides measure 250 miles.

122. I am ashamed to borrow an account of Italy from the Greeks, but Metrodorus[1] of Scepsis states that the river gets its name from the fact that round its source there are many pine-trees of the kind called *padi* in Celtic; in the Ligurian dialect it is called Bodincus, which means 'bottomless'.

138. Such is Italy, sacred to the gods, these are its races and the towns of its peoples. She yields place to no country in the abundance of her mineral-bearing ores; but their exploitation is forbidden by an old decree of the Senate relating to the conservation of Italy.

1. Philosopher and statesman. He lived in the time of Mithridates Eupator, King of Pontus (120–63 BC), who may have been responsible for his death. Metrodorus was celebrated for his phenomenal memory and for his hatred of Rome.

BOOK IV
EUROPE AND BRITAIN

Greece

9. The Peloponnese, formerly called Apia and Pelasgia, is a peninsula, second in fame to no other geographical area. It lies between
the Aegean and the Ionian seas and is like the leaf of a plane-tree in
shape. Its pointed indentations extend the coastline to 563 miles,
according to Isidorus,[1] or nearly twice as much if you add the
coastline of the bays. The narrow strip of land from which the
Peloponnese projects is called the Isthmus. There the Aegean and
Ionian seas make inroads from opposite sides, that is, from the north
and the east. The seas close in on the peninsula leaving a mere 5
miles between the coasts, which are eroded by the onslaught of such
a volume of water from opposite directions. So Greece and the
Peloponnese are joined by a narrow neck of land.

10. The Gulf of Corinth is on one side and the Saronic Gulf on the
other. At one end is Lechaeae, at the other, Cenchreae.[2] The route
around the Peloponnese is long and fraught with danger for those
ships which cannot be transported across the Isthmus on low-loaders
because of their size.[3] King Demetrius, Julius Caesar and the emperors Caligula and Nero all attempted to dig a canal to allow the
passage of ships through the Isthmus, a venture frowned on by the
gods as evidenced by the fate of all three.

11. In the middle of this area that I have called the Isthmus is the
colony of Corinth, formerly Ephyra: its houses cling to the hillside,
and the top of its citadel, the Acrocorinth, from which flows the
spring Peirene, enjoys views of the two opposing seas.

The Peloponnese

23. Hellas, which we call Greece, begins at the narrow part of the
Isthmus. In this region is Attica – in remote antiquity known as

---

1. Isidorus of Charax, a geographer of the first century AD.
2. The harbours of Corinth and the one on the Saronic Gulf respectively.
3. See R. M. Cook, 'The Diolkos', *Journal of Hellenic Studies*, XCIX (1979),
pp. 152 f., and B. R. MacDonald, ibid., CVI (1986), pp. 191–5.

Acte. The harbours of the Piraeus and Phalerum are some 55 miles from the Isthmus and joined by walls to Athens, which is 5 miles distant from them. Athens is a free city and needs no further publicity, its great fame being more than sufficient. In Attica are the springs of Cephisia, Larine, Callirhoë with its nine wells, the mountains of Brilessus,[1] Aegialus, Icarius, Hymettus and Lycabettus, and the place called Ilissus. Some 46 miles from the Piraeus are Cape Sunium and Thoricus.

66. The other islands [of the Cyclades] are Myconus, with Mount Dimastus, 15 miles from Delos, Siphnos, which is 28 miles in circumference, Seriphos, Prepesinthus and Cynthus. By far the most famous of the Cyclades, and situated in the middle of the group, is Delos, renowned for its Temple of Apollo and for its commerce. Legend has it that Delos drifted for a long while and that, down to the time of Marcus Varro, it was the only island not to have felt an earth tremor. Mucianus says that it suffered two earthquakes. Aristotle records that it derived its name from its sudden appearance in the sea. Aglaosthenes calls it the island of Cynthus, others Ortygia, Asteria, Lagia, Chlamydia and Pyrpile (because fire was discovered there). It is 5 miles in circumference, and its highest point is Mount Cynthius.

67. Then there is Paros, famous for its marble, and Naxos, the island of Dionysus, famous for the fertility of its vineyards.

From the Hellespont to the Black Sea

75. The fourth of the major gulfs of Europe begins at the Hellespont and ends at the mouth of the Maeotic Lake. I must include a brief account of the overall shape of the Black Sea so that its regions may be easily known. It is a vast sea lying at the threshold of Asia, kept apart from Europe by the Hellespont. The Black Sea forces a way between the lands, separating Europe from Asia by a narrow channel less than a mile wide, as I have observed. The first part of the straits is called the Hellespont, where King Xerxes[2] of Persia constructed a pontoon bridge and led his army across. From there a narrow channel extends 86 miles to the

---

1. Probably an alternative name for Pentelicus, famous for its marble quarries.

2. Son of Darius, king 486–465 BC. He dug a canal through Mount Athos and bridged the Hellespont by pontoons. Xerxes won victories at Artemisium and Thermopylae (480 BC): the Greeks were forced back to the Isthmus of Corinth. Themistocles, however, won a decisive sea-battle at Salamis (480 BC).

Asiatic city of Priapus; this was the point at which Alexander the Great made his crossing.

76. After that, the sea widens and narrows again. The wide part is called the Sea of Marmara, and the narrow the Bosporus. The place where Xerxes crossed is 500 yards wide: the distance from there to the Hellespont is 240 miles.

77. Next comes the Black Sea, at one time known as the Axenus.[1] It swallows up a large area of land which retreats before it. With a great bend in its coasts the Black Sea curves backwards and stretches out on either side so as to form the exact shape of a Scythian bow. In the middle of the curve, the entrance to the Maeotic Lake joins it. The opening is called the Cimmerian Strait, and is 2½ miles wide. The total distance from the Hellespont is 500 miles, according to Polybius.

The circumference of the Black Sea is 2,150 miles according to Varro and earlier authorities generally. Cornelius Nepos adds 350 miles; Artemidorus says that the circumference is 2,119, Agrippa 2,540, Mucianus 2,425 miles respectively.

The Danube region and south Russia

88. Next come the Ripaean Mountains and the region called Pterophoros[2] because of the continuous fall of feather-like snow, a part of the world condemned by Nature, steeped in dense gloom and occupied only by frost and the freezing lair of the north wind.

89. Behind these mountains and beyond the north wind (if we are to believe this) lives a happy race known as the Hyperboreans who survive to a ripe old age and are famous for marvellous things handed down in stories. It is believed that here are the hinges on which the world turns and the extreme limits of the circuits of the stars. The Hyperboreans enjoy six months of daylight. The sun rises once a year, at midsummer, and sets once, at midwinter. It is a sunny region with an equable climate, free from all harmful gales. The Hyperboreans live in the woods and groves and worship gods both individually and in groups; all disharmony and sorrow are unknown. Death does not come until they have had their fill of life. Setting a banquet, they greet their old age with luxury, and then

---

1. So named from the Greek, *axenos*, 'Inhospitable' Sea: a cold, stormy region, without sheltering islands. Greek sailors normally hopped from island to island, wherever possible, rather than crossing the open sea.

2 From *pterophoros*, 'winged'.

leap into the sea from a certain rock. This method of burial is the most serene.

90. Some authorities locate the Hyperboreans not in Europe, but on the nearest part of the Asian coast. Those who place them in the area where there are only six months of daylight state that they sow in the morning, reap at midday, pick the fruit from the trees at sunset, and spend the night in caves.

91. But there is no room to doubt that this race exists, since so many authorities state that their custom is to send the first-fruits of their harvest to Delos, as offerings to Apollo, whom they specially worship. Virgins used to take these and were for many years held in respect and given hospitality by the local people until, because of a breach of good faith, they began the custom of depositing their offerings at the nearest frontiers of the neighbouring people; these in turn took the offerings to their neighbours and so on until they arrived at Delos. After a while this custom disappeared.

Britain

102. Opposite the Rhine and the Meuse is the island of Britain, well known in Greek and Roman historical records.[1] It is situated to the north-west, facing, but separated by a wide channel from, Germany, Gaul and Spain, which together make up the greater part of Europe. Its native name is Albion, while all the islands, about which I shall speak a little later, are called the Britains. From Gesoriacum on the coast occupied by the Morini the shortest distance is 50 miles. Pytheas and Isidorus record its shoreline as being 4,875 miles in length. Almost thirty years ago Roman armies explored Britain but did not go beyond the vicinity of the forest of Caledonia.[2] Agrippa believes the length of the island to be 800 miles, and its width 300 miles, while Hibernia is the same width but only 600 miles long.

103. Hibernia lies beyond Britain, the shortest crossing – from the Silures – being some 30 miles. None of the remaining islands is said to possess a shoreline of more than 125 miles.

Other islands are the Orcades, Acmodae, Hebudes, Mona and Monapia.

104. The historian Timaeus states that there is an island named Mictis within six days' sailing from Britain, where tin is found and

---

1. As, for example, in Dionysius Periegetes, *A Description of the Known World* (*Periegesis tes oikoumenes*), written in a pseudo-epic style in the time of Hadrian.
2. See generally Tacitus, *Agricola*.

to which the Britons cross in coracles constructed of osier covered with sewn hides.

112. The whole district, from the Pyrenees onwards, is full of gold, silver, iron, lead and tin mines.

119. Opposite Celtiberia are a number of islands that the Greeks call the Cassiterides[1] because of the rich supply of tin found there.

---

1. The 'Tin Islands', first mentioned in Herodotus, III, 115.

BOOK V

THE CONTINENTS OF AFRICA AND ASIA

Africa and the two Mauretanias

1. Africa was called Libya by the Greeks, and the sea in front of it the Libyan Sea; it shares a boundary with Egypt. No other part of the world has fewer bays, and the coast stretches obliquely from the west. The names of its peoples and towns are completely unpronounceable except by the natives.

2. To begin with there are the two countries called Mauretania, which were kingdoms until the time of the Emperor Caligula, but, thanks to his cruelty, were divided into provinces. Beyond the Pillars of Hercules there used to be the towns of Lissa and Cotte; but now there is only Tingi, originally founded by Antaeus and afterwards named Traducta Julia by the Emperor Claudius when he established a colony there. Sixty-two miles from Tingi is Lixus, made a colony by the same emperor, a town about which writers of old recount very many legends.

3. This was the site of Antaeus' palace, where the struggle with Hercules took place, and the location of the Gardens of the Hesperides. A channel flows inland from the sea with a wandering course that, as people nowadays explain, looks like a snake guarding the place. It encompasses an island that is the only part not flooded by the tides, even though the neighbouring area is higher. On this island there is also an altar of Hercules, but nothing else, except wild olive-trees, remains of that famous grove[1] which, according to the legend, bore golden apples.

4. No doubt the outrageously false Greek stories about these snakes and the River Lixus occasion less wonder to people who reflect that we, not very long ago, have recorded stories about them that are little less fantastic. According to these accounts, the city of

1. The scene of one of the Twelve Labours of Hercules, set him by King Eurystheus. According to tradition, the golden apples in the Garden of the Hesperides (beyond the Atlas Mountains) were guarded by a snake. These had been given by Earth to Hera on her marriage to Zeus. Pliny attempts some rationalization of the legend by suggesting that a winding inlet of the sea in this region had led to the idea of the snake.

Lixus is very powerful and greater than Great Carthage, and more-over is on the same meridian as Carthage, an almost immeasurable distance from Tingi. All this, and more, Cornelius Nepos believed with very great relish.

Mount Atlas

6. Mount Atlas has attracted more legends than any other mountain in Africa. Men report that it rises out of the middle of the sands into the sky, rugged and jagged on the side facing the coastline of the ocean to which it has given its name, while on the side facing the hinterland of Africa it is shaded by woods and irrigated by gushing springs. Fruits of all kinds grow spontaneously there in such profu-sion that pleasure is always satisfied.

7. It is said that not a single inhabitant is seen during the day and everything is quiet with a chilling silence like that of the desert; an apprehensiveness that renders one speechless steals over those who approach the mountain, and similarly a fear of the peak soaring above the clouds and reaching almost to the moon. At night Atlas flashes with many fires, so men say, and is filled with the wanton frolics of Goat-Pans[1] and Satyrs and resounds with the music of flutes, pipes, drums and cymbals. Famous authors have published these stories in addition to the exploits of Hercules and Perseus that took place there. Mount Atlas is a tremendous distance away and approached across uncharted terrain.

The exploration of Africa

8. Some notes have survived by the Carthaginian commander Hanno, who, in the heyday of Carthage, was ordered to circum-navigate Africa. The majority of Greek and Roman writers follow Hanno both in their legendary stories and in their accounts of the many settlements founded by him in Africa; neither memory nor trace of these settlements now exists.

9–10. While Scipio Aemilianus[2] was commander in Africa, the historian Polybius was given a fleet to explore that part of the world. Sailing round the coast, Polybius reported that west of Mount Atlas there are forests full of the wild animals that Africa

---

1. Aegipanes.
2. Son of Aemilius Paullus who defeated King Perseus at Pydna (168 BC) in the Third Macedonian War. He was also a grandson, by adoption, of Scipio Africanus.

produces. In the River Bambolus there are many crocodiles and hippopotamuses.[1]

11. The first time Roman forces fought in Mauretania was during the principate of Claudius. King Ptolemy had been put to death by Caligula and his freedman Aedemon was seeking to avenge him; it is generally accepted that our soldiers went as far as Mount Atlas, and at this point the natives fled.

12. There are five Roman colonies in that province. According to widespread reports it might seem to be an accessible region; but, put to the test, this view is found to be almost completely fallacious; persons of rank, although unwilling to track down the truth, are not ashamed to tell falsehoods because they cannot bear to admit their ignorance. Credibility never more readily falls flat on its face than when an authority of weight supports a false assertion. As for myself, I am less surprised that certain matters are unknown to persons of the equestrian order – indeed some now enter the Senate – than that anything should be unknown to Luxury, which is a very great and influential power inasmuch as men scour forests for ivory and citrus-wood and all the rocks of Gaetulia for the murex and for purple.

14. Suetonius Paulinus, consul in my time,[2] was the first Roman commander actually to cross the Atlas Mountains, and he went some miles further.[3] His estimate of their height agrees with that of other authorities, but he further adds that the lower slopes are filled with dense forests of tall trees of an unknown species: they have very tall trunks notable for their sheen and freedom from knots. Their leaves, like those of the cypress except for the heavy scent, are covered with a thin down, from which, with a suitable technique, clothing can be made just like that derived from the silkworm. The summit of Mount Atlas is covered with deep snow, even in summer.

15. Suetonius Paulinus reached there in ten days, and travelled beyond to the River Ger, across deserts of black dust, with

1. Illustrated in Roman wall-paintings and mosaics. See D. Strong, *History of Roman Painting* (Harmondsworth, 1976), p. 36.

2. AD 66.

3. At the end of his reign, Caligula caused a rebellion in Mauretania by putting King Ptolemy to death. Ptolemy was a son of Augustus' nominee Juba. The revolt was suppressed by Suetonius Paulinus, the future governor of Britain. As propraetor (AD 42) he pursued the rebels beyond Mount Atlas to the Sahara. Claudius carried out the plans of his predecessor by making the kingdom into two provinces, namely, Mauretania Caesariensis (Algeria) and Tingitana (Morocco).

projecting rocks in some places that looked as if they had been burnt – a place uninhabitable because of the heat, although it was winter when he experienced it. The Canarii live in the neighbouring forests, which are full of every species of elephant and snake.

Cyrenaica

33. The territory of Cyrene is considered good, to a depth of 15 miles from the coast, for even growing trees, but a further 15 miles inland, for growing only corn; then there is a strip 30 miles wide and 250 miles long, suitable only for silphium.

34. Next after the Nasamones come the Asbytae and Macae; and beyond them, some twelve days' journey from the Greater Syrtes, are the Amantes. They are hemmed in by sand towards the west, but find water without difficulty in wells about 3 feet deep, since this place receives the overflow of water from Mauretania. The Amantes construct their houses of blocks of salt quarried out of their mountains like stone.

The interior of Africa

44. The River Niger has the same nature as the Nile. It produces reeds, papyrus and the same animals, and rises at the same seasons. Some place the Atlas tribe in the middle of the desert and next to them the half-animal Goat-Pans, the Blemmyae, Gamphasantes, Satyrs and Strapfeet.

45. The Atlas tribe is primitive and subhuman, if we believe what we hear; they do not call each other by names. When they observe the rising and setting sun they utter terrible curses against it, as the cause of disaster to themselves and their fields. Nor do they have dreams in their sleep like the rest of mankind. The Cave-dwellers hollow out caves which are their houses;[1] their food is snake meat. They have no voice but make a shrill noise, thus lacking any communication by speech. The Garamantes do not marry but live promiscuously with their women. The Augilae worship only the gods of the lower world. The Gamphasantes wear no clothes, do not fight and do not associate with any foreigner.

46. The Blemmyae are reported as being without heads; their mouth and eyes are attached to their chest. The Satyrs have no

---

1. See Herodotus, IV, 183, Diodorus Siculus, III, 32–3, and Strabo, 775–6 for descriptions of the Cave-dwellers.

human characteristics except their shape. The form of the Goat-Pans is as commonly depicted. The Strapfeet are people with feet like thongs who naturally move by crawling. The Pharusi, formerly from Persia, are said to have been Hercules' companions on his journey to the Gardens of the Hesperides. I cannot think of any more to record about Africa.

Egypt

47–8. Asia is joined to Africa; the distance from the Canopic mouth of the Nile to the entrance of the Black Sea is given by Timosthenes as 2,638 miles. The inhabited country next to Africa is Egypt, which extends southwards into the interior, where the Ethiopians border it to the rear. Two branches of the Nile divide to the right and left, forming the boundaries of Lower Egypt: the Canopic mouth separates it from Africa, and the Pelusiac mouth separates it from Asia, with a space of 170 miles between the two. For this reason some authorities put Egypt among the islands because the Nile divides in such a way as to make a triangular shaped-area of land; consequently many have called Egypt after the Greek letter Delta.[1]

51. The sources of the Nile are uncertain, for it flows through burning deserts for a very great distance and is explored only by unarmed travellers, except in time of war; wars have brought all other countries to light. Its origin, as far as King Juba was able to discover, is in a mountain in Lower Mauretania, not far from the ocean; here it forms a stagnant lake named Nilides. Fish found in this are the *alabeta, coracinus* and *silurus.* A crocodile was brought from the lake by Juba[2] to prove his theory and was placed as an offering in the temple of Isis at Caesarea, where it can be seen today. Moreover, it has been observed that the Nile rises when snow or rain in Mauretania floods it.

53. The Nile separates Africa from Ethiopia, and, even though the river bank is not inhabited by humans, it is full of wild animals and large creatures and gives sustenance to forests. Where it cuts through the middle of Ethiopia it is called Astapus, which in the local language means 'water flowing from the shades'.

54. From time to time the river is dashed against islands and

1. Δ.
2. King of Numidia and, later, of Mauretania also (46–19 BC). Juba was a prolific writer.

spurred on by these obstructions until at last it is shut in by mountains; there its flow is faster than elsewhere and it is borne on rapidly to the place in Ethiopia known as Catadupi. Here, at the First Cataract, because of the crashing sound, it seems not to flow but to race between the rocks that stand in its way. Thereafter, it is gentle and the violence of its waters is broken and subdued; the river is also weary because of the distance it has travelled, and it discharges itself by its many mouths into the Mediterranean. However, when its volume is greatly increased, it spreads out for a certain number of days, flooding the whole of Egypt; in this way it fertilizes the land.

55. People have advanced various explanations for the rising of the Nile, but the most likely are either the backflow caused by the Etesian winds which blow against the current of the river at that time of the year – the sea outside being driven into the mouths of the Nile – or the summer rains in Ethiopia, which are caused by the same winds bringing clouds there from the rest of the world.

58. The Nile's increase is detected by wells that are marked with a scale. An average rise is 24 feet. A larger volume of water, by receding too slowly, stunts growth and detracts from the time available for sowing because the soil is wet, while a smaller volume does not irrigate everywhere and allows too little time because the soil is dry. The greatest rise to date was 27 feet when Claudius was emperor, the smallest, $7\frac{1}{2}$ feet in the year of the Battle of Pharsalus,[1] as if the Nile were trying to avert the murder of Pompey by some portent. When the water has finished rising, the floodgates are opened and they release it. As each piece of land is freed from the onrush of water, it is sown. The Nile is the only river that does not give off vapours.

59. Elephantis is an inhabited island some 4 miles below the Last Cataract and 16 miles above Syene. This is the limit of navigation in Egypt, some 585 miles from Alexandria. Elephantis is the place that Ethiopian ships make for, since they are collapsible and the crew carry them on their shoulders whenever they arrive at the cataracts.

60. As well as other claims to glory in the past, Egypt had the distinction of having had 20,000 cities when Amasis was king[2] and

1. 48 BC.

2. 569–525 BC. He allied himself with Lydia, Samos, Cyrene and, perhaps, Sparta, against the Persians which led to the overthrow of Egypt (525 BC) shortly after his death.

even today it has a very large number, but they are of no consequence.

62. Alexandria,[1] however, is justly worthy of praise; it was built by Alexander the Great on the Mediterranean coast 12 miles from the Canopic mouth towards Africa and next to Lake Mareotis. It was planned by the architect Dinochares,[2] who was famous for a variety of talents. The town is 15 miles wide and, in plan, is the shape of a Macedonian soldier's cloak, with indents in its circumference and projecting corners on the right and left sides. A fifth of the site was set aside for the royal palace.

Lake Mareotis, to the south of the city, sends traffic from the interior by means of a canal from the Canopic mouth of the Nile.

Syria

66–7. The next country on the coast is Syria, formerly the greatest of lands; the whole of the sea off the coast is called the Phoenician Sea. The Phoenician race itself is held in great honour for the invention of the alphabet and for the sciences of astronomy, navigation and military strategy.

Judaea, the Jordan and the Dead Sea

70–1. Beyond Idumaea and Samaria, stretches the wide expanse of Judaea. The River Jordan rises from the Spring of Panias. This is a pleasant stream and winds along as far as the terrain allows, and serves the people who live on its banks; it progresses, seemingly with reluctance, towards the gloomy Dead Sea, by which it is finally swallowed.

72. The Dead Sea produces only bitumen, from which it derives its name.[3] The bodies of animals do not sink in it – even bulls and camels float; and from this arises the report that nothing sinks. It is more than 100 miles long and 75 miles wide at its widest part, but only 6 miles at the narrowest. In the east the Dead Sea is faced by

1. The city, founded by Alexander the Great stood on the neck of land between the sea and Lake Mareotis, with harbours on both sides. Its architect, Deinocrates, laid out Alexandria on the rectangular plan usual in Hellenistic cities. The roads that have been uncovered are Roman. Our knowledge, however, of the Hellenistic city is principally derived from Strabo's account. See further W. W. Tarn and G. T. Griffith, *Hellenistic Civilisation* (London, 1966), pp. 183 ff.

2. Pliny must mean Deinocrates.

3. *Asphaltites lacus.*

Arabia where the Nomads live, in the south by Machaerus, once second only to Jerusalem as a stronghold in Judaea. On the same side there is a hot spring with health-giving properties, Callirrhoë, which shows its reputation by its very name.[1]

73. On the west side of the Dead Sea, away from the coast, where there are harmful vapours, lives the solitary tribe of the Essenes.[2] This tribe is remarkable beyond all others in the whole world, because it has no women, has rejected sexual desires, is without money and has only the company of palm-trees. Day by day the crowd of refugees is renewed by hordes of people tired of life and driven there by the waves of fortune to adopt their customs. Thus through thousands of ages – incredible to relate – the race in which no one is born lives for ever; so fruitful for them is other men's dissatisfaction with life!

Below the Essenes was the town of Engeda, second only to Jerusalem in the fertility of its soil and in its groves of palm-trees, but now, like Jerusalem, another heap of ashes. Then comes Masada, a fortress on a rock, not far from the Dead Sea. This is the extent of Judaea.

74. Next to Judaea, towards Syria, is the region of the Decapolis, so called because of the number of its towns, although not all writers have the same towns in their list; most, however, include Damascus.

Asia Minor

88. Palmyra is a city famous for its location, for the riches of its soil and for its pleasant springs. On all sides its fields are surrounded by sand, and it is, as it were, isolated by Nature from other lands. It has a destiny of its own between the two superpowers, Rome and Parthia.

115. On the coast are Notium and Ephesus built by the Amazons on the slope of Mount Pion, and watered by the River Cayster

---

1. Which means: 'beautiful-flowing'.
2. A religious sect, or brotherhood, which flourished in Palestine from the second century BC to the end of the first century AD. The Essenes were clustered in monastic communities which, generally at least, excluded women. Property was held in common and all details of daily life were regulated by officials. Pliny reckons that they were some 4,000 in number. Like the Pharisees, the Essenes meticulously observed the Law of Moses, the sabbath and ritual purity. They also professed belief in immortality and dire punishment for sin. They denied the resurrection of the body and refused to take part in public life.

which rises in the Cilbian Range and brings down the waters of many streams, and which drains the Pegasaean Marsh formed by the overflow of the River Phyrites. From these rivers comes a quantity of silt that adds to the coastline and has now joined the island of Syrie to the mainland by mud-flats. In the city of Ephesus is the spring called Callippia and the Temple of Diana.

124. In the Troad even now is the small city-state of Scamandria, and 2½ miles from its harbour, Ilium, a town exempt from tribute. Ilium was the setting of the famous epic poem.[1]

Cyprus and other islands

128. Of the islands off the coast of Asia Minor the first is at the Canopic mouth of the Nile, named after Menelaus' helmsman Canopus. The second, called Pharos, joined by a bridge to Alexandria, was a settlement of the dictator Julius Caesar; formerly it was a day's sail from Egypt, and now it guides the course of ships at night from its lighthouse tower. For, because of treacherous shoals, Alexandria can only be approached by three channels, namely those of Steganus, Posideum and Taurus. Then in the Phoenician Sea, off Joppa, lies Paria, the whole of which is a town. This, men say, is the place where Andromeda was exposed to the creature from the sea.[2] Also off Joppa is Aradus; between there and the mainland, according to Mucianus, fresh water is raised by means of a leather pipe from a spring about 75 feet deep that gushes out of the sea-bed.

1. Homer's *Iliad*.
2. Andromeda was rescued by Perseus who used the Gorgon's head to turn the sea monster into stone.

BOOK VI

THE BLACK SEA, INDIA AND THE FAR EAST

1. The Black Sea reaches into Europe and Asia. It was formerly called Axenus because of the inhospitable roughness of its waters, which results from a peculiar jealousy on the part of Nature who there indulges the sea's greed without any limit.

2. The Bosporus derives its name from the fact that the passage can be forded by oxen. There the singing of birds and barking of dogs on one side can be heard on the other, and even the exchange of human speech; indeed conversation is carried on between the two worlds of Asia and Europe, except when the wind bears away the sound.

3. Some authorities give the distance of the Black Sea from the Bosporus to the Maeotic Lake as 1,438 miles but Eratosthenes estimates it as 1,338 miles.

43. Ecbatana, the capital of Media founded by King Seleucus, is 750 miles from Great Seleuceia and 20 miles from the Caspian Gates. The main towns of Media are Phazaca, Aganzaga and Apamea. The reason for the name 'Gates' is that the range is pierced by a narrow, man-made pass 8 miles long, barely wide enough for a single line of waggons to pass through. Rocks overhang on the right and left and give the appearance of having been burnt. The region is waterless for a distance of 28 miles. The narrow passage is obstructed by a stream of salt water from the rocks that makes its way out through the same pass. Moreover, the large number of snakes allows one to pass through only in winter.

The Caspian Sea

51. Alexander the Great said that the water of the Caspian Sea was sweet to drink and Marcus Varro states that drinking water was brought from it for Pompey when he was campaigning in that region during the Mithridatic War. No doubt the volume of the rivers flowing into the Caspian neutralizes the salt.

The Scythians and China

53. After leaving the Caspian Sea we come to the Scythians and to more deserts inhabited by wild beasts, until we reach a mountain range called Tabis, which rises as a cliff over the sea; and, not until we have travelled nearly half the length of the coast, which faces north-east, is that region inhabited.

54. The first human beings we come to are the people called the Chinese, who are well known for a woollen substance[1] obtained from their forests. After soaking the leaves in water they comb off the white down and so give our women the double task of carding the fibres and weaving them together again. So many are the operations involved and so distant is the region of the world from which this material is sought to make it possible for the Roman ladies to show off diaphanous clothes in public. The Chinese are mild in character, but resemble wild animals in that they shun the company of the rest of their fellow men and wait for traders to come to them.

India

56. From the Himalayas onwards there is complete agreement as to the races. First there is the Indian race, who live not only on the Eastern Sea but on the Southern also, which we have called the Indian Ocean.

57. A great many authorities give the length of the coast of India as forty days' and nights' sail and the distance from north to south as 2,850 miles. Agrippa says it is 3,300 miles long and 2,300 miles in width. Posidonius gives a measurement from north-east to south-east, making the whole of India face the west side of Gaul.

58. Its races and cities would be countless, if one wanted to list them all. The troops of Alexander the Great and King Seleucus, and Antiochus who succeeded him, and their admiral Patrocles, who sailed round India even into the Hyrcanian and Caspian Sea, demonstrated this. Likewise, the other Greek authorities who have been entertained by Indian kings, for example Megasthenes[2] and

1. This material was probably cotton made into calico or muslin.
2. An envoy of Seleucus Nicator, founder of the Syrian monarchy. Megasthenes wrote his work on India (*Indika*) as a result of his experience while ambassador to the king of the Prasii.

Dionysius, sent by Philadelphus for that purpose, have reported the strength of these nations.

59. There is, however, no room for precise accuracy in this matter, so divergent and hard to believe are the accounts given. Those who accompanied Alexander the Great have written that in the region of India that he subdued there were 5,000 towns, none with a population of fewer than 2,000, and 9 nations; they also state that India comprises a third of the whole land surface of the world and that its populations are uncountable. The latter statement is very probable since the Indians are virtually the only race that has never left its own territory. From the time of the god Dionysus to Alexander the Great – that is, in a period of 6,451 years and three months – there were 153 kings of India.

The great rivers of India[1]

60. The rivers are enormous in length and it is recorded that Alexander the Great, sailing on the Indus never less than 75 miles a day, could not reach the mouth of that river in under five months. Yet, it is agreed, the Indus is shorter than the Ganges. The Roman author Seneca, who wrote a treatise on India, records 60 rivers and 118 races. It would be an equally laborious task to enumerate all the mountains of India.

65. Some authorities say that the Ganges rises from unknown sources, like the Nile, and irrigates the neighbouring areas in the same way, but others state that its source is in the mountains of Scythia. It flows out in a gentle stream, with an average width of $12\frac{1}{2}$ miles and a width of 8 miles at its narrowest point. The river is nowhere less than 100 feet deep. The last people situated on its banks are the Gangarid Calingae.

66. Their king lives in the city of Pertalis. He commands 60,000 infantry, 1,000 cavalry and 700 elephants, always ready and equipped for war. The peoples of the more civilized Indian races are divided into many classes in their daily life. Some work on the land, others do military service, others export local goods and import goods from abroad; the élite of society and the richest men govern, serve as judges and act as counsellors to the kings. There is a fifth class of persons who devote their efforts to philosophy, which is held in esteem by the Indians and raised to the status of almost a religion.

1. Pliny follows the accounts of Diognetus and Baeton, surveyors on Alexander the Great's expedition.

These men always end their life by a voluntary death on a pyre that they themselves have previously set alight. There is yet a further class: half-wild people, fully engaged in hunting and taming elephants – an occupation from which the classes already mentioned are kept away. These people plough with elephants and are carried by them; elephants are their best cattle and they use them in battle and when defending their frontiers. Strength, age and size are the factors that determine their choice for use in war.

70. To the south of the Ganges the tribes are coloured by the sun's heat, but not burnt like the Ethiopians. The nearer they are to the Indus, the more colour they show. The Indus is located immediately after the Prasii, in whose mountain regions there are said to be pygmies. Artemidorus records the distance from the Ganges to the Indus as 2,100 miles.

The Indus and beyond

71. The Indus, or Sindus, as the natives call it, rises on the east side of a ridge of Mount Caucasus called the Paropanisus; it receives nineteen tributaries, the best known being the Hydaspes, but owing to its gentle course the Indus is nowhere wider than above 6 miles or 75 feet deep. It forms an island of considerable size called Prasiane and another smaller one called Patale.

72. The main river is navigable for a distance of 1,240 miles according to the most conservative authorities, and flows into the ocean after following the sun westwards.

74–5. Next come the Nareae who are hemmed in by the Capitalia Range, the highest mountains in India. The inhabitants on the far side of this range work gold and silver mines over a wide area. Next come the Odonbaeoraes and the Arabastrae, whose fine city, Thorax, is guarded by marshy canals, which crocodiles, very greedy for human flesh, force one to approach by a bridge.

Taprobane[1]

81. Taprobane was long considered to be another world, with the name 'Land of the Antichthones'. The age of Alexander the Great and his achievements proved that it was an island. Onesicritus, an admiral in Alexander's navy, writes that elephants are bred there larger in size and more warlike in spirit than those in India. Megas-

1. Sri Lanka.

thenes says that Taprobane is divided by a river and that the natives are called Aborigines[1] and produce more gold and larger pearls than the Indians. Eratosthenes gives the dimensions of the island as 804 miles in length and 575 miles in breadth;[2] he adds that there are no cities, but 700 villages.

82. Taprobane is in the Eastern Sea and stretches along the side of India from east to west. It was formerly believed to be twenty days' sail from the Prasii, but later, when the voyage was made in boats built of papyrus and with the kind of rigging employed on the Nile, its distance was fixed, with reference to that covered by our ships, as seven days. The sea between Taprobane and the Indian mainland is shallow, not more than 18 feet deep, but in certain channels the depth is such that no anchors can rest on the sea-bed. Therefore, ships with prows at both ends are used so that they do not have to turn round while passing through the narrow part of the channel; their capacity is 3,000 amphorae.

83. The people of Taprobane do not take any observations of the stars while at sea; indeed there the Great Bear is not visible. They carry birds with them, let them go and then follow their course as they head for land. They do not put to sea for more than four months in the year and especially avoid the 100 days from mid-summer, when the sea is stormy.

84. This much has been recorded by early writers. During the principate of Claudius, however, more accurate information became available to us when an embassy came from Taprobane in the following circumstances. Annius Plocamus had obtained a contract from the Treasury to collect the taxes from the Red Sea region. A freedman of his, while sailing round Arabia, was carried by north winds beyond Carmania and fourteen days later reached the harbour of Hippuri in Taprobane, where the king kindly offered him hospitality. In six months he learned the language thoroughly, and in reply to the king's questions told him about the Romans and their emperor.

85. The king heard many things, but was most impressed with Roman honesty in that amongst the money in his guest's possession, the denarii were all equal in weight, although the different likenesses on them indicated that they had been struck by various emperors. He was moved by this particularly to adopt a friendly attitude to the Romans and sent four envoys to Rome, led by one Rachias.

---

1. *Palaegoni*: 'born long ago'.
2. The actual measurements of the island of Sri Lanka are 271 × 137 miles.

From these we discovered the following facts about Taprobane: it contains 500 towns and there is a harbour facing south, next to the town of Palaesimundus, which is the most famous of all places on the island; it has a royal palace and a population of 200,000.

86–7. The nearest promontory in India is called Coliacus, four days' sail away – passing, in mid-voyage, the Island of the Sun. The sea there is a deep green colour and trees grow in it in clumps, the tops of them being grazed by ships' rudders.[1]

88. The envoys also informed us that the side of Taprobane facing India is 1,250 miles long and lies south-east of India; that beyond the Himalayas are the Chinese, who are known to them through trading. Rachias' father had travelled to China; when they arrived on the coast, the Chinese always hurried to the beach to meet them. The people themselves, so the envoys explained, are above average height, have golden-coloured hair, blue eyes and harsh voices, although no conversation was had with them. The rest of what the envoys said agreed with our merchants' accounts, namely that their goods were left on the opposite bank of a river next to those put out for sale; they took these away if the exchange was agreeable. In no other context is hatred of luxury more justified than if the imagination travels to those parts and considers what is obtained from there, the method of trading employed and the reasons for it.

The way of life in Taprobane

89. But not even Taprobane, although banished by Nature beyond our world, is free from our vices. Gold and silver are highly valued there also, and a kind of marble like tortoise-shell and pearls and precious stones are esteemed; indeed the whole accumulated pile of luxury stands higher than ours. The envoys from Taprobane told us that their people had far more wealth than we had, but that we made greater use of ours. No one had a slave, no one slept beyond day-break or took a siesta; their buildings were of moderate height; the price of corn never increased; there were no lawcourts or lawsuits; they worshipped Hercules; the king was elected by the people – someone getting on in years, forbearing in attitude and childless. If the king had a child after his accession, he was forced to abdicate so that the monarchy should not be hereditary.

90. They went on to explain that thirty advisers were assigned to

1. Pliny is describing swamps.

the king by the people and no one could be condemned to death except by a majority vote of these. There was the right of appeal to the people and a jury of seventy was appointed; if they acquitted the accused, the thirty lost their reputation and were held in the deepest disgrace. The king wore the clothes of Father Liber, and the rest Arabian dress.

91. If the king did anything wrong he was condemned to death. No one, however, carried out the sentence; instead, all the people boycotted him and refused to speak to him.

The people spent their holidays hunting; tiger and elephant hunts were the most popular. They were diligent farmers but did not have any call for the vine, although apples were abundant. They also liked fishing, especially for turtles, the shells of which were large and used to roof their homes. A hundred years was regarded as a moderate life-span.

Near Eastern sea routes

121. Babylon is the capital of the Chaldaean peoples and for a long while was outstandingly famous among cities throughout the whole world. For this reason the rest of Mesopotamia and Assyria was called Babylonia. The city has two walls with a circumference of 60 miles, each being 200 feet high and 50 feet wide. The Euphrates flows through the city with marvellous embankments on either side. The Temple of Jupiter Belus[1] still survives; Belus was the discoverer of the science of astronomy.

122. The rest of Babylon has reverted to desert, and its population has been drained by the proximity of Seleuceia, which Nicator founded to replace it slightly less than 90 miles distant, at the point where the Euphrates canal joins the Tigris. Seleuceia is still referred to as Babylonian although nowadays it is a free, independent city which retains Macedonian customs. It is said that the population is 600,000, that the plan of the walls is like the shape of an eagle spreading its wings and, finally, that the region is the most fertile in the whole of the East.

Arabia

160–1. The knight Aelius Gallus is the only person to have carried

---

1. The form Belus is hellenized from Ba'al, Bel sometimes being recognized as a divine title; more often it is taken as the name of an oriental king.

Roman arms into Arabia; for Gaius Caesar, son of Augustus, had only a brief view of the country. The discoveries Gallus reported on his return were that the Nomads live on milk and wild animals' flesh, while the remaining tribes get wine from palm-trees – as the Indians do – and oil from sesame. The Homeritae are the most numerous people; the Minaei have land that grows palm groves and timber in abundance and is rich in flocks; the Cerbani, Agraei and especially the Chatramotitae excel in fighting. The Sabaei are the richest because of the fertility of their forest in producing perfumes, their gold mines, irrigated fields and production of honey and wax.

162. The Arabs wear turbans, or leave their hair uncut; they shave their beards but leave a moustache – others, however, leave the beard also unshaven. Strange to relate, of these countless tribes half live by trade, half by marauding. Overall they are the richest of peoples because very great wealth from Rome and Parthia settles in their coffers when they sell what they catch in the sea or forest, and they buy nothing in return.

Africa and a proposed canal from the Nile to the Red Sea

165. Next comes the Tyro tribe and, on the Red Sea, the harbour of the Daneoi, from which Sesostris, king of Egypt, intended to carry a ship-canal to where the Nile flows into what is known as the Delta; this is a distance of over 60 miles. Later the Persian king Darius had the same idea, and yet again Ptolemy II, who made a trench 100 feet wide, 30 feet deep and about 35 miles long, as far as the Bitter Lakes.

166. Fear of causing a flood stopped him from going further, since it was discovered that the level of the Red Sea is $4\frac{1}{2}$ feet above the level of the land of Egypt. Some do not advance this reason, but say that he was afraid that making an inlet from the sea would pollute the water of the Nile, which is the sole source of drinking water.

Ethiopia

187. It is not at all surprising that the outermost districts of Ethiopia produce animal and human freaks when you consider the capacity and speed of fire in moulding bodies and in carving their outlines. There are certain reports from the interior, on the eastern side, of races without noses and with completely flat faces; in some cases tribes have no upper lip, in others, no tongue.

188. One group has no mouth and no nostrils; these people breathe through a single hole and similarly suck in drink by means of oat straws and also take in grains of wild oats for food. Some use nods and gestures instead of speech. Some even, until the reign of King Ptolemy Lathyrus in Egypt, were unaware of the use of fire. Some writers have actually recorded a race of pygmies living among the marshes in which the Nile rises.

Some islands off the coast of Africa

198. Ephorus,[1] Eudoxus[2] and Timosthenes[3] state that there are a large number of islands scattered over the Eastern Sea. Clitarchus[4] says that one was reported to King Alexander as being so rich in resources that its inhabitants paid a talent of gold for a horse and that another, where a holy mountain had been found, was covered with a dense forest, from the trees of which fell drops of moisture with a wonderfully pleasant smell.

200. Opposite the Horn of the West we are told that there are some islands called the Gorgades, formerly the home of the Gorgons, and which, according to the account of Xenophon of Lampsacus, are two days' sail from the mainland. The Carthaginian general Hanno reached these islands and reported[5] that the women's bodies were covered in hair. Hanno put the skins of two of the women in the Temple of Juno to prove his story and as a curious exhibit; they were on show there until Rome captured Carthage.

1. Ephorus of Cyme (*c.* 405–330 BC), a pupil of Isocrates and the chief source of Diodorus Siculus. He envisaged the history of the Greek peninsula as a unity and was the first to write a complete account from the mythical beginnings down to Philip of Macedon. His work was the basis for many subsequent histories.

2. A Greek traveller from Cyzicus (*c.* 240 BC).

3. A Rhodian who served as an admiral under Ptolemy Philadelphus (*c.* 280 BC).

4. Writer of a history of Alexander the Great's Asiatic expedition, which he had accompanied.

5. In his work *Coasting Voyage* (*Periplous*).

ZOOLOGY

BOOK VII
MAN

Man is the highest species in the order of creation

1. The nature of living creatures in the world is as important as the study of almost any other field, even though the human mind is not able to pursue all aspects of the subject. Pride of place will rightly be given to one for whose benefit Nature appears to have created everything else. Her very many gifts, however, are bestowed at a cruel price, so that we cannot confidently say whether she is a good parent to mankind or a harsh stepmother.

2. Man is the only living creature whom Nature covers with materials derived from others. To the remainder she gives different kinds of coverings – shell, bark, spines, hides, fur, bristles, hair, down, feathers, scales and fleeces. Even tree-trunks she protects from cold and heat by bark, sometimes in a double layer. But only man is cast forth on the day of his birth naked on the bare earth, to the accompaniment of crying and whimpering. No other creature is more given to tears – and that right at the beginning of life. The well-known first smile occurs, at the earliest, only after forty days in any child.

4. The early promise of strength and the first gift of time make him like a four-footed animal. When does man walk? When does he speak? When is his mouth firm enough for solid food? How long does his fontanelle pulsate – a sign that man is the weakest among all living creatures? Then there are the diseases to which he is subject, and the cures devised against these ills that are overcome by new maladies. All other animals know their own natures: some use speed, others swift flight, and yet others swimming. Man, however, knows nothing unless by learning – neither how to speak nor how to walk nor how to eat; in a word, the only thing he knows instinctively is how to weep. And so there have been many people who judged that it would have been better not to have been born, or to have died as soon as possible.

5. Man alone of living creatures has been given grief, on him alone has luxury been bestowed in countless forms and through every single limb – and likewise ambition, greed and a boundless desire for living, superstition, anxiety about burial and even about what there will be after his life ends. No creature's life is more fragile; none has a greater lust for everything; none, a more confused sense of fear or a fiercer anger. To sum up, other creatures spend their days among their own kind. We see them gather together and take their stand against other species; fierce lions do not fight among themselves, serpents do not bite serpents — not even sea-monsters and fish act cruelly, except against different species. But man, I swear, experiences most ills at the hands of his fellow men.

Racial and individual characteristics

6. I shall not now deal with manners and customs, which are innumerable and as many as are the groups of mankind. Yet there are some that I think should not be left out, especially those of people who live furthest from the sea. Among these are some things that I do not doubt will appear fantastic and unbelievable to many. For who believed in the Ethiopians before seeing them? Or what is not considered miraculous when it first comes to our knowledge?

7. How many things are judged impossible before they happen? In truth the power and majesty of Nature at every turn lack credibility if one views these aspects piecemeal and does not embrace them as a whole. Disregarding peacocks, the spotted skins of tigers and panthers, indeed the colouring of countless animals, something which is a small matter to relate but a vast subject to consider is the great many languages that races speak, and their dialects. The variety of speech is such that, in the view of someone belonging to another race, a foreigner is hardly a member of the human species!

8. Although our faces or features contain ten or so characteristics, no two faces exist among all the thousands of human beings that cannot be differentiated — a situation that no form of art could aspire to achieve. In most cases I shall not give an assurance that I believe in my authorities, but shall credit the facts to those who will be quoted on all doubtful matters, assuming that it is acceptable to follow the Greeks with their far greater application and study going back further in time.

9. I have drawn attention to the fact that some Scythian tribes — indeed a large percentage of them — feed on human bodies. This might perhaps seem unbelievable if we did not bear in mind that such monstrous peoples have existed in the central part of the world — namely the Cyclopes and Laestrygones — and in very recent times

Transalpine tribes have practised human sacrifice, which is not far short of eating human flesh.

10. Next to these, towards the north, nor far from where the north wind rises and the cave named after it, which people call earth's door-bolt, the Arimaspi are said to live, a people noted for having one eye in the middle of their forehead. Many authorities, the most distinguished among them Herodotus and Aristeas[1] of Proconnesus, write that there is a continual battle between the Arimaspi and griffins in the vicinity of the latter's mines. The griffin is a type of wild beast with wings, as is commonly reported, which digs gold out of tunnels. The griffins guard the gold and the Arimaspi try to seize it, each with remarkable greed.

11. Beyond the Scythian Cannibals, in a certain large valley in the Himalayas, there is a region called Abarimon where some forest-dwellers live who have their feet turned back behind their legs; they run with extraordinary speed and wander far and wide with the wild animals. Baeton, Alexander the Great's road-surveyor, states that these people could not breathe in another climate, and for that reason none had been brought to the neighbouring kings, or to Alexander himself.

12. According to Isigonus of Nicaea, the Scythian Cannibals, who live ten days' journey to the north beyond the River Borysthenes, drink out of human skulls and use the scalps with the hair attached as napkins to cover their chests. Isigonus also says that certain people are born in Albania with keen-sighted, greyish-green eyes; they are bald from childhood and see more at night than during the day. He further adds that the Sauromatae, who live thirteen days' journey beyond the Borysthenes, always eat every two days.

13. Crates[2] of Pergamum states that there was a race of men near Parium, on the Hellespont, whom he calls Ophiogenes; they used to cure snake-bites by touch and extract the venom from the body by placing their hands on its surface. Varro writes that even now there are a few people in that region whose saliva is an antidote to snake-bites.

1. An epic poet who lived in the time of Croesus and was a legendary servant of Apollo. His story has three features of special interest for Apollonian religion: namely, the inclusion of (a) *ecstasis*, the separation of the soul from the body; (b) the taking on of non-human shape; and (c) the missionary spirit.

2. Crates was the first director of the library at Pergamum and wrote with a philosophical and antiquarian bias on a number of ancient authors. He visited Rome as an envoy of Attalus of Pergamum (probably in 168 BC).

14. There is a similar tribe in Africa called the Psylli after King Psyllus, whose tomb is in the region of the Greater Syrtes, as Agatharchides records. They produce in their bodies a poison deadly to snakes, and its odour puts snakes to sleep. Their custom was to expose children at birth to extremely fierce snakes and to use these snakes to test the faithfulness of their wives since snakes do not flee people born of adulterous blood. This tribe has almost been wiped out by the Nasamones, who now occupy that region. However, a tribe descended from them, who escaped or were away when the battle took place, lives on today in a few places.

15. A similar tribe still exists in Italy, the Marsi, whom men say were descended from the son of Circe, and so naturally possess this power. Yet all men have a poison that is effective against snakes: people say that snakes flee from saliva as though from boiling water, and that if it gets into their throats they die – this is especially the case if the person is fasting. Beyond the Nasamones, but next to their territory, Calliphanes places the Machlyes who are bisexual and assume the role of either sex turn and turn about. Aristotle adds that their right breast is that of a man, their left a woman's.

The evil eye

16. In the same region of Africa, Isigonus[1] and Nymphodorus[2] state there are families practising witchcraft, who by their practices make meadows dry up, trees wither and infants die. Isigonus adds that there are people of the same kind among the Triballi and the Illyrians, who cast spells by a glance and kill those they stare at for a longer time, especially if it is with a look of anger. Adults are more readily susceptible to their evil eye. What is more noteworthy is that each eye has two pupils.

Fire-walking

19. Not far from the city of Rome, in the territory of the Falisci, there are a few families called the Hirpi, who, at the annual sacrifice to Apollo on Mount Soracte, walk over a pile of charred logs and

---

1. A Greek grammarian from Cnidus who became guardian to a young Ptolemy, perhaps Ptolemy IX Soter II of Egypt (c. 141–81 BC). Extracts from his work survive in Photius and Diodorus Siculus.

2. Nymphodorus (fl. c. 335 BC), A Greek from Syracuse, wrote works on Asia and Sicily.

are not burned. For this reason, by a perpetual decree of the Senate, they have exemption from military service and all other public duties. Some people are born with bodily parts that possess special properties; for example, King Pyrrhus' big toe on his right foot cured an inflamed spleen by touch. The story is told that when he was cremated his big toe would not burn along with the rest of his body; it was put in a chest in a temple.

The marvels of India and Ethiopia

21. India and regions of Ethiopia are especially full of wonders. The largest animals grow in India: for example, Indian dogs are bigger than others. The trees are said to be so high that one cannot shoot an arrow over them. The richness of the soil, the temperate climate and the abundance of water make it possible – if one is prepared to believe this – for squadrons of cavalry to hide beneath a single fig-tree. Reeds are of such height that a section between knots will make a canoe to carry three people.

22. It is known that many inhabitants exceed seven feet in height. They never spit, are unaffected by headache, toothache or pain in the eyes, and very rarely experience pain in other parts of the body; they are hardened by the heat of the sun. Their philosophers, whom they call Gymnosophists, remain standing from sunrise to sunset in the burning sun, all the while looking at the sun with fixed gaze, resting first on one foot and then on the other.

23. Megasthenes records that on Mount Nulus there are men with their feet reversed and with eight toes on each foot. On many mountains there are men with dogs' heads who are covered with wild beasts' skins; they bark instead of speaking and live by hunting and fowling, for which they use their nails. He says they were more than 120,000 in number when he published his work. Ctesias writes that in a certain Indian tribe women bear children once only in their lifetime and the children begin to go grey as soon as they are born. He also writes of a tribe of men called the Monocoli who have only one leg and hop with amazing speed. These people are also called the Umbrella-footed, because when the weather is hot they lie on their backs stretched out on the ground and protect themselves by the shade of their feet. The Monocoli are not far away from the Cave-dwellers, and further to the east of these are some people without necks and with eyes in their shoulders.

24. There are also Satyrs[1] in the mountains of eastern India, in the

1. Probably monkeys.

region of the Catarcludi. These are very fast-moving animals, some-times running on all fours, sometimes upright like humans. Because of their speed only the old or the sick are caught. Tauron mentions a forest tribe called Choromandae – a tribe that has no speech but a chilling scream, as well as hairy bodies, grey eyes and teeth like a dog's. Eudoxus says that in southern India men have feet 18 inches long and the women have such small feet that they are nicknamed Sparrow-feet.

The Astomi, Trispithami and Pygmies

25. Among the Nomads of India Megasthenes records a race called Sciritai that has only holes in place of nostrils – like snakes – and has bandy legs. At the extreme boundaries of India, to the east, near the source of the Ganges, he locates the Astomi who have no mouth and whose body is covered in hair. They dress in cotton wool and live only on the air they breathe and the odour they draw in through their nostrils. The Astomi have no food or drink, only the various emanations from roots, flowers and woodland apples that they carry with them on long journeys. Megasthenes says that they can easily be killed by a particularly strong odour.

26. Beyond the Astomi, in the depths of the mountains, so the story goes, live the Trispithami and Pygmies. In height they do not exceed three spans – that is, about 2½ feet. The climate is healthy and always spring-like, sheltered by mountains in the north. Homer has written that the Pygmies were attacked by cranes. The story is that in spring the Pygmies, armed with arrows and riding on the backs of rams and she-goats, go *en masse* down to the sea and eat the cranes' eggs and their young, and this foray takes up three months. If they did not do this they could not protect themselves against the flocks of cranes that would result. The Pygmies' houses are built of mud, feathers and eggshell.

27. Aristotle states that they live in caves, but, apropos the remain-ing details, he agrees with other authorities. According to Isigonus, the Cyrni, an Indian race, live to be a hundred and forty years old – likewise, the long-lived Ethiopians, Chinese and inhabitants of Mount Athos. Because the latter eat snake-flesh, their head and clothes are free from creatures harmful to the body.

Longevity among Indians

28. Onesicritus[1] states that in parts of India where there are no

1. Onesicritus accompanied Nearchus and wrote a history of his Asiatic campaigns. He steered Alexander the Great's ship down the River Jhelum. Onesicritus wrote a

shadows people live to the age of a hundred and thirty. Crates of Pergamum says that the Pandae, who dwell in the valleys, live 200 years and are white-haired when young but black-haired in old age.

30. Artemidorus records that in Taprobane people live very long without any lessening of their physical strength.

Marvels of nature in respect of man

32. In the deserts of Africa ghosts suddenly confront the traveller and vanish in a flash. These and similar kinds of human beings ingenious Nature has made to be playthings for herself and for us, creations at which to marvel. Indeed, who could list the things she does day by day and almost every hour? Let it be sufficient for the revelation of her power to have included races of men among her marvels. I turn now from generalities to a few marvels attested for individuals.

33. The birth of triplets is confirmed by the example of the Horatii and Curiatii. Multiple births above this number are considered ominous events, except in Egypt where the drinking of water from the Nile causes fertility. Recently, on the day of the Emperor Augustus' funeral, a certain woman of the lower classes, named Fausta, gave birth to two boys and two girls at Ostia, which without a shadow of doubt was an omen of the shortage of food that followed. In the Peloponnese we found an example of a woman who produced quintuplets on four occasions, most of whom survived at each birth. Trogus is the authority for a multiple birth of seven children in Egypt. There are people who have the characteristics of both sexes. We call them hermaphrodites, the Greeks *androgyni*. Once considered portents, now they are sources of entertainment. Pompey the Great put representations of famous wonders among the decorations of his theatre: these were made for the purpose with special care by the skill of great artists. Among the representations, we read of Eutyche who was borne to her funeral pyre at Tralles by twenty children – she had given birth thirty times – of Alcippe, who gave birth to an elephant, and of a serving-maid who gave birth to a snake. Claudius Caesar writes that a hippocentaur was born in Thessaly but died on the same day. In his reign I personally saw a hippocentaur brought here from Egypt, preserved in honey. Among these examples of marvels is one where a child at Saguntum immediately returned to the womb, in the year in which that city was destroyed by Hannibal.

historical romance (cf. Xenophon, *Cyropaedia*), with Alexander as a Cynic hero and bringer of culture. Pliny used his work.

Changes of sex

36. That women have changed into men is not a myth. We find in historical records that during the consulship of Publius Licinius Crassus and Gaius Cassius Longinus,[1] a girl at Casinum became a boy before her parents' very eyes and, on the order of the augurs, was taken away to an uninhabited island. Licinius Mucianus records that at Argos he saw a man called Arescon, who, as Arescusa, had married a husband, but had subsequently grown a beard, developed male characteristics and had married a wife. He had also seen a boy at Smyrna who had experienced the same fate. In Africa, I myself saw someone who became a man on his wedding-day.

Pregnancy and birth

37. When twins are born, it is said, neither the mother nor more than one of the two children usually survives. But if the twins are of different sex, it is even more unusual for either to be saved. Girls are born more quickly than boys, and boys are usually carried on the right side, girls on the left. Other animals have a fixed time for mating and reproduction; reproduction in humans occurs all the year around. The period of gestation is not fixed – it is sometimes six months, or seven, or as long as ten months. A baby born before the seventh month is usually stillborn.

39. Only those conceived the day before or the day after the full moon, or when there is no moon, are born in the seventh month. It is common in Egypt for babies to be born in the eighth month – and indeed similarly in Italy – and these survive, contrary to the beliefs of older times. Nine days after conception, headache, giddiness and impaired vision, distaste for food and sickness are symptoms of the formation of the embryo. If the child is male, the pregnant mother has a better colour and an easier delivery, and movement begins in the womb on the fortieth day. If the child is a girl all the symptoms are the opposite: the weight is hard to carry, there is a slight swelling of the legs and groin, and the first movement is on the ninetieth day.

42. But with both sexes the greatest feeling of faintness is experienced when the embryo starts to grow hair, and also at the full moon, which is specially hostile to unborn babies. The way one walks and just about everything that can be mentioned are important

1. 171 BC

in a pregnant woman; so women who eat food that is too salty give birth to children without nails. Delivery can cause death, just as sneezing after intercourse can bring on an abortion.

43–4. From these beginnings tyrants are born, and from these beginnings the proud spirit that destroys. You who put faith in your physical strength and grasp Fortune's gifts and consider yourself not her nursling but her very offspring, you whose mind dwells on empire, who, swelling with some success, believe yourself a god, could you have perished from so trivial a cause? Even today you can perish from an even more insignificant cause: by being bitten by a snake's tiny fang, or even choked by a raisin stone, like the poet Anacreon, or like the praetor Fabius Senator, by a single hair in a drink of milk. Assuredly, only he who is always mindful of the frailty of man will weigh life in a fair balance.

45. A breech delivery, feet first, is against nature and this is the reason why those so delivered are called Agrippa – as having been born with difficulty.[1] Marcus Agrippa, they say, was born in this manner and is almost the only example of a successful person among all those so delivered. Yet he is thought to have paid the penalty that his irregular birth foretold: his youth was rendered unhappy by lameness; he spent a lifetime amid fighting and exposed to death; and unhappiness was caused to the world by all his offspring, but especially by the two Agrippinas, who gave birth to the emperors Gaius Caligula and Domitius Nero, two firebrands to mankind. Agrippina, his mother, writes that Nero, who throughout the whole of his rule was the enemy of mankind, was born feet first. It is Nature's way for a man to be born head first, and man's custom to be carried out for burial feet first.

48. Women are among the few living creatures to have intercourse when pregnant. In the records of doctors and those who are concerned with collecting such data, there was an abortion in which twelve babies were stillborn. But when a small interval separates two conceptions, each may reach term, as was seen in the case of Hercules and his brother Iphicles, and the woman who gave birth to twins, one resembling her husband, the other a lover. Similarly the servant-girl from Proconnesus, who had intercourse twice on the same day and bore one twin like her master and the other like his estate-manager.

1. The name Agrippa is derived from *aegre partus*: 'born with difficulty'.

Characteristics that can be transmitted to children

50. The following knowledge is widely current: deformed children may be born of healthy parents; and healthy children, or children with the same deformity, may be born of deformed parents. Some marks, moles or even scars recur; in some cases a birthmark on the arm appeared in the fourth generation.

51. In Lepidus' family three children were born (but with an interval separating them), I am told, with their eyes covered by a membrane. Other children have resembled their grandfather, and there have been twins, one of which was like his father, the other like his mother. Yet another instance features a child who resembled his brother – like a twin – although he was born a year later. One indubitable example was that of the famous boxer Nicaeus of Byzantium; although his mother was the result of an adulterous union with an Ethiopian, she was no different in complexion from other women, but Nicaeus himself resembled his Ethiopian grandfather.

Unrelated people sometimes look alike

52. Examples of likeness form a vast subject and one in which a great many fortuitous circumstances are believed to exert an influence, including recollections of sights and sounds and impressions received at the very moment of conception.

53. A commoner called Artemo was so like Antiochus, the king of Syria[1] that Laodice, the King's wife, after succeeding in murdering her husband successfully, used Artemo's help to feign her recommendation for elevation to the throne. Pompey the Great had two doubles, almost indistinguishable from him in appearance: a plebeian named Vibius and Publicius a freedman, both of whom reproduced his virtuous countenance and the dignified appearance of his distinguished brow.

54. A similar reason led to Pompey's father being nicknamed Menogenes, this being the name of his cook, although he was already called Strabo because of the appearance of his eyes, which had a squint[2] as was also found in his slave.

57. Some individuals are physically incompatible and incapable

---

1. Antiochus Soter (324–261 BC) ruled the Eastern Seleucid territories from 293/2 BC. Little is known of him except that he was the greatest founder of cities after Alexander the Great.
2. The Latin word *strabo* means 'a squinter'.

of having children, but become fertile when they associate with other partners. Such was the case with Augustus and Livia and others. Also some women produce only daughters or only sons, although for the most part the sexes alternate, as happened for example, twelve times to the mother of the Gracchi and nine times to Germanicus' wife Agrippina. Some women are childless when young; some have only one child during their lifetime.

58. Certain women do not go to full term and, if they overcome this inherent tendency to give birth prematurely by medication, usually produce a daughter. Among a large number of exceptional events connected with the Emperor Augustus, he lived to see his daughter's grandson Marcus Silanus, who was born in the year of his death. Silanus became consul after Nero; he held the province of Asia but was poisoned by Nero.[1]

59. Quintus Metellus Macedonicus[2] left six children and eleven grandchildren, but if we include daughters- and sons-in-law the total number of those who addressed him as 'Father' was twenty-seven.

62. A woman does not bear children after the age of fifty, and the majority reach the menopause at forty. As for men, it is well known that King Masinissa[3] fathered a son, Methimannus, after the age of eighty-six, and Cato the Censor had a son by the daughter of his client Salonius when he was eighty-one.

63. Woman is the only living creature to menstruate.

66. This great disturbance occurs every thirty days and is heavier every three months; in some cases it is more frequent than once a month. Certain women never menstruate and so do not produce children, since this discharge is the material of human procreation.

67. It is said that the easiest time for conception is at the beginning or end of a period. The sure sign of fertility[4] in a woman – so I am told – is if, when the eyes have been treated with medication, traces of this appear in the saliva.

Children

68. Children cut their first teeth after six months – the upper ones

---

1. AD 54. See Tacitus, *Annals*, XIII, 1.

2. Metellus defeated Critolaus (146 BC) when he challenged Roman authority in Greece. He was a severe critic of Tiberius Gracchus and participated in the attack on Gaius Gracchus (122 BC). Metellus died in 115 BC.

3. He was king of Numidia and joined the Romans when they landed in Africa (204 BC) during the Second Punic War.

4. Pliny uses the word *fecunditas*, but possibly means conception here.

usually come first and these milk teeth fall out after the age of six. Some children are born with teeth, two famous examples being Manius Curius,[1] who for this reason received the surname 'Dentatus', and Gnaeus Papirius Carbo.[2] At the time of the Kings this was considered bad luck in the case of female children.

69. Valeria was born with teeth, and soothsayers, when asked about this, prophesied that she would bring disaster to any state to which she was taken. She was taken to Suessa Pometia, then a very flourishing place; the oracle turned out as predicted.

70. Teeth are so resistant to fire that they do not burn with the rest of the body when it is cremated; yet, although they are not overcome by fire, they decay when the saliva is bad. They can be whitened by a certain medication. They are worn down by use, and in some persons they decay long before. Nor are the teeth necessary only for eating food in the interests of nourishment; the front ones control the voice and speech, meeting the impact of the tongue with a sort of harmony and, according to their arrangement and size, cut short, soften or dull the words, and so, when the teeth have decayed, their absence takes away all distinct articulation.

72. I am told that only one person, Zoroaster,[3] laughed on the day he was born, and that his fontanelle pulsated so much that it dislodged hands placed on his head, which was an omen of his future knowledge.

73. It is an established fact that at the age of three a person's height is half what it will finally reach. It is obvious, however, that the whole human race is becoming shorter day by day and that few men are taller than their fathers. When an earthquake split open a mountain in Crete a body more than 70 feet tall was discovered, which some people thought was that of Orion,[4] others of Otus.[5] According to historical records, the body of Orestes, which had been exhumed[6] on the orders of an oracle, was over 10 feet tall.

1. Manius Curius Dentatus celebrated a triumph after defeating King Pyrrhus of Epirus at Beneventum (275 BC). He died in 270 BC. Pliny's dates are incorrect.

2. A supporter of Tiberius Gracchus, on whose agrarian commission he served (129 BC).

3. Zoroaster was supposed to have instructed Pythagoras in Babylon and to have inspired the Chaldaean doctrines of astrology and magic.

4. In mythology, Orion was a great hunter. He became the constellation named after him.

5. Otus, with his brother Ephialtes, threatened to wage war on the gods. They tried to pile Mount Ossa on Mount Olympus, and Mount Pelion on Mount Ossa, so that they could reach the gods' dwelling; Zeus killed them with a thunderbolt.

6. By the Spartans in 554 BC. See Herodotus, I, 65 ff.

Indeed, nearly a thousand years ago, Homer the famous poet never stopped lamenting that mortals were shorter than in the old days.

74. The tallest person seen in our age was a man called Gabbara, brought from Arabia when Claudius was emperor; he was just under 10 feet tall.

75. During the principate of Augustus, the smallest man was called Conopas: he was about $2\frac{1}{2}$ feet tall. The smallest woman was Andromeda, a freedwoman of Julia Augusta. Marcus Varro states that the Roman knights Manius Maximus and Marcus Tullius were about 3 feet tall, and I myself have seen their bodies preserved in coffins.

76. I have found in historical records that at Salamis the son of Euthymenes grew to a height of $4\frac{1}{2}$ feet in his third year: he walked slowly and his senses were dull, but he was already sexually mature and had a deep voice. However, he died from a sudden attack of paralysis soon afterwards. I myself saw all these features, except sexual maturity, in a son of a Roman knight, Cornelius Tacitus, who was in charge of finances in Belgic Gaul. The Greeks call these cases 'freaks';[1] we have no name for them.

Unusual attributes and examples of outstanding strength, sight, voice and memory

79. The story is told that Crassus, the grandfather of the Crassus killed at Carrhae in Parthia, never laughed, and for this reason was called Agelastus[2] and there are many examples of men who never cried. Socrates, who was celebrated for his philosophy, always maintained the same expression, neither joyful nor troubled. Sometimes this disposition of mind turns into a sort of rigidity and a hard unyielding nature, and eliminates human emotions.

80. The Greeks, who have known many men of this mould called them 'unfeeling',[3] and − surprisingly − they were mainly founders of schools of philosophy: Diogenes the Cynic, Pyrrho, Heraclitus and Timo. The last one indeed went so far as to hate the whole human race. But these small indications of nature are known to vary in many people; for example, Drusus' daughter Antonia never used to spit, and Pomponius, an ex-consul and poet, never belched.

---

1. *ektrapeloi.*
2. The Greek word *agelastos* means 'not-laughing', 'grave'.
3. *apatheis.*

81. In his account of examples of exceptional strength Varro writes that Tritanus, famous in the gladiators' school for his use of Samnite weapons, was a man of slight build but outstanding strength; and his son, a soldier of Pompey the Great, had muscles that made a trellis-like pattern all over his body – even on his arms and hands. He challenged an enemy to single combat, defeated him, although unarmed, and finally caught hold of him with one finger and carried him off to camp.

82. Vinnius Valens, who saw service as an officer in the Emperor Augustus' praetorian guard, used to hold up carts loaded with wineskins until the latter were emptied. He also used to catch hold of wagons with one hand and bring them to a stop by leaning against the beasts of burden as they struggled forward, and he accomplished other amazing feats, which can be seen carved on his tombstone.

83. The same Varro says that Rusticelius, who was nicknamed Hercules, used to lift up his mule. Fufius Salvius would climb up a ladder carrying 200-pound weights fixed to his feet, and with the same weight in his hands and on his shoulders. When Milo the athlete stood his ground no one could budge him, and when he gripped an apple no one could straighten his fingers.

84. Phidippides ran the 130 miles from Athens to Sparta in two days[1] – a mighty feat, until Anystis the Spartan runner and Alexander the Great's messenger Philonides ran the 150 miles from Sicyon to Elis in a single day. Indeed nowadays some men, as we know, can manage 160 miles in the Circus, and recently, during the consulships of Fonteius and Vipstanus, a boy aged eight ran 75 miles in the hours between noon and evening. The full significance of his marvellous achievement can only really be grasped if one bears in mind that Tiberius Nero completed the longest known journey within a twenty-four-hour period (200 miles) when he raced to Germany to visit his sick brother Drusus.

85. There are examples of keen eyesight that surpass one's wildest belief. Cicero states that a copy of Homer's *Iliad*, written on parchment, was kept in a nutshell. He also mentions a man who could see a distance of 135 miles: Marcus Varro gives his name as Strabo and states that during the Punic Wars he used to stand on the headland of Lilybaeum in Sicily and would tell even the number of ships leaving harbour at Carthage. Callicrates made such small ivory

1. He was sent to ask for help against the Persian invaders (490 BC); see Herodotus, VI, 105.

representations of ants and other creatures that no one else could see their parts. A certain Myrmecides was famous for the same kind of feat: he made a four-horse chariot out of ivory so small that a fly's wings could cover it, and also a ship that a small bee could hide with its wings.

86. There is one marvellous example of the relay of sound over a distance. The battle in which Sybaris was destroyed was heard at Olympia on the actual day it took place.[1] The messengers of the victory over the Cimbri,[2] and the brothers Castor and Pollux who brought news to Rome of the victory over Perseus,[3] on the very day it happened, were visions and prophetic warnings sent by the gods.

87. Fate is full of disasters, and there have been innumerable examples of physical endurance. The most famous example among women is that of the prostitute Leaena who, although tortured, did not betray the tyrannicides Harmodius and Aristogiton.[4] Among men there is the example of Anaxarchus,[5] who, when tortured for a similar reason, bit off his tongue – the potential instrument of betrayal – and spat it in the tyrant's face.

88. With regard to memory, a most essential tool in life, it is not easy to say who was the most exceptional since so many men have gained fame for it. King Cyrus[6] could name all the soldiers in his army, and Lucius Scipio, the whole Roman people. Cineas, King Pyrrhus'[7] ambassador, could put names to all the senators and knights at Rome the day after he arrived in the city. Mithridates, who was king of twenty-two peoples, gave judgements in as many languages in an assembly, addressing each group without an interpreter.

89. In Greece a certain Charmadas repeated the contents of any volumes in libraries that anyone asked him to recall, just as if he was reading them. Finally, a system of memorizing was invented by the lyric poet Simonides, and the finishing touches were put to this by

1. This actually refers to a battle fought at the River Sagrada in south Italy in which the Locrian settlers defeated Crotona (560 BC).

2. Marius defeated the Cimbrians at Campus Raudius (101 BC).

3. Aemilius Paullus defeated King Perseus at Pydna (168 BC).

4. They allegedly killed Hipparchus at Athens (514 BC), although there is no evidence to support this claim.

5. A philosopher at the court of Alexander the Great, who was put to death by Nicocreon, king of Salamis in Cyprus.

6. Cyrus reigned 559–529 BC and was the founder of the Achaemenid (Persian) Empire.

7. Pyrrhus (319–272 BC) was King of Epirus and last of the military adventurers that the age of Alexander the Great had bred in profusion among the Greeks.

Metrodorus of Scepsis so that anything heard could be repeated verbatim.

90. No other human function is equally susceptible to upset. It reacts to injuries from diseases and mishaps, or to the fear of them, sometimes with the loss of a specific aspect of memory, at other times with complete loss of memory. One person struck by a stone forgot solely how to read and write. Another who fell from a very high roof forgot his mother, relatives and acquaintances. Yet another forgot even his own servants. The orator Messala Corvinus could not remember his own name. Lapses occur when the body is sound, but resting. As sleep gradually steals over one, it restricts the memory and causes the inactive mind to wonder where it is.

Caesar and Pompey the Great

91. I consider the dictator Julius Caesar the most outstanding person in respect of his mental vigour. In this I am not talking about courage and resolve, nor about a loftiness that embraces all things under the sky, but I am thinking about his intrinsic vigour and quickness of mind, winged, as it were, with fire. I am told that he used to read or write and dictate or listen simultaneously. He used to dictate to his secretaries four letters at once when dealing with very important matters or, if not busy with anything else, seven letters at a time.

92. Caesar also fought fifty pitched battles and was the only general to surpass the record of Marcus Marcellus,[1] who had fought thirty-nine. But I would not credit to his glory the fact that in addition to winning victories against fellow citizens, he killed 1,192,000 men in battle − a very great wrong against the human race, even though this had been forced upon him. He himself admitted as much by not publishing details of the slaughter resulting from the Civil Wars.

93. With more justice would Pompey the Great be credited with the 846 ships he captured from the pirates.[2] Furthermore, let Caesar

1. Consul in 222 BC and conqueror of Syracuse (211 BC) in the Second Punic War, in spite of Archimedes' engineering skill which was put to its defence.
2. In the first century BC pirates were the scourge of the Mediterranean. After the punitive expeditions of Quintus Metellus against those operating from Crete (68–67 BC), the island became a Roman province. The threat of famine at Rome, however, drove the Tribal Assembly to reconstitute the *imperium infinitum* of Antonius (67 BC) and to entrust this unlimited authority to Pompey; with 270 warships and 100,000 legionary infantry he cleared the Mediterranean of pirates.

enjoy his own peculiar reputation for clemency, in which he surpassed all men, exercising this virtue even to his own hindrance. Caesar also provided an example of high-mindedness that cannot be paralleled.

94. To count under this heading the shows he put on and the wealth he lavishly expended, or the magnificence of his public works, would be to illustrate characteristics of someone addicted to extravagance. But when, however, the dispatch-cases of Pompey the Great were captured at Pharsalus[1] and those of Scipio at Thapsus, Caesar acting out of the highest moral principles, burnt them and did not read them; this showed the true and incomparable loftiness of an unconquered spirit.

95. In truth it concerns the glory not merely of one man but of the whole Roman Empire to detail at this point all the names of Pompey the Great's victories and triumphs; they equal not only the brilliance of the deeds of Alexander the Great, but almost even those of Hercules and Bacchus.

96. So, after he had recovered Sicily, from which he emerged as a champion of the Republic and a member of Sulla's party, and when the whole of Africa had been conquered and brought under our control, Pompey acquired, as a trophy, the title of 'The Great'. Although only a knight, he rode back in a triumphal chariot, something no one had ever done before. Immediately he crossed to the West, and set up trophies in the Pyrenees; adding to his victories, he brought under our control 876 towns from the Alps to the frontiers of Further Spain. With great magnanimity he did not trumpet his victory over Sertorius,[2] and when he had put an end to the Civil War, which was stirring up all foreign policy, he led triumphal chariots into Rome as a knight – having twice been commander before ever serving in the ranks.

98. Pompey freed the sea-coast from pirates and restored the command of the sea to the Roman people. He celebrated a triumph over Asia, Pontus, Armenia, Paphlagonia, Cappadocia, Cilicia,

---

1. Pompey engaged Caesar at the Battle of Pharsalus (48 BC). The Pompeians were defeated and fled to Egypt, where Pompey was murdered by order of the young King Ptolemy Dionysius.

2. Quintus Sertorius raised a rebellion in Spain among the Lusitanians (80 BC) that led to nine years of fighting. A law passed in 73 or 72 BC granting a pardon to Lepidus' former associates, undermined the authority of Sertorius over his Italian officers. In 72 BC one of these refugees, Marcus Perpenna, murdered Sertorius and usurped his command. Pompey soon defeated Perpenna (71 BC).

Syria, the Scythians, the Jews and Albanians, Iberia, Crete, the Bastarnae, and, in addition to these, over King Mithridates and Tigranes.

99. The high-water mark of his fame, as he himself said in an assembly when talking about his career, was finding Asia Minor the furthest of the provinces, and making it a central one. Yet, if anyone similarly wishes to review the achievements of Caesar, who appeared greater than Pompey, he must mention the whole world, which one must agree to be a never-ending task.

Outstanding achievements, culminating with those of Marcus Sergius, the great-grandfather of the conspirator Catiline

104. Although these men performed huge feats of courage, fortune played the most important part. No one – at least in my view – can rightly put any man above Marcus Sergius, although his great grandson Catiline brings shame upon his name.[1] In his second campaign Sergius lost his right hand. In two campaigns he was wounded twenty-three times and, because of this, had no use in either hand or either foot: only his spirit remained intact. Sergius served in many subsequent campaigns, although disabled. He was captured twice by Hannibal – no ordinary enemy – and twice he escaped from captivity although kept in chains and shackles every day for twenty months. He fought on four occasions with only his left hand, while two horses he was riding were stabbed beneath him.

105. He had a right hand made of iron for him and, going into battle with this bound to his arm, raised the siege of Cremona, saved Placentia and captured twelve enemy camps in Gaul. All these exploits were confirmed by the speech he made during his term as praetor when his colleagues tried to keep him away from the sacrifices, as one who was disabled. What heaps of wreaths he would have accumulated if he had confronted a different enemy!

106. For it is of the greatest consequence on what occasions a

---

1. Catiline (108–62 BC) was a member of an old impoverished patrician family which had served Sulla. A candidate for the consulship (63 BC), he staked his chances on a programme for the general cancellation of debts (*novae tabulae*), hoping thus to get the votes of the impoverished peasant farmers. The knights and Cicero opposed him so Catiline plotted with disgruntled Sullan colonists, and others, to march on Rome. The Senate accepted Cicero's evidence of the conspiracy, passed on to him by the Allobroges; most of the conspirators were summarily executed, but Catiline escaped to be killed in battle the following year.

man's bravery happens to be displayed. What civic wreaths were awarded as a result of Trebia or Ticinus or Trasumenus? What crown was won at Cannae?[1] From which place was escape the highest exploit of courage? Assuredly, others have defeated men, but Sergius overcame fortune also.

107. Who could make a select list of famous people who have displayed exceptional ability across many disciplines and such a very great variety of subjects and a treatments of subjects? – unless perhaps it is agreed that no more successful example than Homer, the Greek epic poet, has existed, whether the criterion is the presentation or the content of his work. Once there was a golden perfume-case valuable for its pearls and precious stones; friends had pointed out the various uses to which it might be put, but Alexander the Great – for a soldier grimy from the fray has no use for perfume – rejoined: 'No thanks, but let this box guard the books of Homer so that the most valuable achievement of man's mind may be preserved in the richest possible container.'

109. Alexander the Great ordered that the house of the poet Pindar should be spared when he sacked Thebes. He believed that the philosopher Aristotle's native land was his too, and combined that evidence of kindness with the very great fame of his deeds. Apollo exposed the murderers of the poet Archilochus at Delphi. Bacchus, on the death of Sophocles the leading tragic poet, ordered his burial even though the Lacedaemonians were besieging the city's walls. Their king, Lysander, was often warned in his sleep to allow the burial of the man who had been Bacchus' favourite. Lysander asked what people had died at Athens, and immediately understood whom the god meant and granted a truce for the funeral.

110. Dionysius the tyrant, who was otherwise naturally cruel and proud, sent a ship decorated with garlands to meet Plato the 'high priest' of philosophy.[2] When he came ashore, he met him in person in a chariot with four white horses. Isocrates sold a speech for 20 talents. The foremost Athenian orator Aeschines read his speech *Against Ctesiphon* to the Rhodians and also Demosthenes' *On the Crown*, which had driven him into exile at Rhodes. They greeted this with admiration, but he said that they would have admired it

1. A series of disastrous defeats sustained by the Roman army at the beginning of the Second Punic War (218–216 BC). At Cannae one consul died, but the other, Varro, escaped with part of his forces.

2. Plato visited the younger Dionysius, tyrant of Syracuse, soon after 367 BC and again on a later occasion.

even more if they had heard the orator himself. Because of Demosthenes' own misfortune he had become a strong witness for his enemy's case.

111. The Athenians drove Thucydides into exile when he was a general, but recalled him in the role of historian.[1] They admired the eloquence of a man whose courage they had condemned. The kings of Egypt and Macedon offered convincing proof of Menander's achievement in comedy when they sent a fleet of ambassadors for him, but stronger evidence came from Menander himself when he preferred literary success to a royal fortune.

112. The leading men of Rome, also, have borne witness to the skill of foreigners. At the end of the Mithridatic War, Gnaeus Pompey, when about to enter the house of the famous professor of philosophy Posidonius, forbade his lictor to knock on the door in the usual manner, and he to whom East and West submitted lowered the fasces to the door of learning. After the celebrated embassy of three leading philosophers was sent from Athens and Cato the Censor had listened to Carneades, he proposed that the envoys should be sent away as soon as possible, because when Carneades was expounding his arguments it was hard to determine where the truth lay.[2]

113. What a great change in *mores*! On other occasions that famous man always proposed that the Greeks should be expelled from Italy. But his great-grandson, Cato of Utica, brought back a philosopher from his military tribunate and another from his mission to Cyprus. Of the two Catos, the former is noted for having driven out, the latter for having introduced, the same language.

Romans of exceptional intellect

114. Let us now take stock of famous Romans. The elder Africanus[3] ordered a statue of Quintus Ennius to be placed on his own tomb and for that famous name to be read over his last ashes, together with the epitaph written for the poet. The late Emperor

1. As commander of the Athenian fleet, Thucydides arrived too late to save Amphipolis from being captured by the Spartan general Brasidas. As a result Thucydides went into voluntary exile to avoid being tried. He was recalled soon after the restoration of democracy and a general amnesty (403 BC).

2. Carneades (*c.* 214–128 BC) was an Academic philosopher at Athens. He was an ambassador to Rome in 155 BC.

3. In the Second Punic War Scipio Africanus Major (236–184 BC) carried the war into Africa and finally defeated the Carthaginians at Zama (202 BC).

Augustus forbade the burning of Virgil's poems, overriding the modesty expressed in his will. In this way the poet enjoyed a greater tribute to his work than if he had commended his own works himself.

115. In the library at Rome founded by Asinius Pollio[1] – the first in the world put together from the spoils of war – the only statue of a living person set up was that of Marcus Varro. That this high honour was bestowed by an eminent orator and citizen on only one of the large number of men of genius who were then alive was, in my opinion, no less a distinction than when Pompey the Great gave this same man a naval crown after the war against the pirates.

116. There are countless Roman examples, if one were disposed to pursue them, since this single race has produced more distinguished men than other countries in every field of activity. But what excuse could I have for not mentioning you, Marcus Tullius? Or by what mark of distinction can I proclaim your outstanding excellence? By what else than the most outstanding evidence of that whole nation's decree, selecting from your entire life only the achievements of your consulship?

117. Thanks to your speech, the tribes renounced the agrarian law that is their very life-support. On your advice they forgave Roscius,[2] the proposer of the law relating to the theatre, and bore with equanimity the slight of discrimination implied by the allocation of seats. Your entreaty made the children of men who had been proscribed ashamed to stand for office. Catiline fled from your genius. You proscribed Mark Antony. Hail first person to be called 'Father of your Country', the first civilian to win a triumph and a laurel wreath for his speeches, parent of eloquence and of Latium's literature. As the dictator Caesar, your former enemy, wrote of you, winner of a greater laurel wreath than any gained from a triumph, inasmuch as it is greater to have advanced the frontiers of the Roman genius than those of the Roman Empire.

Outstanding men of science and art

123. Countless people have become eminent in the knowledge of the different sciences, and it is proper, to touch upon these when we

1. The orator, poet and historian (first century BC) who launched Virgil on his career.

2. Roscius passed a law (67 BC) reserving for the knights the first fourteen rows of seats in the theatre behind the patricians.

are culling the flower of mankind: in astronomy, Berosus,[1] to whom the Athenians in the name of the state set up a statue with a gilded tongue because of his divinely inspired predictions; in philosophy, Apollodorus,[2] honoured by the Amphictyons of Greece; in medicine, Hippocrates,[3] who foretold a plague that was coming from Illyria, and sent his students round the cities to help; for this service Greece decreed him the honours it had bestowed on Hercules. At the Megalensian Festival King Ptolemy rewarded Cleombrotus of Ceos with 100 talents for his medical knowledge, after he had saved King Antiochus'[4] life.

124. Critobulus enjoyed a great reputation for extracting an arrow from Philip's eye,[5] and for treating his loss of sight without any disfigurement to his face. Asclepiades[6] of Prusa has a very great reputation for the following reasons: he founded a new school of medicine; despised the ambassadors and promises of King Mithridates; discovered a means by which wine might cure the sick; brought a man back from the grave and saved him. But, most of all, he is renowned because he made a bet with Fortune that he should not be considered a doctor if ever he himself should be ill in any way, and he won his bet because he lost his life as a very old man by falling down stairs.

125. Marcus Marcellus gave Archimedes[7] striking proof of his

1. Berosus (*fl. c.* 290 BC) was a priest of Bel who wrote a history of Babylon in three books, dedicated to Antiochus I. The value of his work lay in its transmission of Babylonian history and astronomy.

2. Apollodorus, (third century BC), who came from Alexandria, was a physician and scientist. His chief work, *Poisonous Creatures (peri ton therion)*, was a primary source for all later pharmacologists.

3. Born on Cos (*c.* 460–*c.* 367 BC), Hippocrates was the most famous physician in antiquity; he is regarded as the founder of scientific medicine.

4. Antiochus I, Soter (324–261 BC), king of Syria.

5. During the siege of Methone (354 BC).

6. Physician to King Prusias of Bithynia. Asclepiades came to Rome (*c.* 50 BC) to practise. Opposed to the theory of humours and of the healing powers of Nature, he explains health as the unhindered movement of the bodily corpuscles; diseases, as their inhibited movement. His therapy consisted in diet rather than drugs.

7. The famous mathematician and inventor from Syracuse (287–212 BC). Archimedes constructed devices for use against the Romans at the siege of Syracuse. He understood hydrostatics and the principle of levers: 'Give me a place to stand on and I will move the earth.' He measured the volume of Hieron's crown by its displacement of water, thereby allowing its density (and so purity) to be determined. Archimedes also invented a helical screw for raising water. In the field of mathematics he worked out the value of π and made many other discoveries relating to conoids, spirals, the centres of gravity of planes and the quadrature of the parabola.

knowledge of geometry and mechanics when at the capture of Syracuse he ordered that Archimedes alone should be spared from any violence – but the ignorance of a soldier thwarted Marcellus' order. Among those who have also won praise are Chersiphron, from Cnossus, because he built the famous Temple of Diana at Ephesus; Philo who made a dockyard at Athens for 400 ships; Ctesibius who discovered the theory of the pneumatic pump and invented hydraulic engines;[1] and Dinochares, Alexander's surveyor during the foundation of Alexandria in Egypt. Also, Alexander the Great decreed that no one other than Apelles[2] should paint his portrait, only Pyrgoteles should carve his statue in stone, and only Lysippus[3] should cast a statue of him in bronze. There are many famous examples of these arts.

Who is the happiest of all men?

130. Undoubtedly the one race of outstanding virtue in the whole world is the Roman. But what man has enjoyed the greatest happiness is not a question that can be decided by human judgement. Different people have different ideas of prosperity, each according to his own temperament. If we make a fair judgement and come to a decision, ignoring all Fortune's ostentation, no mortal is happy. Fortune deals lavishly and comes to an indulgent arrangement with the man who can rightly be said to be not unhappy. On the other hand, there is always a fear that Fortune may grow tired, and, once this is entertained, happiness has no sure foundation.

131. What of the proverb that no mortal is happy all the time? Would that as many men as possible would disagree with this view rather than accept it as the statement of a prophet! Mortality is

One of the most interesting of his innovations was the introduction of a notation in which 100,000,000 was used as a base (as we use 10) for the expression of extremely large numbers.

1. Ctesibius, a mechanical engineer from Alexandria (born *c.* 250 BC), was also, according to Vitruvius (*On Architecture*, IV, 8, 2 ff.), one of the first to investigate the principles of construction of water-clocks.

2. The famous painter from Colophon (and later Ephesus) of the fourth century BC, who died on the island of Cos. He was sometimes called Coan because of his *Aphrodite* of Cos.

3. Lysippus (second half of the fourth century BC) worked mainly in bronze, and no fewer than 1,500 statues have been credited to him. Of surviving statues the most famous are the *Agias* (Delphi) and the *Apoxyomenus* (in the Vatican Museum).

delusory and ingenious in deceiving itself, and makes its calculations in the manner of the Thracian tribe that puts stone counters of different colours in an urn, according to each day's experience, and on the last day separates and counts them and thus makes a pronouncement about each individual.

132. What are we to deduce from the fact that the very day praised by the whiteness of the counter held the source of misfortune? How many men has the acquisition of power overthrown? How many men have possessions ruined and plunged into the worst distress?

The mutability of fortune: the chequered career of Augustus

134. Now we come to examples of changing Fortune, which are innumerable. For what great joys does she bring except after disasters, or what immense disasters except after enormous joys?

147–9. Even in the case of the late Emperor Augustus, whom all mankind names in the list of happy men, if all things are carefully weighed up, great reversals of the human lot may be discovered. For example, his uncle[1] denied him the office of Master of the Horse and preferred his opponent Lepidus;[2] the proscriptions engendered much hatred, and he was connected in the triumvirate with the most wicked citizens – nor did he have an equal share, for Antony had the more important role. After the Battle of Philippi[3] he was sick and a fugitive and spent three days in a marsh, ill and in spite of, as Agrippa and Maecenas state, his swollen condition due to dropsy. He was shipwrecked off Sicily and also had another spell of hiding in a cave there; he requested Proculeius to kill him in the naval retreat when a unit of the enemy was already pressing hard. Then there was the anxiety of the contest at Perusia; the uncertainties of the Battle of Actium;[4] then again his fall from a tower in the Pannonian Wars, so many mutinies among his troops, so many critical illnesses, his suspicion of Marcellus' vows, the disgrace of Agrippa's banishment, so many plots against his life and the charge of causing the death of his children.[5] His grief included bereavement,

1. Julius Caesar.
2. In 46 BC.
3. 42 BC.
4. 31 BC.
5. Lucius and Gaius, sons of Julia and Agrippa, whom he adopted as his sons. Their removal was perhaps effected by Livia. Augustus, however, was suspected of complicity, in order to ensure the succession of Tiberius.

his daughter's adultery and the revelation of her plots against her father's life, the insolent withdrawal of his stepson Nero,[1] another adultery involving his granddaughter. But this was not all. There was also a long succession of misfortunes that included lack of money for the army, the revolt of Illyria, the enlistment of slaves, the shortage of young men, plague at Rome, famine in Italy, determination to commit suicide, and death more than half-achieved by fasting for four days.

150. Then there was the disaster of Varus[2] and the foul affront to his dignity, the disowning of Postumus Agrippa after he had been adopted as heir, and the feeling of loss when he had been banished; then, his suspicion against Fabius and the betrayal of secrets, and after this the intrigues of his wife and Tiberius – the final cause for disquiet during his last years.

The vicissitudes of fortune

165. Other examples of the fickleness of man's fortune are as follows: Homer tells how men of such different destinies as Hector and Polydamas were born in the same night.[3] Marcus Caelius Rufus and Gaius Licinius Calvus, both orators but with very different success, were born on the same day in the consulship of Gaius Marius and Gnaeus Carbo.[4] This happens daily throughout the whole world, even to people born at the same hours: masters and slaves, kings and poor men, can share their moment of birth.

166. Lucius Publius Cornelius Rufus, who was consul with Manius Curius, lost his sight in his sleep while dreaming that this was happening to him. In contrast, Jason of Pherae[5] who was ill with a tumour and given up by his doctors, sought death in battle but, wounded in the chest, found himself cured by his enemy. In the battle against the Allobroges and Arverni on the River Isara on

1. Tiberius Claudius Nero, who became the Emperor Tiberius. He was the son of Livia by her first marriage, and so stepson of Augustus. In 6 BC he went into voluntary exile in Rhodes, where he remained for seven years studying philosophy.

2. Quintilius Varus and his army were annihilated in Germany (AD 9) by rebels under Arminius, a chieftain of the Cherusci who had served in the Roman forces.

3. See *Iliad*, XVIII, 249 ff. Polydamas was the son of Panthus, and noteworthy for his wise advice to Hector which the latter rejected to his cost.

4. 82 BC.

5. Assassinated in 370 BC. Jason had been well on his way to attaining, in east Greece, the supreme position enjoyed by Dionysius in the west. Had he survived, Thessaly might indeed have accomplished the role reserved for Macedonia.

8 August when 130,000 enemy soldiers were killed, the consul Quintus Fabius Maximus was cured of a quartan ague in combat.[1]

The shortness of active life and signs of impending death

167. Whatever gift of Nature is bestowed upon us is uncertain and fragile, indeed grudging and brief, even for those who have been very fortunate – certainly when viewed in the context of the whole of time. What then of the fact that in view of our nightly rest, each of us is alive for only half his life, and an equal part of our existence is spent in a state that is like death, or, if one is an insomniac, like punishment. Nor are we counting the years of infancy, which lack sensation, nor old age, which survives to experience torture, nor the many kinds of dangers, diseases, fears and anxieties when death is so often invoked that it becomes the most frequent.

168. Indeed Nature has given men nothing better than life's brevity. The senses grow dull, the limbs become numb, sight, hearing, even the teeth and alimentary tracts die before we do, and yet this period of degeneration is reckoned as part of life. It is therefore miraculous – and this is the only example found – that the musician Xenophilus lived to be a hundred and five years old without any physical disability.

169. The decay of intelligence is, in a way, a disease. For Nature has imposed certain conditions even for diseases. A quartan fever never begins in midwinter or in the winter months, and some people are immune to it over the age of sixty, while others, especially women, are not likely to succumb to it after puberty. Old men are least susceptible to disease. For diseases attack whole nations and particular classes: sometimes slaves, sometimes the ranks of the nobility, sometimes other social strata. It has been observed that plague always travels from southern parts to the west and almost never in another direction: plagues never occur in winter nor do they last more than a three-month period.

171. The signs of impending death are as follows: madness; laughter; disorder of the mind; playing with fringes and making folds in a bedspread; taking no notice of persons trying to wake one; and incontinence. The most unmistakable signs are the appearance of the eyes and nostrils; continually lying on one's back; an irregular or slow pulse; and other symptoms noted by Hippocrates, the leading light in medicine. Although the signs of death are countless, there is

---

1. 121 BC. Quintus Fabius Maximus defeated the Arverni, when the bridges they had built across the Rhône broke under the weight of their retreating armies.

none that provides the assurance of good health. Indeed Cato the Censor made the pronouncement – as if from some oracle – that indications of senility in youth are a sign of premature death.

172. There are so many diseases that Pherecydes the Syrian died with a mass of parasitic creatures bursting out of his body. Some people have a perpetual fever, like Gaius Maecenas, who never had an hour's sleep during the last three years of his life. The poet Antipater[1] of Sidon had an attack of fever every year on his birthday, and died from this condition when he reached a fairly advanced age.

173. The former consul Aviola came back to life on his funeral pyre, but because the flames were very powerful he could not be helped and so was burnt alive. It is on record – and several authorities repeat this fact – that the one-time praetor Lucius Lamia died in similar circumstances, but that Aelius Tubero, who had held the office of praetor, was saved from the pyre. This is the lot of mortals: we come into the world to face these and similar surprises meted out by Fortune, so that man can place no trust even in death.

The activities of the soul outside its bodily home

174. Among other examples we find that Hermotimus of Clazomenae used to leave his body and wander about, reporting many things from afar that only an observer on the spot could know; meanwhile his body was only half alive. This continued until some enemies, called the Cantharidae, burnt his body and took away the sheath, as it were, from his soul when it returned. The soul of Aristaeus, also, was seen at Proconnesus flying out of his mouth in the form of a raven; this subsequently gave rise to the invention of many stories.

175. My reaction to the story of Epimenides of Cnossus is similar. When a lad, Epimenides became tired while travelling in the heat and slept in a cave for fifty-seven years. When he awoke, just as if it were the day after, he was surprised at the appearance of things and how they had changed. Old age came upon him within the same number of days as he had slept years – yet he lived to the age of one hundred and fifty-seven years.

The revival of people pronounced dead

176. Varro, when one of the twenty officials dividing out land at

1. *fl. c.* 100 BC.

Capua, cites the incident of a man who was being carried out on a bier for burial but returned home on foot.

177. He also adds marvellous happenings which it would be appropriate to describe in full. There were two brothers named Corfidius, both knights. The elder appeared to have died. When his will was read, the younger brother was declared his heir and began arranging the funeral. Meanwhile, the brother who seemed to be dead clapped his hands and summoned his servants. He told them that he had come from his younger brother, who had entrusted his daughter to him, and had also shown him the place where, unknown to anyone else, he had buried some gold in a hole in the ground. The younger brother had also asked that the arrangements he had made for the elder's funeral should be used for him. While the elder brother was telling the story, the younger brother's servants hurriedly brought the news that their master had died and that the gold had been discovered in the place he had said.

178. Life is full of such predictions but they should not be collected, since more often than not they are false, as I shall show by a remarkable example. In the Sicilian War[1] the bravest man in Caesar's fleet, Gabienus, was captured by Sextus Pompeius, by whose order his throat was cut and almost severed. He lay on the shore for a whole day. As evening drew in, he asked the assembled crowd, amid groans and entreaties, to get Pompeius to come to him, or to send one of his trusted men, since he had come back from the dead and had some news to give him.

179. Pompeius sent several of his friends. Gabienus told them that the gods of the lower world approved of Pompeius' cause – the party that was in the right – so that the ultimate result would be what he wished. Gabienus went on to say that he had been ordered to convey this news and that proof of the truth of this message would be that he would die as soon as he had completed his commission. So it turned out. There are examples of people appearing after burial, but my subject treats the works of Nature, not supernatural happenings.

Sudden death

180. Sudden deaths (the supreme happiness in life) are miraculous and frequent: I shall show that these are natural. Verrius has put on record a great many, but I shall be moderate and quote only a

---

1. Fought between Octavian and Sextus Pompeius (38–36 BC).

selection. People who have died of joy include Sophocles and Dionysius, the tyrant of Sicily, each of whom died after learning the news that his tragedy had won first prize. Then there was the mother who saw her son return safely from Cannae in spite of a false report of his death, and Diodorus, the professor of logic, who died of shame because he could not immediately solve a problem that Stilpo had asked him as a joke.

182. Also, an ambassador who had pleaded the cause of the people of Rhodes before the Senate amid great admiration died suddenly on the steps of the Senate House as he was leaving. Gnaeus Baebius Tamphilus, who had been a praetor, died after he had asked his slave the time. Aulus Pompeius expired on the Capitol after paying his respects to the gods.

183. Marcus Terentius Corax died suddenly while writing a letter in the Forum, and last year a Roman knight, while whispering something in the ear of an ex-consul, fell dead just in front of the ivory statue of Apollo in the Forum of Augustus. Most remarkable of all, Gaius Julius, a doctor, died by running a probe through his eye while applying ointment.

184. Cornelius Gallus,[1] who had been praetor, and Titus Hetereius, a Roman knight, both died while having sexual intercourse with women. The most wished-for example of a peaceful death is the one recorded by writers of former times, namely that of Marcus Ofilius Hilarus.

185. He was a comedy actor who had achieved a high degree of favour in the eyes of the people, and on his birthday was giving a party. When dinner was served he asked for a hot drink in a mug, and at the same time picked up the mask he had been wearing that day and, looking at it, put the wreath from his own head on this. He remained in this position and became stiff without anyone noticing, until the person lying nearest to him warned him that his drink was getting cold.

186. These are happy examples; on the other hand there are innumerable unhappy ones. Lucius Domitius, a man of a very famous family, who was defeated at Massilia and captured by Caesar at Corfinium,[2] took poison because he was disenchanted with life but then made every effort to save himself. It is found in official records that at the funeral of Felix, the charioteer of the Reds, one of the punters who had backed him threw himself on the pyre – a

1. Gallus (c. 69–26 BC), poet and politician, was a friend of Augustus and Virgil.
2. 49 BC.

pathetic story – and the opposition tried to prevent this incident being added to the charioteer's glory by claiming that the man had merely passed out from the effects of the large amount of perfume. Not long before, the body of Marcus Lepidus, a man of a very high-born family, who died because of the stress occasioned by his divorce, was thrown off his pyre by the violence of the flames, and, because his body could not be put back on account of the intense heat, he was burnt naked by having other bundles of brushwood put next to him.

Cremation

187. Cremation was not the original practice at Rome since the dead used to be buried. But, when it became known that the bodies of those who had fallen in wars in far-off places were being dis-interred, cremation was brought in. However, many families adhered to the time-honoured ritual, as in the case of the Cornelii, no one of whom was cremated before the dictator Sulla – and he had expressed his wish to be cremated because he feared reprisals for having dug up the corpse of Gaius Marius.

Belief in an after-life

188. There is some confusion concerning the spirits of the departed after burial. All men are in the same state from their last day forward as they were before their first day, and neither body nor mind has any more sensation after death than it had before birth. But wishful thinking prolongs itself into the future and falsely invents for itself a life that continues beyond death, sometimes by giving the soul immortality or a change of shape, sometimes by according feeling to those below, worshipping spirits and deifying one who has already ceased to be even a man. We do this as if man's method of breathing differs in some way from that of other animals, or as if we could not find many animals that live longer but for which no one prophesies a similar immortality!

189. If we consider the soul separately, of what is it formed? What is its material? Where is its power of thought? How does it hear or touch? What use does it get from these senses, or what good can it experience without them? Further, what is its abode and how great is the crowd of souls or shadows from so many ages past? These imaginings are characteristic of childish gibberish and of mortal men greedy for everlasting life. Similar also is the vanity about preserving the bodies of men[1] and Democritus' promise of our coming to life again; but he did not come back on earth!

190. A plague on this mad idea that life is renewed by death! What respite is there ever for new generations if the soul keeps sensation in the upper world and the ghost in the lower? To be sure this sweet but naïve view destroys Nature's particular boon – namely death – and doubles the sorrow of one about to die by the thought and sorrow to come. For if it is pleasant to live, for whom can it be pleasant to have finished living? How much easier and much surer a foundation it is for each person to trust in himself, and for us to gain our pattern of future freedom from care from our experience of it before birth!

The discovery of arts and sciences and technological advances

191. It seems not inappropriate, before leaving our discussion of man's nature, to point out what different people have discovered. Bacchus introduced buying and selling, the crown, the royal emblem and the triumphal processions. Ceres discovered corn; previously men had lived on acorns. She also invented milling and the making of flour in Attica (or, according to some authorities, in Sicily). This was the reason that Ceres was judged to be a goddess. Ceres was also the first to give laws – or, as others think, it was Rhadamanthus.[2]

192. I think that the Assyrians have always had writing, but others, such as Gellius, hold the view that it was invented by Mercury in Egypt; yet others credit the discovery of writing to Syria. Both sets of authorities believe that Cadmus imported an alphabet of sixteen letters into Greece from Phoenicia and that Palamedes added four letters at the time of the Trojan War, namely, Z, Ψ, Φ, X; and that subsequently the Greek lyric poet Simonides added a further four – A, Ξ, Ω, Θ – the sounds of which are also recognized in the Roman alphabet. Aristotle states that in ancient times the alphabet had eighteen letters and that Epicharmus, rather than Palamedes, added Ψ and Z.

193. Anticlides records that someone called Menos invented the alphabet in Egypt 15,000 years before Phoroneus, the earliest king of Greece, and tries to prove this from the monuments. On the other hand, Epigenes, a weighty authority and one of the leading lights, informs us that the Babylonians have had astronomical obser-

---

1. A reference to embalming which was practised in Egypt.
2. Rhadamanthus, son of Zeus and Europa, was famed for his justice and impartiality, and became a judge in Hades.

vations inscribed on baked bricks for 730,000 years; but those authorities who give the shortest span of time, namely Berosus and Critodemus, say 490,000 years. From these remarks it is evident that the use of an alphabet goes back a long way. The Pelasgians introduced the alphabet into Latium.

194. The brothers Euryalus and Hyperbius first introduced brick kilns and houses into Athens; before, caves served as houses. Gellius is of the opinion that Toxius, son of Uranus, was the inventor of building with clay, having taken as his example the construction of swallows' nests. Cecrops named the first town Cecropia after himself, and this is now the Acropolis at Athens. Some allege that Argos had been founded earlier by King Phoroneus and certain authorities say Sicyon as well, but the Egyptians state that Diospolis was founded in their country long before.

195-6. Cinyra, son of Agriopa, invented tiles and copper-mining on the island of Cyprus; he was also the inventor of tongs, the hammer, crowbar and anvil. Danaus, who came from Egypt to Greece to the region that used to be called 'Dry Argos', invented wells, and Cadmus invented stone-quarrying, at Thebes, or, according to Theophrastus, in Phoenicia. Thrason introduced walls, the Cyclopes towers, according to Aristotle, but the latter invention is credited to the Tirynthians by Theophrastus. The Egyptians pioneered weaving, the Lydians the dyeing of wool at Sardis, while the use of the spindle in making wool was devised by Closter, son of Arachne. Arachne herself discovered linen and nets. Nicias of Megara invented fulling, and Tychius of Boeotia shoemaking. The Egyptians assert that medicine was discovered among them, but according to other authorities the discovery came about through the agency of Arabus, son of Babylon and Apollo. The science of herbs and drugs was discovered by Chiron, son of Saturn and Philyra.

197. Aristotle thinks that the Scythian Lydus showed people how to smelt and work copper, but Theophrastus attributes this to Delas from Phrygia. Some say that the Chalybes discovered the art of working in bronze, others that it was the Cyclopes. Hesiod attributes the forging of iron to Crete – to the Dactyli of Mount Ida. Erichthonius of Athens, or according to others, Aeacus, discovered silver. The mining and smelting of gold was invented by Cadmus the Phoenician at Mount Pangaeus, or according to others by Thoas, or Aeacus, in Panchaia,[1] or by Sun, son of Oceanus, to whom Gellius also attributes the discovery of medicine derived from mined

1. An imaginary island east of Arabia, rich in precious stones, incense and myrrh.

substances. Tin was first imported by Midacritus from the Cassiterides.

198–9. The working of iron was invented by the Cyclopes, potteries by Coroebus of Athens, the potter's wheel by Anacharsis the Scythian, or according to others by Hyperbius of Corinth. Daedalus invented woodworking, together with the saw, axe, plumb-line, gimlet, glue and isinglass, but the square, plummet, lathe and lever were the invention of Theodorus of Samos, and we owe weights and measures to Pheidon of Argos. Pyrodes produced fire from flint, and Prometheus preserved it in a fennel stalk. The Phrygians constructed a vehicle with four wheels while the Phoenicians developed trade. Eumolpus of Athens was responsible for the cultivation of vines and trees. Staphylus, son of Silenus, discovered the dilution of wine with water. Aristaeus of Athens manufactured oil and oil-presses, and likewise produced honey. Buzyges, an Athenian, developed the ox-plough, or some say it was Triptolemus. The Egyptians introduced monarchy, the Athenians, after Theseus, democracy.

200. The first tyrant was Phalaris[1] of Akragas. The Spartans invented slavery, and murder trials were first held in the Areopagus at Athens.

Timekeeping devices

213. Fabius Vestalis records that the first sundial was set up eleven years before the war with Pyrrhus,[2] at the Temple of Quirinus, by Lucius Papirius Cursor when dedicating this temple which had been vowed by his father. Fabius Vestalis does not indicate the principle of the sundial, or the name of its maker, or where it originated, or the name of his authority for this assertion.

214. Marcus Varro asserts that the first state sundial was set up on a column along the Rostra, during the First Punic War after Catina in Sicily had been captured by the consul Manius Valerius Messala, and that it was brought from Sicily thirty years later than the date assigned to Papirius' sundial, namely, AUC 491.[3] The lines on the face did not match up with the hours, yet the Romans followed its time for ninety-nine years until Quintus Marcius Philippus, who was censor with Lucius Paulus, set up a more carefully calibrated

1. Reigned *c.* 570–554 BC.
2. In the early years of the third century BC.
3. 263 BC.

one next to it; this gift was one of the most welcome products of Philippus' activities.

215. Even then, however, in cloudy weather the hours were uncertain until five years later when Scipio Nasica, the colleague of Laenas, instituted the first water-clock, which divided the hours of night and day equally; he dedicated this timekeeping device in a roofed building in AUC 595.[1] For so long a time the hours of daylight had remained undivided for the Roman people.

BOOK VIII

LAND ANIMALS

The elephant

1. The elephant is the largest land animal and is closest to man as regards intelligence, because it understands the language of its native land, is obedient to commands, remembers the duties that it has been taught, and has a desire for affection and honour. Indeed the elephant has qualities rarely apparent even in man, namely honesty, good sense, justice, and also respect for the stars, sun and moon.

3. Elephants are also credited with an understanding of another's religion, since, when they are on the point of going across the seas, they do not go aboard the ship before being coaxed by their mahout with a sworn assurance about their return. They have been seen, when exhausted by sickness – since diseases assail even those huge bodies – lying on their backs and throwing grass towards the sky as though beseeching Earth to answer their prayers. Indeed, as an example of their docility, they do homage to their king, kneeling before him and offering him garlands. The Indians use a smaller species for ploughing: these they call 'mongrel' elephants.

4. Elephants were harnessed together in Rome for the first time, and drew Pompey the Great's chariot in his African triumph; just as it is recorded that they had been employed on a former occasion, when Bacchus celebrated his triumph after conquering India. Procillus remarks that at Pompey's triumph the yoked elephants were unable to go out through the gates. Some elephants, at the gladiatorial show staged by Germanicus Caesar, even performed clumsy gyrations like dancers.

5. It was a common trick for them to throw weapons through the air – the wind did not deflect them – and to engage in gladiatorial contests with each other, or to play together in light-hearted war-dances. Later, elephants even walked tightropes, four at a time, carrying in a litter a woman pretending to be in labour. Or they walked between couches to take their places in dining-rooms crowded with people, picking their way gingerly to avoid lurching into anyone who was drinking.

6. It is a known fact that one elephant, somewhat slow-witted in understanding orders, was often beaten with a lash and was discovered at night practising what he had to do. It is amazing that elephants can even climb up ropes in front of them, but more so that they can come down again when the rope is sloping! Mucianus, who held the consulship on three occasions, is the authority for the statement that one elephant learnt the shapes of Greek letters and used to write in Greek: 'I myself wrote this and dedicated these spoils taken from the Celts.'

Mucianus adds that he had seen elephants at Puteoli, when made to disembark at the end of their voyage, turn round and walk backwards to try to deceive themselves about their estimate of the distance because they were frightened by the length of the gangplank stretching out from the land.

7. Elephants themselves are aware that their instruments of protection are a valuable commodity that are sought as plunder. Juba calls these 'horns', whilst Herodotus, a much earlier source, more appropriately refers to them as teeth, following the commonly accepted term.[1] When these fall out because of some accident or old age, the elephants bury them. Only the tusk is of ivory, otherwise, in these animals too, the skeleton is ordinary bone. Recently, however, because ivory is scarce – outside India an ample supply is rarely found – the elephant's bones have begun to be cut into layers. So much of the ivory in our world has been yielded up to the demands of luxury.

8. A young elephant is identifiable by the whiteness of its tusks. Elephants exercise the greatest care with their tusks. They use the point of one for fighting, but sparingly to prevent it being blunted, while the other is used for digging out roots and moving large masses. When surrounded by hunters, elephants station those of their number with smaller tusks at the front, so that fighting them is not considered such a challenge. Afterwards, when tired out, they break off their tusks by beating them against trees, and ransom themselves by parting with their booty.

9. It is strange that most animals know why they are hunted, and almost all of them know what to be on their guard against. An elephant that accidentally encounters a man wandering across its path in some remote place is mild and quiet and even, it is said, points out the way. Yet the very same animal, when it notices a man's footprint, trembles in fear of an ambush before catching sight of the man himself: he stops to pick up the scent, looks about him

1. III, 97.

and trumpets in anger and does not tread on the footprint but digs it up and passes it to the next elephant, and that one to the one following, and so on to the last, with a similar message. Then the column wheels round, retires and forms a line of battle. So much does the scent persist to be smelt by all of the elephants.

11. Elephants always move in herds. The oldest is the leader, the next in age brings up the rear. When about to cross a river, they send the smallest ahead to prevent the depth being increased by the footsteps of the larger animals wearing away the river bed. Antipater says that two elephants employed by King Antiochus for military uses were known by their names. Indeed elephants know their own names. Although Cato removed the names of the commanders from his *Annals*, he certainly records that the bravest elephant to fight in the Carthaginian army was called 'the Syrian', and had one broken tusk.

12. Antiochus was once trying to ford a river, but his elephant Ajax jibbed, although at other times it was always at the head of the column. So Antiochus announced that the elephant that crossed would have the leading place. Patroclus dared and Antiochus gave him silver trappings (a source of the greatest pleasure for elephants) and all the other privileges of a leader. Ajax, in disgrace, preferred to die of starvation rather than face dishonour. For their sense of shame is remarkable, and when defeated an elephant shuns the voice of its conqueror and offers earth and sacred leaves.

13. Elephants mate in secret because of their modesty, the male when five years old, the female when ten. This happens for two years in an elephant's life on five days in each year, so men say, and no more.

14. It is no cause for surprise that animals with memory also show affection. Juba records that an elephant recognized, many years later in old age, a man who had been its mahout when young. The same author also cites an example of a sort of insight into justice: King Bocchus tied to stakes thirty elephants he had resolved to punish, and exposed them to the same number of other elephants, while men ran among the latter to provoke them to attack; but they could not be made the instruments of another's cruelty.

16. Elephants appeared for the first time in Italy during the war with King Pyrrhus, and they got the name 'Lucanian oxen' because they were seen in Lucania in AUC 474.[1] They first appeared in Rome, however, five years later, in a triumph. A large number of elephants were captured from the Carthaginians in Sicily by the

1. 280 BC.

victory of the pontifex Lucilius Metellus in AUC 502:[1] there were 142, or, as some authorities state, 140, and they were ferried across on rafts which Metellus had made by putting a layer of planks on rows of wine-jars secured together.

17. Verrius records that these elephants fought in the Circus and were killed by javelins, because the Romans were at a loss what to do with them, since they had decided not to look after them or give them to local kings. Lucius Piso says that the elephants were simply led into the Circus, and, in order to increase the contempt for them, were driven round it by men carrying spears tipped with a ball. The authorities who do not think that they were killed fail to explain their subsequent fate.

18. There was a famous contest between a Roman and an elephant when Hannibal compelled Roman prisoners to fight one another. He matched a survivor against an elephant and agreed to let him go free if he killed the animal. The prisoner engaged the elephant single-handed and, to the great indignation of the Carthaginians, killed it. Hannibal, realizing that the account of this contest would bring contempt for the beasts, sent horsemen to kill the man as he left the arena.

20. In Pompey's second consulship,[2] when the temple of Venus Victrix was dedicated, twenty elephants (some say seventeen) fought in the Circus against Gaetulians armed with throwing-spears. One elephant put up a fantastic fight and, although its feet were badly wounded, crawled on its knees against the attacking bands. It snatched away their shields and hurled them into the air. The spectators enjoyed the curving trajectory of their descent – as if they were being juggled by a skilled performer and not thrown by a raging beast. There was also an extraordinary incident with a second elephant when it was killed by a single blow: a javelin struck under its eye and penetrated the vital parts of its head.

21. All the elephants, en masse, tried to break out through the iron railings that enclosed them, much to the discomfiture of the spectators. (For this reason, when Caesar as dictator was intending to stage a similar spectacle, he surrounded the arena with a moat filled with water. The Emperor Nero did away with this moat when he added seats for the knights.) But when Pompey's elephants had given up hope of escape, they played on the sympathy of the crowd, entreating them with indescribable gestures. They moaned,

1. 252 BC.
2. 55 BC.

as if wailing, and caused the spectators such distress that, forgetting Pompey and his lavish display specially devised to honour them, they rose in a body, in tears, and heaped dire curses on Pompey, the effects of which he soon suffered.

22. Elephants also fought on the side of Julius Caesar the dictator in his third consulship:[1] 20 were matched against 500 infantry and, on a second occasion, a further 20, equipped with howdahs, each carrying 60 men, fought with the same number of infantry as previously and an equal number of cavalry. Later, during the principates of Claudius and Nero, elephants fought men in single combat; for gladiators this was the high point of their career.

24. Captured elephants are very quickly tamed by barley juice. In India they are rounded up by a mahout, who, riding a tame elephant, either catches a wild one on its own, or separates one from the herd and beats it so that when it is exhausted he can mount it and control it in the same way as the tame one. Africans employ covered pits to trap elephants. When one strays into a pit the rest of the herd immediately heap branches together, roll down rocks and build ramps, using every effort to drag it out. Previously, in order to tame elephants, the kings herded them with the aid of horsemen into a trench constructed by hand and deceptive in its length. Enclosed within its banks the beasts were overcome by starvation. The proof of submission was when an elephant gently took a branch that a man held out to it. Nowadays hunters pierce their feet, which are very soft, with javelins in order to obtain their tusks.

27. Elephants, once tamed, are used in war and carry howdahs full of armed soldiers on their backs. In the East they make a major contribution to warfare, scattering battle-lines and trampling down armed men. Yet these beasts are terrified by the slightest shrill sound made by a pig. When wounded and frightened, they always yield ground and cause no less destruction to their own side. African elephants fear their Indian counterparts and do not dare to look at them, for the Indian species is larger.[2]

28. People commonly think that elephants carry their unborn offspring for ten years, but Aristotle says two years: they produce only one at a time, and they live for 200, in some cases, 300 years. They reach maturity at sixty. They particularly enjoy rivers and roam around streams, although unable to swim because of the size

1. 46 BC.
2. This statement is incorrect.

of their bodies and because they cannot bear cold; this is their greatest weakness.

29. Elephants eat with their mouth, but breathe, drink and smell with their trunk, not inappropriately called their 'hand'. They hate mice most of all living creatures, and if they see one even touch the food put in their stall they back away from it. They experience extreme pain if, when drinking, they swallow a leech (which I observe has now begun to be commonly called a 'bloodsucker'). This fixes itself in the windpipe and inflicts unbearable pain.

31. Their tusks command a high price and the ivory is excellent for images of the gods. Our extravagant life-style has found another reason for singing the praises of the elephant, namely the taste of the hard skin of the trunk, sought after for no other reason than that one seems to be eating ivory itself. Large tusks are seen in temples. Polybius, on the authority of Prince Gulusa,[1] also records that in remote parts of Africa having a common border with Ethiopia tusks are substitutes for doorposts in houses, and that in these and in cattle stalls, partitions are made with elephants' tusks in place of stakes.

32. India produces the biggest elephants, as well as snakes that continually fight them. The snakes are of such a size that they easily surround the elephants in coils, and tie them up with a twisted knot. In this struggle both die, for the defeated elephant falls and its weight crushes the snake coiled round it.

Snakes

36. Megasthenes writes that in India snakes grow to such a size as to be able to swallow whole stags and bulls. Metrodorus says that round about the River Rhyndacus, in Pontus, they catch and swallow birds flying over them though they are high and moving at speed.

37. There is the well-known story of the snake 120 feet long that was killed in the River Bagradas during the Punic Wars by the Roman general Regulus:[2] he used catapults and ballistae as if he

1. Livy, XLII, 23.
2. Atilius Regulus, consul in 256 BC, set out for Africa with a fleet of 230 galleys. The Romans defeated the Carthaginian fleet off Cape Ecnomus. In 249 BC, after two successive victories, in one of which Regulus had been captured, the Carthaginians negotiated with the Romans about an exchange of prisoners. Regulus was sent on parole to Rome to facilitate a settlement. Out of patriotism he broke his trust by warning the Senate against any deal with the enemy.

were storming a town. Its skin and jaw-bones remained in a temple in Rome right down to the Numantine War.[1] Snakes in Italy called 'boas' lend credence to these stories: they reach a great size, so that, during the principate of the Emperor Claudius, a whole child was seen in the stomach of one killed on the Vatican Hill.

The zoological researches of Aristotle

44. King Alexander the Great had a burning desire to acquire a knowledge of zoology, and delegated research in this field to Aristotle, a man of supreme authority in every branch of science. Orders were given to some thousands of people throughout the whole of Asia Minor and Greece – people who made their living by hunting, catching birds, and fishing, as well as those in charge of warrens, herds, apiaries, fish-ponds and aviaries: they were to see that he was informed about any creature born in any region. The result of his inquiries from such people led to the publication, in nearly fifty volumes, of his famous work *On Animals*. I ask my readers to be favourably disposed to my presentation of this information – together with facts of which Aristotle was unaware – while making their brief excursion under my direction into all the works of Nature.

Lions

46. Aristotle writes that there are two species of lion: one is thickset and short, with a curly mane, and is more timid than the other, which is long, straight-haired and makes light of wounds. The smell of lions is strong and no less so their breath. Lions rarely drink; they feed on alternate days, and when full occasionally abstain from food for three days. They devour whole what they can while feeding, and when their stomach will not accept an excess of food they put their claws into their throats and drag it out so that if they have to run away they may not go away over-full.

48. Only the lion among wild animals shows mercy to suppliants; it spares those bent down before it, and, when angry, turns its rage on men rather than on women, and only attacks children when desperately hungry. Juba believes that lions understand the meaning of prayers. He was told of an attack by a pride of lions in the forests on a woman of Gaetulia who was captured but escaped; they were

1. 142–133 BC.

pacified by a speech in which she dared to say that she was a woman, a fugitive, a weakling, a suppliant to the most noble of all animals, who ruled all other beasts – prey unworthy of their glory.

49. The lion's tail gives a clue to his state of mind; the ears serve the same function as in horses. For Nature has endowed all the noblest beasts with these means of expressing themselves. So the lion's tail is still when he is calm, and moves gently when he wishes to cajole, which is rarely. Indeed, his anger is more frequently displayed: at its onset his tail lashes the earth, and, as it increases, his back, as if to goad him on. The lion's strength is in his chest. Black gore flows from every wound, whether the injuries result from claw or tooth. When lions have eaten their fill they are harmless.

50. The lion's noble spirit is most discernible in dangers: he sneers at weapons and protects himself for a long time by fearsome threats only – it is as if he protests that he will be acting against his will. And then he leaps forward, not as if forced by danger but rather as roused to anger by madness. A further sign of his noble spirit is that however large the number of dogs and hunters bearing down upon him in level plains and open ground, he gives ground contemptuously and stops every now and again. But when he reaches the undergrowth and forest, he races off very swiftly, as if the place hides his disgrace. He bounds forward when in pursuit, but not when running away.

51. If wounded, he memorizes his attacker with a marvellous power of observation and singles him out in however large a crowd you care to imagine. He seizes anyone who hurls a weapon at him, yet does not wound him, and whirls him round and throws him to the ground, but without inflicting any injury.

53. Quintus Scaevola, son of Publius, when curule aedile, was the first to put on a fight between several lions at Rome. Lucius Sulla, however, later to become dictator,[1] in his praetorship[2] was the first to mount a fight involving 100 maned lions. After Sulla, Pompey the Great displayed 600 lions in the Circus (including 315 with manes), and Caesar, when dictator, some 400.

54. The capture of lions was formerly a difficult undertaking mainly achieved by means of covered pits. During the principate of the Emperor Claudius chance taught a Gaetulian shepherd a method

---

1. Sulla (138–78 BC) was appointed dictator in 81 BC to 'restore the Republic' (reipublicae constituendae), a position that he held until 79 BC, when he resigned and withdrew to a country estate in Campania.
2. 93 BC.

almost to be ashamed of with a wild beast of this kind. As the lion charged he threw his cloak at it, an act that was immediately transferred to the arena as the basis of a show. The lion's great ferocity subsides in a scarcely believable way when its head is covered by even a light garment, with the result that it is completely overcome without a fight. Indeed all his strength is in his eyes, so it is not surprising that when Lysimachus, on Alexander's order, was shut up in a cage with a lion he strangled it.

55. Hanno, one of the most famous of Carthaginians, was the first man recorded as having dared to handle a lion and to show it when tamed. He was brought to trial on the grounds that he was regarded as a person likely to persuade others to do anything because of his artful character, and that their freedom was ill-entrusted to one to whom even ferocity had surrendered so devastatingly.

56. There are also instances of mercy shown by lions, which came about by chance. The Syracusan Mentor met a lion in Syria that rolled over on its back in the attitude of a suppliant. Struck with fear he tried to run away, but the lion confronted him everywhere he went and followed his footsteps like someone cajoling him. He noticed a swollen wound in its foot, and by pulling out a thorn freed the lion from its severe pain. A painting at Syracuse bears witness to this incident.

57. Similarly, a native of Samos called Elpis, on landing from a ship in Africa, saw a lion near the coast with its jaws open in a threatening way. He took flight up a tree, calling on Bacchus, since the principal time for prayer is when there is no room for hope. The lion did not bar his way – although he could have done – as he ran away, but, lying down by the tree, began to ask for pity with the gaping jaws that had terrified Elpis. A bone had stuck in its teeth when he had eaten too ravenously, and was torturing him not only because of the pain of the splinter, but also because he was unable to take food.

58. Finally Elpis came down from the tree and pulled out the bone as the lion offered his mouth inclined at the most appropriate angle. The story goes that as long as the ship stayed on the coast the lion showed its gratitude by taking Elpis what he caught. For this reason Elpis dedicated a temple to Bacchus on Samos, and from the incident the Greeks call it the 'Temple of Dionysus with his mouth open'.

Camels, giraffes and other animals

67–8. The East provides pasture for camels among its herds of cattle. There are two different types: the Bactrian and the Arabian.

The former have two humps on the back, the latter one, with a second hump beneath the chest, on which they can lean. Both resemble oxen in that they do not have teeth in the upper jaw. Both do service as beasts of burden and as cavalry in battles. The Bactrian and Arabian camels differ in size and in strength. A camel never travels beyond its normal daily mileage or carries more than a prescribed load. They are not so fast as horses for which they have an inborn hatred. Camels can endure thirst for four days and, when they have an opportunity to drink, fill themselves to make up for the time they have gone without and for their future needs; they stir up the water by trampling in it, otherwise they do not enjoy their drink. Camels live for fifty years, some even for a hundred, although even camels are liable to contract rabies. A method has been invented for spaying female camels to make them ready for war. When mating is denied them they become stronger.

69. Some likeness to camels is found in two other animals. The Ethiopians call one *nabun*: this has a neck like a horse, feet and legs like an ox and a head resembling a camel's; it is tawny in colour, with white spots, for which reason it is called a spotted camel[1] or giraffe. The giraffe was first seen at Rome at the Circus games given by Caesar when dictator.

The legendary manticore, basilisk and werewolf

75. Ctesias writes that in India is born a creature that he calls the manticore: this has a triple row of teeth like a comb, the face and ears of a man, grey eyes, a blood-red colour, a lion's body, and inflicts stings with its tail like a scorpion. The manticore has a voice that sounds like a pan-pipe combined with a trumpet, achieves great speed and is especially keen on human flesh.

78. The basilisk is found in Cyrenaica and is not more than a foot in length; it is adorned with a bright white spot on its head like a diadem. It puts all snakes to flight by its hissing and does not move forward with many winding coils, like other snakes, but travels along with its middle sticking up. It destroys bushes not only by its touch but also by its breath, and it burns grass and splits rocks. Its power makes it a threat to other creatures. It is believed that once one was killed with a spear by a man on horseback and its destructive power rose through the spear and killed both the rider and his horse. Kings have often wished to see a basilisk once dead beyond a

1. *camelopardus.*

shadow of doubt. For such a fantastic creature the venom of weasels is fatal – thus does Nature determine that nothing is without its match. Men throw basilisks into weasels' dens, which are easily recognized by the putrefaction of the ground. The weasels kill them by their foul smell and then die themselves. Nature's fight is over.

80. In Italy people believe that to see wolves is dangerous, and that if a wolf looks at a man first it renders him momentarily speechless. In Africa and Egypt wolves are inactive and small, but in colder regions they are fierce and cruel. I am obliged to consider – and with confidence – that the assertion that men are turned into wolves and back to themselves again is false, otherwise we must also believe in all the other things that over so many generations we have discovered to be fabulous. I must nevertheless show the origin of this belief, which is so ingrained in popular lore that werewolves are regarded as people under a curse.

81. According to Euanthes, a writer well respected among Greek authors, the Arcadians say that someone chosen by lot, out of the clan of a certain Anthus, is led out to a marsh in that region. Having hung his clothes on an oak-tree, he swims across the water and goes to a deserted place. There he is changed into a wolf and associates with other wolves for nine years. If he has avoided contact with a human during that period, he returns to the same marsh, swims across it and regains his shape with nine years' age added to his former appearance. Euanthes also affirms the more incredible fact that he gets back into the same clothes.

82. It is astonishing how far Greek gullibility will go. There is no occurrence so fabulously shameless that it lacks a witness.

Crocodiles

89. The crocodile dwells in the Nile: it is a four-footed evil thing, as dangerous on land as in the river. It is the only land creature without a tongue and the only one that bites by pressing with its movable upper jaw. It is frightening also in another way, namely because of its row of teeth set close together like a comb. Its length is generally more than thirty feet. It lays as many eggs as geese and, by a sort of clairvoyance, hatches them always beyond the line to which the Nile is likely to flood in any year when at full flood. Nor does any other creature grow to a greater size from such a small beginning. It is armed with talons and with a hide invincible against all blows. It spends its days on land, its nights in water, in both cases for considerations of warmth.

90. When a crocodile has gorged itself on fish and is asleep with its mouth full of food, a small bird (called *trochilus* in Egypt, but in Italy 'king of birds') tempts the creature to open up its mouth for the sake of a meal: it hops in, and first of all cleans the crocodile's mouth, then its teeth and its throat also, which gapes open as wide as possible for the pleasure of this scraping. The weasel watches for the crocodile to fall asleep while enjoying this, and like a spear pierces its throat and eats away its stomach.

The hippopotamus

95. The Nile produces a creature even mightier than the crocodile: the hippopotamus. It has hooves like those of oxen; a horse's back, mane and neighing sound; a turned-up snout; a boar's tail and curved tusks, although less damaging; and an impenetrable hide used for shields and helmets, unless soaked in water. The hippopotamus feeds on crops, marking out an area in advance for each day, so men say, and it makes its footprints lead out of a field so that no traps are prepared for its return.

96. Marcus Scaurus was the first to put a hippopotamus, together with five crocodiles, on show at Rome when he staged the games during his aedileship;[1] they were kept in an artificial lake. The hippopotamus stands out as a teacher in one branch of medicine. For when it lumbers ashore after excessive eating – in which it indulges all the time – to look for recently cut rushes, and sees a very sharp stalk, it presses its body on to it and pierces a vein in its leg, and so, by losing blood, lightens its body, which would otherwise become ill. Then it covers up the wound again with mud.

Hedgehogs

133. Hedgehogs prepare food for the winter. They fix fallen apples on their spines by rolling on them and, with an extra one in their mouth, carry them to hollow trees. They predict a change in the wind from north to south by burying themselves in their dens. When they sense a predator they draw together mouth, feet and all the lower part of their body which has sparse and harmless down on it, and roll up into the shape of a ball so that no part can be seized except the spines.

134. In a hopeless situation they urinate over themselves; this

1. 58 BC.

causes their hide to disintegrate and damages their spines; when this happens they know they are caught. Therefore, the skill is to hunt them immediately they have urinated. Then the hide is of special quality; otherwise, it is ruined and brittle, while the spines disintegrate and fall out, even if the hedgehog scurries away to freedom and survives. On this account the hedgehog wets itself with this harmful liquid only in the last resort, since even animals hate to poison themselves, and it waits until the last moment, with the result that usually it is captured before this. Once the hedgehog is caught the ball can be unrolled by sprinkling hot water over it. The hedgehog is hung by one of its back feet and left to die by starvation. Otherwise it is not possible to kill it and save the hide.

135. The animal itself is not, as most people think, of no benefit to man, since, if it had no spines, the softness of cattle's hides would have been given to mortals to no purpose. Garments made out of this skin are smooth. Even in this, fraud makes great profit out of monopoly: nothing has attracted more senatorial decrees, and all emperors have been subjected to complaints from the provinces.

Dogs

142. Many of our domesticated animals are worth learning about, and the most faithful to man, bar none, are the dog and the horse. We are told about a dog that fought against robbers in defence of his master, and, although exhausted by wounds, did not leave his corpse but drove off birds and beasts of prey. Another dog in Epirus recognized in a crowd the man who had struck down his master and by biting him and barking forced him to admit the crime. The king of the Garamentes was led back from exile by 200 dogs who fought those trying to oppose them.

143. The Colophonians and the people of Castabulum had troops of dogs. These fought in the front rank, never shirked battle, and were very loyal; nor did they need pay. After the Cimbrians were killed, their dogs defended their houses, which were placed on wagons. When Jason of Lycia was murdered, his dog refused to take food and starved to death. A dog that Duris called Hyrcanus threw himself into the flames of Lysimachus' blazing pyre.

146. Only dogs know their master and recognize a stranger if he arrives unexpectedly. They alone recognize their own names and the voice of members of the family. Dogs remember the way to places, however far away, and no animal has a better memory, except man.

147. A vicious attack by a dog can be lessened if one sits on the ground. Every day of our lives we find very many other qualities in dogs, but it is in hunting that their skill and the acuteness of their senses are particularly remarkable. A dog explores and follows tracks, dragging the handler who accompanies it by its lead towards their quarry. When it sights the quarry, how silent and secret, but how significant, an indication its tail gives, then its muzzle. So, even when they are worn out by old age and blind and weak, men carry dogs in their arms, waiting for winds and scents and pointing their muzzles towards the lairs.

Alexander the Great's famous dog

149. When Alexander the Great made his expedition to India, the king of Albania gave him a dog of unusually large size as a present. Alexander was pleased with its appearance and ordered bears, then boars, and finally hinds, to be released for coursing, but the dog lay motionless and wilfully took no notice of them. Alexander, a man of noble spirit, was annoyed at this laziness in so large a dog and ordered him to be put down. Common talk brought this news to the king. So he sent a second dog to Alexander, adding that he should not try to test it on small beasts, but on a lion or an elephant.

150. He had owned only two of these dogs, he said, and if this one were killed there would be none left. Alexander did not delay and immediately saw a lion torn to pieces. Afterwards he ordered an elephant to be brought in and he derived more pleasure than from any other show. For the dog's hair bristled all over his body, and first it barked thunderously, and then, leaping about, attacked the elephant's limbs on this side and that; it fought skilfully, attacking and drawing back as was necessary, until the elephant, ceaselessly turning round and round, fell to the ground with a thud.

Horses

154. Alexander the Great had the good luck to possess a horse of great rarity. People called it Bucephalus, either because of its fierce appearance or because of the famous brand of a bull's head on its shoulder. The story goes that he was purchased for 16 talents from the herd of Philonicus, who came from Pharsalus; Alexander was still a young boy and impressed by the horse's beauty. When Bucephalus was wearing the king's saddle, he would allow only Alexander to ride him, although at other times he would let all and

sundry. Bucephalus is famous for a memorable action in battle: although wounded in the attack on Thebes, he did not allow Alexander to change over to another horse. There were many incidents of this kind, and so Alexander led the funeral procession when Bucephalus died, and built a city round his tomb; this city was named after the horse.

155. Julius Caesar the dictator had a horse that is also said to have refused to let anyone else ride it. This horse had front feet like a man's, as in the statue standing in front of the Temple of Venus Genetrix. The late Emperor Augustus made a funeral mound for a horse which was the subject of a poem by Germanicus Caesar. At Agrigentum the tombs of many horses have pyramids above them. Juba records that Semiramis was deeply in love with a horse to the point of having sexual intercourse with it.

156. Scythian cavalry is famed on account of its horses. A chieftain was killed after being challenged to fight, and when the victor came to despoil his body the Scythian's horse kicked him and bit him to death.

157. Horses are very easy to manage; the cavalry of the entire army of Sybaris used to appear performing movements to the sound of a band. The Sybarite horses had premonitions about battles and mourned their lost masters.

159. The intelligence of horses defies description. Mounted javelin-throwers experience their responsiveness when they sway their bodies to help in difficult manoeuvres. Horses also gather weapons lying on the ground and give them to their riders. Horses yoked to chariots in the Circus show beyond doubt that they understand encouragement and praise.

160. In the races in the Circus, which were part of the Secular Games put on by Claudius Caesar,[1] a charioteer of the Whites named Corax was thrown out of his chariot at the start; his team were in front and hung on to the lead by crowding their opponents and throwing everything into the race against their rivals – which they would have had to do under a skilled driver – but because they were ashamed for human skill to be surpassed by horses, on completing the prescribed course they stopped dead in their tracks at the chalk finishing-line.

162. The Sarmatians prepare their horses in advance for a long journey by withholding their fodder the day before and only allowing them a little water. This way they ride 150 miles non-stop.

1. AD 47.

Some horses live fifty years, but mares are less long-lived. Mares stop growing at five years, stallions at six. The appearance of the ideal horse has been most beautifully described by the poet Virgil,[1] but I have treated this matter in my book *On the Use of the Spear as a Cavalry Weapon*.[2] I observe that almost everyone agrees about this. But a different type is sought for the Circus. Thus, although a two-year-old may be broken in for other service, a horse is not accepted for racing in the Circus before it is five years old.

Bullfights, sacrifices, and the worship of the ox in Egypt

181. Bulls are noble in appearance, with a grim brow and shaggy ears; while their horns are threatening and seem to spoil for a fight. Their main threat lies in their front feet. A bull stands blazing with fury, bending each front foot in turn and pawing the sand so that it hits its stomach. The bull is the only animal that goads itself into a rage like this. I have seen bulls stand in a two-wheeled chariot, like charioteers, racing at full tilt. The Thessalians started the sport of killing bulls by racing a horse alongside and twisting the neck by the horn. The dictator Julius Caesar was the first to stage such a contest at Rome.[3]

183. Bulls are the most prized sacrificial victims and the most extravagant way of appeasing the gods. Only in the bull, amongst long-tailed animals, is the tail not in proportion from birth; instead it grows until it hangs down to its feet. Among the strange happenings of former times, often an ox spoke; whenever this was reported to the Senate, it met in the open air.

184. In Egypt an ox, which they call Apis, is worshipped as a god. A white spot on his right side, in the form of a crescent moon, is a distinctive marking. Apis also has a knot under his tongue, which the Egyptians call a 'beetle'. Sacred law does not allow the ox to live beyond a fixed number of years: the Egyptians kill it by drowning it in the priests' fountain and, while lamenting this, they seek out a replacement. Until they find one they continue in mourning with their heads shaved. The search, however, never takes long.

185. When Apis' successor is found, he is led to Memphis by 100 priests. Apis has two shrines, which the Egyptians call 'bedrooms'. These provide the basis of omens for the people. When Apis enters

1. *Georgics*, III, 72 ff.
2. *De iaculatione equestri.*
3. 45 BC.

one, this is a favourable sign; when he enters the other, it presages terrible happenings. He gives responses to individuals who seek his advice by taking food from their hands. He turned aside from Germanicus Caesar's hand and, not long after, Germanicus was murdered.[1] For the rest of the time Apis lives in isolation. When he rushes into gatherings, he goes along with lictors who clear a path; crowds of boys escort him singing a song in his honour. He seems to understand, and to want to be worshipped. The crowds suddenly become frenzied and prophesy the future.

186. Once a year Apis is put to a cow with distinctive features, although these differ from his own features. The tradition is that she is always found and killed on the same day. At Memphis there is a place on the Nile called Phiale because of its configuration: here every year the people throw a broad flat dish – a *phiale* – of gold and silver into the Nile on the seven days that they celebrate as Apis' birthdays. For some strange reason crocodiles do not attack on those days, but after midday on the eighth their aggressive instinct returns.

Sheep and wool

189. There are two main kinds of sheep: those protected with a coat, and ordinary farm sheep, which are soft. The farm sheep are fastidious about their pasture, while those with coats feed on brambles. The best coats are those made from the wool of Arabian sheep.

190. Apulian wool is the most highly praised and the kind that is called in Italy 'of the Greek flock', and elsewhere 'Italian' wool. Wool from Milesian sheep holds third place. Apulian fleeces are of short hair and used only for rain-proof cloaks. Those that come from Tarentum and Canusium enjoy the highest reputation, like fleeces from Laodice in Asia Minor derived from the same breed. White fleeces in the region of the River Padus are second to none – no other fleece, to date, has fetched more than 100 sesterces a pound.

191. Sheep are not shorn everywhere, and the practice of plucking out the wool survives in some places. There are many different colours but no specific names for wool, so they are named by reference to their place of origin. Spain has the best black fleeces, Pollentia, near the Alps, the best white, Asia Minor and Baetica the

1. In Egypt (AD 49), allegedly by Piso, the governor of Syria.

best red, which the locals call 'Erythrean', and Canusium has tawny fleeces. Tarentum has a dark colour peculiar to itself. All fleeces when fresh have a medicinal property. Fleeces from Istria and Liburnia are closer in texture to hair than wool, and so are unsuitable for garments with a nap – similarly the fleeces that Salacia in Lusitania uses to advantage with its check pattern. There is a similar wool in the vicinity of Piscinae, in the province of Gallia Narbonensis, and also in Egypt; this is used for darning clothes worn threadbare, and makes them last a long time. Also, coarse fibre from shaggy fleeces has long been popular in floor coverings. Homer is the authority for its use since ancient times.[1] Gauls and Parthians employ different methods for dyeing fleeces.

192. Clothing is made from felted fleeces, and with the addition of vinegar these withstand steel and even fire and the latest method of cleaning. Indeed fleeces obtained from the copper vessels of polishers are used to stuff cushions. This, I believe, is a Gallic custom – at any rate these cushions are nowadays distinguished by Gallic names.

193. I could not easily say when this practice began, for in olden times bedding was filled with straw just as palliasses are in the army today. My father could remember rough woollen topcoats, and I can recall coats that were shaggy inside and outside, as well as shaggy bands of woollen material used as cummerbunds. Now tunics made out of broad strips after the manner of a rough woollen coat are coming into fashion for the first time.

Embroidery and dyeing of woollen cloth

194. Marcus Varro tells us, on his own authority, that the wool on the distaff and spindle belonging to Tanaquil, also called Gaia Caecilia,[2] is still intact in the Temple of Sancus. Also, a royal toga, with folds, worn by Servius Tullius, is in the Temple of Fortune. This is the reason for young brides being accompanied by a decorated distaff and spindle with thread; Tanaquil was the first to weave a straight tunic such as novices and newly married brides wear with the plain white toga.

195. Originally a pleated robe was the mark of *haute couture*.

1. *Odyssey*, IV, 298.
2. A woman of noble family, from Tarquinii. According to the Roman version of the story, she married Tarquinius Priscus and secured the succession for his son-in-law Servius Tullius.

Then the robe with a spotted pattern became *démodé*. Fenestella writes that togas of Phrygian wool with a smooth surface began to be in vogue in the last years of the late Emperor Augustus. Togas closely woven with poppy fibres go back further, and are already alluded to by the poet Lucilius in the case of Torquatus. The toga with a purple border had its origin in Etruria. I understand that kings used robes of state. Embroidered robes were already in existence in Homer's time[1] and are the origin of those worn at triumphs.

196. The Phrygians introduced embroidery with the needle, and for this reason embroidered robes are called 'Phrygian'. Also in Asia Minor, King Attalus invented weaving with gold, the origin of the term 'Attalic' robes. Babylon in particular made famous the weaving of different colours, and gave this process its name. Alexandria introduced damask, a material woven from very many threads, and Gaul invented check patterns. Metellus Scipio includes among the charges laid against Capito that Babylonian throw-over covers for couches were sold for 800,000 sesterces, when not long ago in Nero's principate these cost 4 million.

197. The state robes of Servius Tullius, which covered the statue dedicated by him of Fortune, lasted until the death of Sejanus,[2] and it is extraordinary that they had not disintegrated or suffered damage from moths in the course of some 560 years. I have, before now, seen the fleeces even of living animals dyed purple, scarlet and crimson as if luxury had compelled them to be born like that.

Pliny gives a brief but entertaining description of apes

215. The types of apes that are closest to humans in shape are distinguished from one another by their tails. Apes are extraordinarily cunning characters. People say that they smear themselves with bird droppings and in imitation of hunters put on nooses, set to catch them, as if they are shoes. Mucianus says that apes with tails have played draughts and can distinguish real nuts from imitations made from wax. They are sad when the moon wanes and worship the new moon with great glee. Other quadrupeds also are afraid of eclipses.

216. Apes are notably fond of their young. Domesticated monkeys carry their new-born young about. They show them off

1. In *Odyssey*, III, 125, Helen is embroidering a battle scene.
2. AD 31.

to everyone and are pleased to have them fussed over, and look like persons who understand they are being congratulated. So in many cases they smother their young by hugging them. The baboon is naturally fiercer, just as the orang-utan is very gentle. Ethiopian apes are almost completely different; they are bearded and have a tail that is wide and flat at its base. This animal is said to be unable to live in any other climate but that of Ethiopia, its birthplace.

BOOK IX

CREATURES OF THE SEA

1–4. I shall now speak about creatures of the seas, rivers and ponds. There are very many aquatic animals that are even larger than land animals. The obvious reason for this is the rich nature of water. The greatest number of animals and the largest in size are in the Indian Ocean, among them whales covering somewhat less than 3 acres and sharks more than 150 feet long. In fact, lobsters in the sea grow to almost 6 feet in length, and eels in the River Ganges to nearly 300 feet.

6. There is a very large peninsula in the Red Sea called Cadara; its projection forms a vast bay that took King Ptolemy twelve days and nights to row across when there was not a breath of wind. In this undisturbed place certain animals grow to a vast motionless mass.

7. Alexander the Great's admirals of the fleet have recorded that the Gedrosi, who live by the River Arabis, make the doorways of their houses from the monsters' jaws and use their bones as roof-beams; many bones have been found that were almost 60 feet in length. Also in that country vast sheep-like creatures come ashore from the sea and, after grazing on the roots of bushes, go back again. There are some with the heads of horses, asses and bulls, which feed on the crops.

8. In the Indian Ocean the largest animals are the shark and the whale. In the Cantabrian Sea the largest is the sperm whale, which rears itself up like a huge column higher than a ship's sail and throws up a sort of spray. The largest animal in the Gulf of Gades is the tree-like polypus, which spreads out such large branches that it is believed never to have passed through the Pillars of Hercules. There are some creatures called 'Wheels' because they resemble the shape of a wheel and are distinguished by four spokes, with their 'hubs' centred on two eyes, one on either side.

Tritons

9. Ambassadors from Olisipo, sent on a mission with this purpose in view, reported to the Emperor Tiberius that a Triton, whose appearance is well known, had been seen and heard playing on a shell in a certain cave.

10. I have illustrious knights as authority for the assertion that a Triton has been seen by them in the Gulf of Gades, perfectly resembling a man in his physical appearance. They say that he climbs aboard ships during the night, and that the side of the ship on which he sits is weighed right down, and, if he should happen to stay there an unduly long time, the ship is submerged.

11. The bones of this monster, to which Andromeda was said to have been exposed, were brought by Marcus Scaurus from Joppa in Judaea during his aedileship and shown at Rome among the rest of the amazing items displayed. The monster was over 40 feet long, and the height of its ribs was greater than that of Indian elephants, while its spine was 1½ feet thick.

Whales

12. Whales can be found even in our seas. People say that they are not seen in the Gulf of Gades before winter, but during the summer they lie in a certain large, calm bay, where they take extraordinary delight in breeding. Killer whales, creatures that are the enemies of other species and whose appearance can only be described as a huge mass of flesh with fearsome teeth, are well aware of this.

14. A killer whale was seen in Ostia harbour and was attacked by the Emperor Claudius. It had come when he was building the harbour, tempted by the wreck of a cargo of hides imported from Gaul. Eating its fill of these for many days, the whale had made a furrow in the shallow sea-bed and the waves had banked up the sand to such a height that it was absolutely unable to turn round; its back stuck out of the water like a capsized boat.

15. Claudius gave orders for a number of nets to be stretched across the entrances of the harbour and, setting out in person with the praetorian cohorts, provided the Roman people with a show. The soldiers hurled spears from ships against the creature as it leapt up, and I saw one of the ships sink after being filled with water from the spouting of the whale.

16. Whales' mouths are in their foreheads, and so when they are swimming on the surface of the sea they blow out clouds of water into the air. It is generally admitted that a very few other sea-animals breathe; they are those that have a lung amongst their internal organs, for it is thought that no creature can breathe without a lung. Those who hold this view believe that fish, which have gills, do not breathe air in and out, and that many other species, even those without gills, do not breathe. I observe that Aristotle was of this opinion and convinced people of its truth by his learned research.[1]

17. I shall not pretend that I do not immediately accept this view, since it is possible that animals may have other means of breathing instead of lungs, if Nature is of such a mind, just as many have different fluid in place of blood.

18. Certainly, there are additional facts that impel me to believe that all creatures in the water breathe as can be seen from the bubbles made on the surface of the water by air rising from below.

Dolphins

20. The fastest of all animals, not only among creatures of the sea, is the dolphin. It is quicker than a bird and swifter than a javelin and, if its mouth were not below its snout, almost in the middle of its stomach, no fish would escape its speed. But Nature's foresight adds an element of delay, because dolphins cannot snatch their prey except by turning on to their backs.

23. The dolphin's tongue, unlike that of aquatic animals generally, moves and is short and wide, somewhat similar to a pig's tongue. For speech dolphins moan like a human being. They have an arched back and turned-up snout.

24. Dolphins are not afraid of humans as something alien, but come to meet vessels at sea and play and leap around them; they try to race ships and overtake them even when they are under full sail.

26. In recent years a dolphin at Hippo Diarrhytus on the coast of Africa used to eat out of men's hands and allow itself to be stroked and play with the swimmers; it carried men on its back. Flavianus, the governor of Africa, covered it all over with perfume, but because it was not used to the smell the dolphin drifted in a stupor, and avoided men for some months as if it had been driven away by this insult. Later, however, the dolphin returned and was regarded with the same wonder.[2]

29. In the province of Gallia Narbonensis and the region of Nemausus there is a marsh called Latera where dolphins and men co-operate to catch fish. At a fixed season a huge number of mullet rushes through the narrow mouth of the marsh into the sea, after

1. *Historia Animalium*, VIII, 2.
2. Compare Pliny the Younger's version of this story in *Letters*, IX, 33.

watching for a turn of the tide that prevents nets from being stretched across the channel.

30. When the fishermen see this and a crowd collects, as they know the time and are keen on this sport, the whole population shouts as loudly as it can from the shore, calling on 'Snubnose' for the finale of the show. The dolphins soon hear their wish and at once hurry to the spot to help.

31. Their battle-line appears and immediately takes up position where the fray is to commence. They put themselves between the open sea and the shore and drive the mullet into shallow water. Then the fishermen set their nets and lift the fish out of the water with two-pronged spears. The speed of some of the mullet enables them to leap over the barriers, but the dolphins still catch them. But satisfied for the moment with killing them, the dolphins put off their supper until complete victory has been achieved.

32. The battle hots up. The dolphins press on very bravely and do not mind being caught in the nets, and, fearing that this may encourage the enemy's flight, they slip out gently between boats, nets and swimming fishermen to avoid opening up ways of escape. None tries to get away by leaping out of the water, which in other circumstances they like doing, unless the nets are placed beneath them.

Tunny-fish

44. The tunny is especially large. I know of a tunny weighing more than 800 pounds, with a tail a little more than a yard wide. Fish of similar size are also found in rivers: the catfish in the Nile, the pike in the Rhine, and the sturgeon in the River Padus – a fish that grows so fat from inactivity that it sometimes reaches almost half a ton and is dragged from the water only by teams of oxen. A very small fish called the anchovy kills this creature by biting a certain vein in its throat; it presses its attack with extraordinary eagerness.

47. Male tunny-fish, which do not have a fin under their belly, enter the Black Sea in shoals from the Mediterranean in springtime, and do not spawn elsewhere. Tunny-fish are cut into pieces; the neck and belly are highly prized, and likewise the collar-bone, provided it is fresh; but, even when fresh, it causes severe indigestion. All the rest of the tunny, with the fleshy part intact, is preserved in salt. These pieces are called 'black oaks' from their resemblance to splinters of oak.

50. No creatures harmful to fish enter the Black Sea except seals and small dolphins. The tunny-fish enter by the right shore and

leave by the left. This is thought to happen because they can see more with the right eye, although naturally poor-sighted in both eyes. In the channel of the Bosporus by which the Sea of Marmara and the Black Sea are joined and Europe is separated from Asia, there is a rock of extraordinary brightness near Chalcedon, on the Asiatic side, which shines through the water from the sea-bed to the surface.

51. Always alarmed by the sudden sight of this shining rock, the tunny make headlong for the promontory of Byzantium in a shoal. The promontory is called the Golden Horn because of this rock. So all the catch is landed at Byzantium and there is a great dearth of fish at Chalcedon, owing to the intervening channel of somewhat less than 1,000 yards. Tunny-fish are often seen from the stern of ships as they sail along; the fish accompany the ships in a fascinating manner for several hours at a stretch and for some miles, not even deterred by having tridents thrown repeatedly at them. Some authorities call tunny that do this 'pilot-fish'.

Lampreys

76. In northern Gaul all lampreys have seven spots on their right jaw arranged like the constellation of the Great Bear; these spots have a golden colour while the fish is still alive, but disappear when it dies. Vedius Pollio,[1] a knight of Rome and member of the Council under the late Emperor Augustus, used this creature as a means of proving his cruelty: he threw condemned slaves into tanks of lampreys, not because land-beasts were unable to carry out the punishment, but because only with the lamprey could he watch a man being completely torn to pieces as it happened.

Cuttlefish, squid and octopuses

83. Now I will speak about bloodless fish. Of these there are three kinds: those called 'soft' fish, those covered with thin shells, and, finally, fish encased in hard shells. The first category includes the cuttlefish, squid, octopus and others of this species. They have their head between their feet and belly and have eight small feet. In the squid and cuttlefish, however, two of the feet are long and rough, and they use these to take food into their mouths, and as anchors to

1. A friend of Augustus who built the villa of Pausilipum, near Naples. When he died (15 BC) he left a large part of his property to Augustus.

keep them steady in waves; the other feet are tentacles for hunting. Both sexes, when they sense imminent capture, pour out an inky fluid, which they have in place of blood, and hide themselves by darkening the water.

85. There are many kinds of octopus, the ones on land being larger than those in the sea. Octopuses swim with their head on one side; while they are alive the head is hard, as if inflated. They stick to things by suckers spread over their tentacles. They hold on by lying flat out so that they cannot be torn away. Octopuses do not attach themselves to the sea-bed and have less holding power when large. They alone of the soft fish leave the water to go on to dry land, providing that it is rough, since they hate smooth surfaces.

86. Octopuses feed on the flesh of shellfish, which they break by squeezing them in their tentacles. Consequently their lairs can be pin-pointed by the broken shells lying in front of them. And although in other respects the octopus is stupid, for instance swimming towards a man's hand, in domestic matters it has a kind of intelligence. It gathers everything indoors and then, after eating the flesh, puts out the shells and catches little fish which are attracted to these.

87. Octopuses change colour to match their surroundings, especially when frightened.

90. I must not fail to mention the facts about octopuses discovered when Lucius Lucullus was governor of Baetica, data published by one of his entourage, Trebius Niger. He states that they are very greedy for shellfish and that these close up at a touch and cut off the tentacles, thus obtaining food from the creatures trying to plunder them.

91. Trebius Niger also adds that no animal is more savage in killing a man in the water: it struggles with him, embracing him with its tentacles, swallows at him with its many suckers and pulls him apart; it attacks shipwrecked men or men who are diving. If, however, the octopus turns over, its strength diminishes; for when these creatures are on their backs they collapse inwardly. The other facts recounted by Niger may appear rather bizarre.

92. At Carteia an octopus used to come from the open sea into the uncovered tanks of the fish-farms and there forage for salted fish. All sea-creatures are powerfully attracted to the smell of salted fish, and for this reason baskets used in catching fish are smeared with salted fish. In view of the continual theft, the overseers became exceedingly angry. Fences were put up to obstruct the octopus, but it used to climb these by means of a tree. It could be caught only by employing dogs with a keen scent. These surrounded the octopus as

it was returning at night and roused the overseers, who were terrified by its strange appearance. Its size was unheard of, and likewise its colour; it was smeared with brine and had a dreadful smell. Who would have expected to find an octopus there, or to recognize it against such a background? They seemed to be locked in a struggle with something out of this world, for it nauseated the dogs with its terrible breath, lashed them with the ends of its tentacles, and then struck them with its stronger arms, which it used in the manner of clubs. After great trouble, it was dispatched with the aid of many tridents.

93. Lucullus was shown the octopus's head, which was as big as a jar and had a capacity of well over 90 gallons. In the words of Trebius himself, its cluster of tentacles, round which one could scarcely put both arms, were knotted like clubs and almost 30 feet long; it had suckers, or cups, like 3-gallon basins, and teeth in keeping with its size. The octopus's remains, which were kept as a curiosity, weighed 700 pounds.

Crabs

97–8. Types of crab include the common crab, the crayfish, the spider-crab, the Heraclean crab, lion-crabs and other lesser species. The common crab differs from others in respect of its tail; in Phoenicia it is called the horse-crab and moves so fast that it cannot be overtaken. Crabs have a long life. They have eight feet, all curved at an angle, the front foot being double in the female, and single in the male. Crabs have two claws with serrated nippers; the upper half of the forepart of these moves, the lower half is fixed. In all species the right claw is the larger.

99. In a state of panic crabs can go backwards with equal speed. They fight one another like rams, charging with their claws raised. Crabs provide a cure for snake-bite. It is said that when crabs die as the sun passes through the sign of Cancer, their bodies are transformed into scorpions during the drought.

The decay of morality is caused by the produce of the sea

104. But why do I mention these trivial matters when shellfish are the prime cause of the decline of morals and the adoption of an extravagant life-style? Indeed, of the whole realm of Nature the sea is in many ways the most harmful to the stomach, with its great variety of dishes and tasty fish.

105. But the foregoing pale into insignificance beside the purple-fish, purple robes and pearls. As if it were not enough for the produce of the seas to be stuffed down our throats, it is also worn on the hands, in the ears, on the head and all over the body of women and men alike! What has the sea to do with clothing, the waters and waves to do with wool? The sea receives us in a proper way only when we are without clothes. There may well be a strong alliance between the sea and our stomach, but what connection is there with our backs? Are we not satisfied by feeding on dangerous things without also being clothed by them? Do we get most bodily pleasure from luxuries that cost human life?

Pearls and pearl-diving

106. The Indian Ocean is our main source of pearls, the most prized of all jewels. To get pearls men – including the Indians – go to the islands, which are very few in number. The most productive are Taprobane and Stoidis, together with the Indian promontory of Perimula. Specially praised are the pearls from the islands around Arabia and in the Persian Gulf and Red Sea.

107. Pearls come from shells much like oyster shells. When stimulated by the season for procreation, they open up, as it were, and are impregnated with dew, so the story goes. Then these pregnant shells give birth, and their offsprings are pearls of a quality corresponding to the quality of the dew they have received. If it was pure, the pearls are brilliant, if cloudy, a dirty colour. For it is certain that pearls are conceived from the sky, with which they have more connection than with the sea, and so get from it their cloudiness, or, if the morning is bright, a clear colour.

108. If the pearls are well fed, they duly increase in size. In the event of lightning, the shells close up and the pearls shrink to a size corresponding to the oysters' failure to take in food. But if it also thunders, the oysters take fright and close up so suddenly that they produce what are called 'wind' pearls, which are inflated and have a light and empty form: these are the oysters' miscarriages. Indeed healthy pearls have a skin of so many thicknesses that it could quite appropriately be compared to a hardening of the surface.

109–10. A large pearl is soft in the water but becomes hard immediately it is taken out. Whenever a shell sees a hand, it closes up and hides its wealth since it knows that this is the object of the search. If a hand is inserted, the shell cuts it off with its sharp edge – no other punishment is more just. It is armed with other penalties also, for most shells occur among rocks, and even in deep water sharks are found with them. Nevertheless, these things do not safeguard pearls from women's ears!

111. Some authorities state that groups of shells, like swarms of bees, have an especially large, old shell as their leader – one marvellously skilful at looking out for dangers – and that divers deliberately seek these shells, since, when they are caught, the rest wander aimlessly and are easily trapped in nets. Then they are heavily salted in earthenware pots; the salt eats away all the flesh, and the nuclei, as it were, of their bodies, namely the individual pearls, sink to the bottom.

112. There is no doubt that pearls are worn away by use and that lack of care causes them to change colour. Their value lies in their brilliance, size, roundness, smoothness and weight – all such uncommon qualities that no two pearls are found exactly alike. This is why Roman luxury has given them the name *uniones*[1] – a term not in the vocabulary of the Greeks. Indeed foreigners who discovered this fact call pearls *margaritae*.

113. There is a great variation in their brilliance. Pearls found in the Red Sea are bright, while those in the Indian Ocean are like flakes of mica and exceed others in size. The longer ones have their own intrinsic charm. The greatest praise is for pearls to be called alum-coloured.

114. Now even poor people desire pearls, saying that a pearl is like an attendant for a lady when she is out and about.

117. I have seen Lollia Paulina, the consort of Gaius – not at an important, solemn ceremonial feast, but at an ordinary dinner – celebrating her betrothal covered with alternating emeralds and pearls, which glittered all over her head, hair, ears, neck and fingers, to the value of 40 million sesterces. Paulina herself was ready, at the drop of a hat, to give written proof of her ownership of the gems. They were not presents from an over-generous emperor but ancestral possessions, obtained with spoils from the provinces.

118. They were the proceeds of extortion and the reason why Marcus Lollius disgraced himself by accepting gifts from kings throughout the whole of the East and was deprived of his contacts by the Emperor Gaius Tiberius, son of Augustus; he took poison so that his granddaughter could be seen in the lamplight covered with 40 million sesterces!

119. Two specimen pearls were the largest of all time. Cleopatra,

1. 'Specimen' pearls.

the last queen of Egypt,[1] owned both, which she had inherited from the kings of the East. When Antony was taking his fill every day at choice banquets, Cleopatra, headstrong woman as she was, with a haughty contempt, poured scorn on all his elegance and splendour. When Antony asked how it could be more magnificent, Cleopatra replied that she would spend 10 million sesterces on one banquet.

120. Antony wanted to know how this could be achieved, although he did not believe it possible. So bets were laid and next day, when the dispute was to be settled, Cleopatra put before Antony a banquet sufficiently splendid for the day not to be wasted, but one such as was served every day; he laughed and complained at its limited budget. She explained that this was just the starter and that the banquet proper would complete the account – her dinner alone would cost 10 million sesterces. Cleopatra ordered the main course. Following her instructions, the servants set before her only a single vessel of vinegar, the acidity of which can dissolve pearls.[2]

121. On her ears she was wearing that remarkable and truly unique work of Nature. Antony waited breathlessly to see what on earth she was going to do. Cleopatra took off one ear-ring, dropped the pearl in the vinegar and, when it had dissolved, swallowed it. Lucius Plancus, who was umpire for the wager, put his hand on the other pearl when she prepared to destroy it in a similar way and declared that Antony had been defeated – an omen that was to be fulfilled. The story is told that when Cleopatra, who had won this substantial wager, was captured, the second pearl was cut in two so that half a dinner might adorn each ear of the statue of Venus in the Pantheon at Rome.

123. Fenestella states that pearls came into general use at Rome after Alexandria had been brought under Roman rule,[3] but that small, cheap pearls were first introduced about the time of Sulla.[4] This is clearly incorrect since Aelius Stilo says that the term 'pearls'

1. Cleopatra was the eldest daughter of Ptolemy Auletes. On the death of her father she shared the throne with her younger brother Ptolemy whom she married. She was queen of Egypt from 51–30 BC. She supported Caesar by whom she had a son, Caesarion (47 BC). After Caesar's death she met Antony in Cilicia and joined his cause. Subsequently married to Antony, she committed suicide after their defeat at Actium (31 BC).

2. No form of acetic acid can actually dissolve pearls. Cleopatra no doubt swallowed the pearl (undissolved) and subsequently recovered it in the natural course of events.

3. 47 BC.

4. 81–79 BC.

was given to large *margaritae* at about the time of the Jugurthine Wars.[1]

The purple-fish, murex and the use of purple robes

125. Purple-fish live at most seven years. They hide, like the murex, for thirty days at the rising of Sirius. They gather together in spring and by rubbing together produce a wax-like slime. The murex does the same but has in the middle of its throat the famous purple flower sought after for dyeing robes.

126. There is a white vein with a very small amount of liquid in it; from it is obtained that well-known dye which shines faintly with a deep rosy colour, but the rest of the body is unproductive. Men try to catch the murex alive because it discharges its juice when it dies. They obtain the juice from the larger purple-fish by removing the shell; they crush the smaller ones together with their shell, which is the only way to make them yield their juice.

127. The best-quality Asiatic purple is found at Tyre, the best African at Meninx and on the coast of Gaetulia, and the best European in the region of Sparta. The lictors clear a path for it, and purple indicates the dignity of boyhood. It distinguishes senators from knights, and is summoned to appease the gods. Purple makes all clothes bright and, in the case of triumphal robes, is mixed with gold. Therefore, the obsession with purple may be excused.

The manufacture of the purple dye

133. The vein already mentioned is removed, and to this, salt has to be added in the proportion of about one pint for every 100 pounds. It should be left to dissolve for three days, since, the fresher the salt, the stronger it is. The mixture is then heated in a lead pot, with about seven gallons of water to every fifty pounds, and kept at a moderate temperature by a pipe connected to a furnace some distance away. This skims off the flesh which will have adhered to the veins, and after about nine days the cauldron is filtered and a washed fleece is dipped by way of a trial. Then the dyers heat the liquid until they feel confident of the result. A red colour is inferior to black.

134. The fleece is soaked for five hours, carded, and again dipped until it absorbs all the dye.

---

1. 112–106 BC.

136. I observe that purple has always been in vogue at Rome, but Romulus used it only for his cloak. There is general agreement that the first of the Kings to have an edged toga and broad purple stripe was Tullus Hostilius, after the defeat of the Etruscans.

137. Cornelius Nepos, who died during the principate of the late Emperor Augustus, says: 'When I was a young man, the violet-coloured purple dye was in fashion and cost about 100 denarii a pound. Not much later it was the reddish purple from Tarentum. This was followed by the double-dyed Tyrian variety, which could not be bought for less than about 1,000 denarii a pound. Publius Lentulus Spinther, who was curule aedile in Cicero's consulship,[1] was the first to use this in a toga bordered with purple, but met with disapproval. Yet who at the present time does not employ such purple for covering dining-room couches?' Material dipped twice used to be called 'double-dyed' and this was considered an extravagantly expensive luxury, whereas nowadays almost all the more popular purple materials are so dyed.

Sponges and diving for sponges

148. I am told that there are three types of sponge: one that is thick, very hard and rough, called *tragos*; a less thick, softer variety known as *manos*, and the 'Achilles' sponge, which is thin and close-textured; paintbrushes are made from the latter. All sponges grow on rocks and feed on shells, fish and mud. These creatures apparently possess intelligence because, when they sense the presence of a sponge-diver, they contract and are torn away from the rocks with much more difficulty. They do the same when waves beat against them. The very small shells found inside them clearly show that they live on food of this kind. People say that in the vicinity of Torone they can be fed on such shellfish even when they have been torn away, and that other sponges grow on the rocks from the roots left behind.

151. The great number of sharks that swim amongst sponges attack divers and pose a serious danger. Also, the divers speak of a sort of 'cloud' that thickens above their heads – this 'creature' is like a flat-fish – pressing on them and preventing them from getting back to the surface. Divers therefore carry sharp, pointed spikes attached to cords, since the 'clouds' will not withdraw unless they are pierced through by the spikes. This account, I believe, bears the hallmark of the fear of darkness, for no one has any hard information

1. 63 BC.

about such a creature as a 'cloud' or 'fog', which are the names given to this harmful thing.

152. Divers have fierce encounters with sharks, which make for their groin, heels and all the pale parts of their body. The only safe course is to turn on the sharks and frighten them. For sharks fear men just as much as men fear them, which means that in deep water they have an even chance. When the diver reaches the surface the situation is critical for him, because he loses his means of attack as he tries to get out of the water, and his safety is completely dependent on his shipmates. These pull on a rope tied to his shoulders. He keeps up the struggle and tugs on the rope with his left hand, as a danger signal, while his right hand holds his knife and is busy fighting.

153. Most times the men pull the diver in gently. But when he gets to the boat they see him devoured unless they haul him aboard quickly. And often divers are snatched from the grasp of their mates even as they are being hauled aboard if they do not help them by making themselves into a ball. Some thrust three-pronged gaffs at the sharks, but they are cunning enough to dive beneath the ship and so fight in safety. Divers, therefore, concentrate all their attention on keeping a look out for this danger. The most reliable guarantee of freedom from danger is a sighting of flat-fish, which are never seen where there are sharks; for this reason divers call these 'holy' fish.

Oyster-beds and fish-farms

168–9. Sergius Orata was the first to lay down oyster-beds – in the Bay of Baiae, at the time of the orator Lucius Crassus, before the Marsian War.[1] His reason was not gluttony but monetary greed. He also obtained good financial returns from his great ingenuity as the inventor of the shower-bath, and then from selling country houses fitted out with showers. He was the first to rate Lucrine oysters as having the best flavour, because some kinds of fish are of better quality in different places: for example, the pike in the Tiber between the two bridges; the turbot at Ravenna; the lamprey in Sicily; the sturgeon at Rhodes; and similarly other kinds, not to make this culinary review exhausting. The shores of Britain were not yet under our control when Orata made the Lucrine oysters famous. Afterwards it seemed worth the trouble to obtain oysters

1. 91–88 BC.

from the foot of Italy, from Brundisium, and, to obviate any quarrel between the two delicacies, a plan has been devised to compensate for the hunger caused by the long journey, by feeding the oysters from Brundisium in the Lucrine bed.

170. In the same period the elder Licinius Murena[1] invented farms for other kinds of fish. Then Philip and Hortensius became famous by following Murena's example. Lucullus drove a cutting through a mountain near Neapolis and let in the sea: the channel he excavated cost more than a country house. This was the reason why Pompey the Great used to call him 'Xerxes in a Roman toga'.[2]

1. Propraetor in Asia, 84 BC.
2. Xerxes cut a canal through Mount Athos for his fleet prior to the invasion of Greece.

BOOK X
BIRDS

The ostrich and eagle

1. The largest species of bird, which can almost be classified as an animal, is the ostrich from Africa or Ethiopia. This is taller and faster than a mounted rider and its wings are used only to assist it in running. Despite these wings, the ostrich cannot fly and does not leave the ground.

6. The eagle is the most admired and also the strongest of birds known to us; there are six kinds. The first two kinds are the black and the hare-eagle.

7. The third kind, the *morphnos*, which Homer calls the dusky eagle,[1] has a clever idea for breaking the shells of tortoises which it has seized – namely dropping them from a height. Such an accident killed Aeschylus the dramatist, who was trying to avoid a disaster of this kind which had been prophesied by the fates, so the story goes, by relying for his safety on being in the open air.[2]

8. The fourth kind is the hawk-eagle, also called the mountain stork, which is like a vulture in respect of its very small wings, but larger in the rest of its parts; it is, however, unwarlike and untypical of its species. It always has a ravenous hunger and mournful cry. The hawk-eagle alone makes off with the dead bodies of its prey; other eagles alight when they have killed their quarry. The fifth kind is called the 'true eagle' since it is a genuine and pure-bred species; it is of medium size, reddish in colour and rarely seen. There remains the sea-eagle, which has the keenest eyesight; this hovers at a great height and, when it spots a fish in the sea, swoops down headlong towards it, strikes the water with its breast and seizes its prey.

12. The first three eagles and the fifth species have the 'eagle stone' (which some authorities call *gagates*) built into their nests – a stone that is efficacious for many cures and loses nothing of its

---

1. *Iliad*, XXIV, 316.

2. He had been told he would be killed by the fall of a roof; the prophecy was fulfilled by the shell or 'roof' of the tortoise.

properties through exposure to fire. The stone is large and contains another which rattles as if in a jar when shaken. But only stones taken from a nest have a curative power. Eagles build their eyries in rocks and trees.

14. A pair of eagles requires a large tract of country over which to hunt in order to get enough food. Thus eagles mark out territories for themselves and do not look for prey in a neighbour's area. On making a catch they do not carry this away immediately but place it on the ground and only fly off with it when the weight has been tested.

15. They die from hunger, not from old age or sickness, as the upper part of their beaks grows so large that its inward curve does not allow it to be opened. They are active and fly about from midday onwards, but earlier in the day they perch idly until the market-places fill up with crowds of people. It is said that the eagle is the only creature never killed by a thunderbolt, and for this reason it is generally thought to be Jupiter's armour-bearer.

The eagle as the standard of the Roman Legions

16. Originally the legions had the eagle, wolf, Minotaur, horse and boar as their badges, which went before the respective columns. Subsequently eagles alone began to be carried into battle, while the other emblems were left in camp. Marius, in his second consulship,[1] dispensed with the other emblems and assigned the eagle to the legions as their special badge. From that time it was observed that there was scarcely ever a legionary winter camp without a pair of eagles in the immediate vicinity.

17. The eagle is not satisfied with one adversary and its fiercest struggle is with a large snake: the outcome is in the balance, although the fight takes place in the air. The snake with evil greed tries to get the eagle's eggs, and the eagle seizes it wherever seen. The snake hampers the eagle's wings, coiling itself round them a number of times in such a way that it falls to the ground with the bird.

Cocks and geese

46. Almost as proud as peacocks are our night-watchmen, which Nature designed for interrupting sleep and waking men to work. They have a knowledge of the stars and mark out each three-hour

1. 104 BC.

interval during the day with song, go to sleep at sunset and, in the fourth watch of the night, recall us to our duties and toil. They prevent the sunrise from creeping up on us unnoticed, and announce the arrival of the day with song and the song itself with the beating of their wings.

47. Cocks rule the roost and exercise royal sway in whatever house they live. In the case of fighting cocks this superiority comes from challenging one another; they seem to know that the spurs on their legs are for this purpose. Often the fight does not end until both cocks are dead. The winner sings a victory song and proclaims himself champion; the defeated cock, if not killed, hides in silence and is scarcely able to endure his servile position. Ordinary cocks strut with comparable pride, their necks uplifted and combs held high. Alone of the birds they frequently look up towards the sky, rearing their sickle-shaped tails. Thus cocks are a source of terror even to lions, the noblest of wild animals.

48. Some cocks are born only for continual fighting and battles. Those from Rhodes and Tanagra have bestowed fame on their native lands. The fighting-cocks from Melos and Chalcidice are second in renown. The Roman purple bestows its great honour on this bird, which clearly deserves it.

49. Cocks give or withhold the most favourable omens. Day by day they rule our magistrates and shut or open their own homes to them. They dispatch or hold back the fasces and order or forbid battles, providing the auspices of all our victories won throughout the world. Cocks hold very great power over the government of the world, being, thanks to their entrails and innards, as acceptable to the gods as the most costly victims.

50. Every year at Pergamum there is a public show of cocks fighting like gladiators. In the public records there is an account of how during the consulship of Marcus Lepidus and Quintus Catulus[1] a farmyard cock spoke at the country house of Galerius in the district of Ariminum – a unique happening, as far as I am aware.

51. The goose also is a careful watchman, as its defence of the Capitol bears witness,[2] during the time when our fortunes were betrayed by the silence of our dogs. For this reason it is a matter of priority for the censors to place contracts for the supply of food for the geese.

1. 78 BC.
2. In 390 BC, when Rome was captured by the Gauls, Manlius the ex-consul was awakened by the cackling of the geese in the Temple of Juno just in time to save the Capitol from the Gauls who were storming it.

52. Our countrymen are wise who know the goose by the good quality of its liver, which is distended by forcibly feeding the bird with food; on removal the liver is further increased by being soaked in sweetened milk. Nor is it an idle matter for inquiry who discovered so great a benefit. Was it Scipio Metellus, the consul, or his contemporary the knight Marcus Seius?

Swallows

70. The swallow is the only bird without hooked talons that is carnivorous. During the winter months swallows migrate but only go to places near by, making for sunny recesses in the mountains where they have been found completely bare of feathers. They do not, it is said, enter under the roofs of houses in Thebes because that famous city has been captured so often; nor do they go under the roofs of Bizyes, in Thrace, because of the crimes of Tereus.

71. A knight from Volterra, named Caecina, the owner of a four-horse racing chariot, used to catch swallows and take them to Rome. He used to set them loose to take news of a success to his friends since the birds homed in on their nests; they would have the winning colour smeared on them. Fabius Pictor writes in his *Annals* that, when a Roman garrison was under siege by the Ligurians, a swallow, removed from her chicks, was brought to him so that he could indicate by knots in a thread tied to her foot how many days later help would arrive and when a sortie should be made.

Nightingales and talking birds

81. Nightingales are not the least remarkable of birds. They sing continuously for fifteen days and nights as the buds burst into leaf. They have a remarkably loud voice and persistent flow of breath for so diminutive a body. Then there is the perfect knowledge of music present in one bird: the nightingale modulates its sound, at one time drawing out a long, sustained note with one continuous breath, now varying it by adjusting the breath, now making it staccato by checking it; or it suddenly lowers the note or, when it pleases, makes the sound soprano, mezzo or baritone.

82. In short that little throat contains everything that human skill has devised in the complicated mechanism of the flute, so that there can be no doubt that this sweet sound of Stesichorus was foretold by the omen of the nightingale as it sang perched on his infant lips.

The birds have several songs each; they are not all the same, and every bird sings its own repertoire.

83. Nightingales compete with each other and the contest is clearly a lively one. The defeated bird often dies, her breath failing with her song. Younger birds practise singing and are given verses to imitate. The pupil listens with rapt attention and repeats them, and the two nightingales keep silence turn and turn about. An improvement in the one under instruction and a critical approach on the part of the teacher can be observed.

84. Nightingales therefore command the sort of prices paid for slaves, and indeed greater prices than for armour-bearers in former times. I know a bird – a white one, an extremely rare variety – which was sold for 600,000 sesterces, to be presented to Agrippina the wife of the Emperor Claudius.

Carrier-pigeons and pigeon-fanciers

110. Pigeons have acted as messengers in important matters; at the siege of Mutina, Decimus Brutus sent dispatches to the consuls' camps tied to their feet. What use to Antony were his rampart and watchful besieging force and even nets stretched across the river, when the messenger travelled through the sky? Many go to absurd extremes as pigeon-fanciers. They build pigeon-lofts above their roofs for these birds and tell stories about the high breeding and pedigrees of individual birds, for which there is now a long-standing precedent. Before Pompey's civil war,[1] Lucius Axius, a Roman knight, offered pigeons for sale at 400 denarii a pair, as Marcus Varro records. The largest birds, which are believed to have come from Campania, have brought fame to their native land.

Parrots, parakeets and magpies

117. Most remarkable of all, birds imitate the human voice. Parrots indeed even talk. India sends us the parakeet, called *Siptax*[2] in the Indian language. The parakeet's body is green but distinguished by a red collar round the neck. It greets its master and repeats words it hears, being especially full of fun when given wine. The parakeet's head and beak are equally hard, and are beaten with an iron rod when it is being taught to speak, for otherwise it does not feel the

---

1. 49 BC.
2. A local form of the Roman word *psittacus*.

blows. After flying it lands on its beak and leans its weight on this to compensate for the weakness of its feet.

118. Less famous because it does not come from afar, yet a bird which talks more distinctly, is a certain kind of magpie. These birds love to utter certain words which they not only learn but are fond of and ponder carefully; they do not conceal their obsession. It is generally held that if they are defeated by the difficulty of a word, they die, and that if they do not hear the same words again and again, their memory fails, and that when they are searching for a word, they cheer up miraculously if they hear it. Their shape is uncommon although not especially remarkable. The magpie enjoys sufficient distinction from being able to imitate the human voice.

Men say that only those magpies that feed on acorns can learn to speak, and, among these, those with five claws on their feet learn more readily, and that this is within the first two years of their life. All the birds in each species that imitates human speech have rather broad tongues – although this is a feature of almost all birds.

120. Agrippina owned a thrush that imitated what men said, but this was hitherto unique. When I was writing down these examples, the young Britannicus and Nero had a starling and also nightingales that had been taught to speak Greek and Latin and, moreover, practised assiduously and spoke new words every day in ever longer phrases. Birds are taught in private in a place where no other voice disturbs them. The teacher sits by them to repeat over and over again the words he wants stored up, and coaxes the birds with food as a reward.

Ravens

121. Let our thanks be given to the raven as is its due. During the principate of Tiberius a young raven, from a brood hatched on top of the Temple of Castor and Pollux, flew down to a shoemaker's shop near by, where it was welcome to the owner because of religious considerations. It soon learnt how to talk, and every morning flew to the Rostra facing the Forum and greeted Tiberius by name, then Germanicus and Drusus Caesar and, after that, the people of Rome as they passed by; finally it returned to the shop. The raven was remarkable in that it performed this duty for several years.

122. The tenant of the next shoemaker's shop killed the bird, either out of rivalry or in a sudden fit of anger because he claimed that some droppings had spotted his shoes. This aroused such an uproar among the general public that the man was driven out of the district and subsequently lynched, while the bird's funeral was celebrated with great pomp. The draped bier was carried on the shoulders of two Ethiopians, preceded by a flautist; there were all kinds of floral tributes along the way to the pyre, which had been constructed on the right hand side of the Appian Way at the second milestone, on what is called Rediculus' Plain.

123. The Roman people considered the bird's intelligence a sufficiently good reason for a funeral procession and for the punishment of a Roman citizen. Yet in Rome many leading men had no funeral rites at all, while no one avenged the death of Scipio Aemilianus after he had destroyed Carthage and Numantia.

Aviaries

141. The knight Marcus Laenius Strabo was the first to introduce aviaries containing birds of all kinds, at Brundisium. Thanks to him we began imprisoning creatures to which Nature had assigned the sky.

Human reproduction

171. Man is the only animal whose first experience of mating is accompanied by regret; this is indeed an augury for life derived from a regrettable origin. All other animals have fixed seasons during the year for mating, but man, as has been stated, has intercourse at any hour of the day or night. All other animals derive satisfaction from having mated; man gets almost none.

Deviant behaviour

172. Messalina, the wife of Claudius Caesar, thinking it would be a royal triumph, chose to compete against a certain young servant girl who was a most notorious prostitute, and, over a twenty-four-hour period, beat her record by having sex with twenty-five men. In the human race males have devised every kind of sexual deviation – crimes against Nature; women, for their part, have invented abortion. How much more shameful are we in our sex life than wild animals! Hesiod records that men are keener on sexual intercourse in winter, and women in summer.

BOOK XI
INSECTS

Bees

11–12. Among all insect species the pride of place is reserved for bees,[1] and they are particularly admired – rightly so, because they alone have been created for man's benefit. Bees collect honey: the sweetest, finest, most health-promoting liquid. They make combs and wax useful for a thousand purposes, put up with hard toil and construct building works. Bees have a government; they pursue individual schemes but have collective leaders. What is especially astonishing, they have manners more advanced than those of other animals, whether wild or tame. Nature is so great that from a tiny, ghost-like creature she has made something incomparable. What sinews or muscles can we compare with the enormous efficiency and industry shown by bees? What men, in heaven's name, can we set alongside these insects which are superior to men when it comes to reasoning? For they recognize only what is in the common interest. Disregarding the question of breath, let us agree that they have blood; yet what a tiny amount there must be in these very small creatures. After such considerations let us evaluate their natural intelligence.

13. Bees hibernate – for how could they raise the strength to face frost, snow and the north wind's blasts? Indeed all insects hibernate, but not for so long a time as those that hide in our house-walls and get warm sooner than the others. As for bees, either the course of the seasons or climate has changed or earlier authorities were wrong. Bees go into hibernation after the Pleiades have set and are not seen again until they have risen. Thus they do not emerge until the spring, as writers have said, nor does anyone in Italy think about hives until the beans flower.

14. The bees go out to work and toil, and, weather permitting, no day is lost through idleness. First, they build combs and mould wax; in this way they build their new homes and cells. Then they produce offspring, and subsequently honey, wax from flowers,

---

1. See generally Virgil, *Georgics*, IV.

bee-glue from the droplets of gum-trees, sap, glue and resin from the willow, elm and reed.

The hive

15. They smear the whole inside of the hive first with these substances, as if with a kind of plaster, and then with other rather bitter juices to protect themselves against the greed of other small creatures, for they know that they are about to make something which may be an object of envy. They also build fairly wide gateways round the hive with the same substances.

16. Experts call the first foundations 'gum', the second layer 'pitch-wax', the third 'bee-glue'. The latter is obtained from the mild gum of vines and poplars, thickened by the addition of pollen; though not as yet wax, it strengthens the combs. With this substance they seal off all places where cold can get into the combs, or where they are vulnerable to damage. Bee-glue has a heavy scent and even nowadays is used by the majority of people in place of *galbanum*.[1]

The sources of honey

17. Apart from these substances, the bees collect *erithace*, which some call 'ambrosia', others 'bee-bread'. This provides food for the bees while they work, and is often found stored in the hollows of combs.

18. Bees make their wax from the pollen of all trees and plants except sorrel and echinopus, which are kinds of herbs. Esparto grass is wrongly excluded, because much of the honey in Spain originating from places where broom grows tastes of this plant. I also think that olives are wrongly excluded, since it is an established fact that the most swarms are produced where olives grow. No harm is suffered by any fruit. Bees do not settle on dead flowers, let alone dead bodies.

19. Bees work within a radius of sixty paces, and when nearby flowers have been fully exploited, they send scouts to more distant pastures. If they are caught by the onset of night on such a foray, the bees camp out flat on their backs to protect their wings from the dew. It should come as no surprise that Aristomachus of Soli and Philiscus of Thasos were both obsessively devoted to bees. This was the former's sole occupation for fifty-eight years, while the

1. The resinous sap of a Syrian plant used for medicinal purposes and in perfumes.

latter kept bees in the wilderness and was given the name 'Wild Man'. Both have written about bees.

The organization of bees

20. The work of bees is wonderfully organized on the following plan: they post a guard at their gates, in the manner of a military camp; they sleep until dawn, when one bee wakes them with a double or triple buzz, like a sort of reveille blown by a bugle. Then the bees fly out in a body if the day is likely to be mild. It is a sign of winds and rain if they remain indoors, and so men regard this inactivity as one of the methods of forecasting the weather. The column of bees goes to its work, some bringing back pollen with their feet, others water in their mouth and droplets clinging to the fur that covers their body.

21. While the young ones go to work and collect the substances mentioned, the older ones work in the hive. Those collecting pollen load their hind legs with their front feet – which are naturally rough for this reason – and load their front feet with their beaks. Once fully laden, they return sagging under the weight of their burden.

22. Three or four other bees receive each of them and take their load from them. Duties in the hive are allocated: some build, others polish, others bring up material, yet others prepare food from what is brought up to them. Bees do not feed separately, to prevent any unfairness in respect of work, food or time. Bees begin by building the vaulting of the hive: they bring down a web, as it were, from the top of a loom; a path either side of the work areas provides a separate means of entrance and exit.

Honeycombs

23. The combs hang from the upper part of the hive and adhere slightly to the sides but do not reach the floor. Sometimes they are rectangular, sometimes round, according to the requirements of the hive, and occasionally there are both kinds, when two swarms, although friendly, have different customs. Bees prop up combs that are in danger of falling; the walls between the pillars are arched from ground level to allow access for repairs.

24. The first three rows or so are left empty, so that there may not be anything ready to hand to tempt a thief, the last rows being filled up with as much honey as possible. For this reason combs are

taken from the back of the hive. Working bees catch favourable breezes. If a storm blows up, they balance themselves with the weight of a little pebble gripped by their feet. Some state that the stone is placed on the bee's shoulders. When the wind is against them, the bees fly close to the ground, avoiding brambles.

Bees' discipline

25. The bees have a wonderful way of supervising their work-load: they note the idleness of slackers, reprove them and later even punish them with death. Their hygiene is amazing: everything is moved out of the way and no refuse is left in their work areas. Indeed the droppings of those working in the hive are heaped up in one place so that the bees do not have to go too far away. They carry out the droppings on stormy days when they have to interrupt work.

26. As evening draws in, the buzzing inside the hive diminishes until one bee flies round, as though giving the order for 'lights out', and makes the same loud buzzing with which reveille was sounded, just as if the hive were a military camp. Then suddenly all becomes silent. They build homes first for ordinary bees and then for the queens.[1] If a large yield of honey is anticipated, they add shared quarters for the drones: these are the smallest cells; the worker-bees' cells are larger.

Drones[2]

27. Drones have no stings, being, as it were, imperfect bees: they are the latest offspring of those bees who are exhausted and now have served their time – a late brood, so to speak, of servants to the true bees, who therefore order them about and drive them out first to work, punishing without mercy those who lag behind. The drones help the bees not only in work but in breeding since their horde contributes much to the warmth of the hive.

28. Certainly, the greater the number of drones, the greater the supply of swarms. When the honey begins to mature the bees drive

1. I have used the term 'queen' throughout, although Pliny refers to the leader of the hive as 'king'.

2. Drones are the male bees which impregnate the queens and then remain idle. When the honey harvest begins to fail, in autumn, they are killed by the worker-bees, which are female, but have no part in the reproductive cycle.

away the drones, and, outnumbering them many to one, attack and kill them. The drone is seen only in the spring. If a drone has its wing pulled off and is then thrown back into the hive, it, in turn, removes the wings from other drones.

The queen bee's quarters

29. Bees build spacious, magnificent, secluded palaces in the bottom of the hive for those who are destined to be their rulers. These have a protuberance, and, if this is squeezed, no young are born. All cells are hexagonal, each side being the work of one of the bee's six feet. None of their tasks is done at a set time, but they jump to their duties on fine days. They fill their cells with honey within one or at most two days.

Honey and the management of bees

30. Honey comes out of air and is chiefly formed in the time just before dawn, at the rising of the stars – especially when Sirius shines forth; it is never made before the rising of the Pleiades.

32. The finest-quality honey is obtained from the calyx of the best flowers. This happens at Hymettus and Hybla in the regions of Attica and Sicily respectively, which are sunny places, and also on the island of Calydna. Initially it is honey diluted with water; in the first few days it ferments like new wine and purifies itself; on the twentieth day it thickens and is covered with a fine skin formed from the foam of fermentation.

33. In some places honeycombs are formed that are remarkable for their wax, as in Sicily and among the Paeligni, while in other places they are noted for the quantity of their honey, as in Crete, Cyprus and Africa. In yet other places they are outstanding for their size, as in northern countries; a comb has been seen, in Germany, nearly eight feet long and black in the hollow part.

34. There are three kinds of honey, whatever the region. There is 'spring honey', whose comb is made from pollen; for this reason it is called 'flower-honey'. Some people say that this should not be touched, so that progeny may leave the hive strengthened by the plentiful supply of food. Others, however, leave less of this than of any other kind for the bees, because, as they say, a plentiful supply follows at the rising of the great stars and at the solstice when thyme and vines begin to flower – the principal material used for the cells.

35. Careful management is necessary in taking away the combs, since lack of food causes the bees to despair and die, or fly away, while on the other hand an abundant supply brings idleness, and then the bees feed on honey, not bee-bread. So the more careful beekeepers leave a fifteenth part of this harvest for the bees. The day for commencement, according to a sort of natural law – if only men would observe this – is the thirtieth after the leading-out of the swarm. The harvest is usually within the month of May.

36. The second kind of honey is 'summer honey', which the Greeks call 'ripe honey', because it is made in the best season when Sirius shines forth brightly, about thirty days after midsummer.

40. Cassius Dionysius is of the opinion that a tenth part of the summer honey should be left to the bees if the hives are full; if they are not, a proportionate amount should be left for them; and if they are empty, it should not be touched at all.

41. A third kind – the least valued honey – is the wild variety called 'heath-honey'. It is collected after the first rains of autumn when only heather is in flower in the woods, and for this reason it is similar to sandy honey.

42. Common sense suggests that two-thirds of this honey should be left for the bees, and always the parts of the combs that contain bee-bread. During the sixty days from midwinter to the rising of Arcturus the bees hibernate without food. In the warmer stretch, from the rising of Arcturus to the spring equinox, they are awake but still remain in the hive and fall back on the food put by for this time. In Italy they do the same after the rising of the Pleiades and hibernate until that time.

44. Some people, when removing the honey, weigh the hives and put aside the amount to be left. There is a sense of justice binding us even in the case of bees, and men say that if this partnership is violated the hives perish. Therefore one of the first rules is that people must wash themselves and be clean before they take honey.

45. When honey is being removed, it is very helpful to drive off the bees with smoke to prevent them getting angry or greedily eating the honey themselves. Beekeepers use fairly thick smoke to rouse them from inactivity to get on with their tasks, but too much smoke kills them.

The queen bee

51. More queens are produced so that they may not be in short supply. But afterwards, when the offspring of these have begun to mature, by a unanimous vote they kill the inferior ones to avoid dividing the columns. There are two kinds of queen; the better, red; the inferior, black, or mottled. All are always recognizable by their shape and twice as big as the other bees; their wings are shorter, their legs straight and their bearing more lofty; they have a spot on their forehead which shines white in a sort of fillet. The queens also differ greatly from the ordinary bees by their bright appearance.

52. One may as well ask whether Hercules was one person and how many Bacchuses there were and about all the other matters buried under the dust of antiquity. There is no agreement among pundits about whether the queen bee alone does not have a sting and is armed only with the dignity of her office, or whether Nature has given her a sting but merely denied her its use. This much is certain, the queen does not sting.

53. The general bee population shows a remarkable obedience to her. When she sets out, the whole swarm goes too, and groups round her, encircling and protecting her, and not allowing anyone to see her. At other times, when the bees are at work, she goes round all the works inside, like someone offering encouragement, while she alone is free from any duties. Certain followers and 'officials' surround her, as constant guardians of her authority.

54. She only goes abroad when the swarm is about to migrate. This is known about long in advance because a buzzing sound goes on for some days in the hive, a sign of their preparations while they are choosing a suitable day. If anyone were to cut off one of the queen's wings the swarm would not go away. When they have set out, each bee wants to be near the queen and delights to be seen on duty. When the queen is tired they support her on their shoulders, and when she is completely exhausted, they carry her bodily. Any bee that is parted from the swarm through weariness, or is accidentally left behind, follows on by picking up the scent. Wherever the queen alights, this becomes the camp for all.

Portents provided by bees

55. Bees provide signs of future events both private and public, when a cluster of them hangs down in houses and temples – portents that have often been presaged by momentous events. They settled on the mouth of Plato when he was a young child and foretold the charm of his very pleasing eloquence. They settled in Drusus' camp at the time of our great victory at Arbalo: indeed augurs, who

always think the presence of bees is a bad omen, are not invariably correct.

Further observations about bees

56. If their leader is captured, it is a set-back for the whole column, and if the queen is lost, there is a general breakaway and migration to other leaders. In any case bees are unable to exist without a queen. But when queens become too numerous they reluctantly kill them, and for preference destroy their homes as they are born. If the yield of honey falls desperately low, then they drive away even the drones.

57. I notice that there is some doubt about drones: some people regard them as belonging to a species of their own, like robber bees, which are the largest of the drones but black in colour with a wide stomach; these are so called because they secretly consume the honey. It is an established fact that the drones are killed by the bees; at any rate they do not have a queen like bees. It is a debatable point, however, whether they are born without a sting.

58. More bees are born when the spring is damp, but the supply of honey is better in a dry spring. If some hives are short of food, the bees attack neighbouring ones in pursuit of plunder. The defenders draw up in line of battle, and if the beekeeper is present, he is not attacked by the side that thinks itself favoured by him. They also fight for a variety of other reasons and the two leaders draw up opposing lines of battle. The greatest cause of contention occurs in the collection of pollen, and each side calls out its own supporters. The fight can be broken up by throwing dust on it, or by smoke, while a reconciliation can be effected by milk or water sweetened with honey.

59. There are wild bees and bees found in woods; they have a bristling look and are much more easily stirred up, yet are noteworthy for their industry and application. There are two kinds of domesticated bees: the best is short and speckled and of a compact, round shape; the inferior kind is long and looks like a wasp, while the worst is hairy.

Bee-stings

60. Nature has given bees a sting joined to the stomach; it can be used only once. Some authorities think that bees die immediately they have implanted their sting. Others think that they die only if it

is driven in so far that some of the bee's intestine follows, that afterwards they become drones and do not produce honey, as though their strength had been drained, depriving them of the powers to help and harm alike.

The renewal of the swarm

68–9. Bees like the sound of bronze when it is struck, and assemble when summoned by this. It is clear, therefore, that they have a sense of hearing. When they have completed their work and reared their offspring, although they have discharged all their duty they have a solemn ritual. The bees range abroad in the open and soar on high, describing circles as they fly, and then at length they return to their food. Their longest life-span, if they successfully weather enemy attacks and accidents, is seven years. It is said that hives have never lasted intact beyond ten years. There are some who think that dead bees revive if they are sheltered indoors in winter, then exposed to the heat of the spring sunshine and kept warm by hot ash from fig-wood.

70. When bees have died, some think that they can be restored – inasmuch as Nature changes some things from one state to another – by covering them with the stomach of oxen newly killed, and with mud, or, as Virgil states, with the carcasses of bullocks.[1] Similarly wasps and hornets are brought to life by the bodies of horses, and beetles by asses.

The silk-moth and silk production

75. Another species of insect is the silk-moth which is a native of Assyria. It is larger than the insects already mentioned.[2] Silk-moths make their nests of mud, which looks like salt, attached to stone; they are so hard that they can scarcely be pierced by javelins. In the nests they make wax combs on a larger scale than bees and produce a bigger larva.

76. Silk-moths have an additional stage in their generation. A very big larva first changes into a caterpillar with two antennae, then becomes what is termed a chrysalis, from which comes a larva which in six months turns into a silkworm. The silkworms weave webs like spiders and these are used for *haute couture* dresses for

1. *Georgics*, IV, 280 ff.
2. i.e. bees, wasps and hornets.

women, the material being called silk. The technique of unravelling the cocoons and weaving the thread was first invented on Cos by a woman named Pamphile, the daughter of Plateas. She has the inalienable distinction of having devised a way of making women's clothing 'see through'.

77. Silk-moths, so they say, are produced on Cos, where a vapour from the ground breathes life into the flowers – from the cypress, terebinth, ash and oak – that have been beaten down by the rain. First, small butterflies without down are produced; these cannot endure the cold so they grow shaggy hair and equip themselves with thick coats to combat winter, scraping together down from the leaves with their rough feet. They compact this into fleeces, card it with their claws and draw it out into the woof, thinned out as if by a comb, and then they wrap this round their body.

78. Then they are taken away, put in earthenware containers and reared on bran in a warm atmosphere. Underneath their coats a peculiar kind of feather grows, and when they are covered by these they are taken out for further treatment. The tufts of wool are plucked out and softened by moisture and subsequently thinned out into threads by means of a rush spindle. Even men have not been ashamed to adopt silk clothing in summer because of its lightness. Our habits have become so bizarre since the time we used to wear leather cuirasses that even a toga is considered an undue weight. However, we have left Assyrian silk dresses to the women – so far!

Man compared with other animals

130. Of all animals man has the most hair on his head: this is equally true of both sexes – at any rate among peoples that do not cut their hair. Man is the only species to suffer baldness, except in the case of animals actually born without hair. Only the hair of man and horses turns grey; in the former it always begins at the forehead, then spreads to the back of the head.

132. There are instances of double crowns, but only in human beings. The bones of the skull, in man, are flat and thin, without marrow; they are serrated and interlock like combs. All animals that have blood possess a brain, as do those marine creatures that I have called soft species: for example, the octopus, although it is bloodless. Man, however, has the largest brain in proportion to his size and the most moist one; it is the coldest of his organs, being covered by two membranes above and below; if either is fractured death ensues. The brain of a man is larger than that of a woman.

134. In all human beings the brain has no blood or veins, and in some cases is without fatty tissue. The experts assert that it is different from marrow because it hardens when boiled. In the middle of the brain of all creatures there are very small bones. Only in man does the brain pulsate in infancy; the fontanelle does not become firm until the child begins to talk. The brain is the seat of the mind's government.

136. Only man has ears that do not move and this is the origin of the nickname 'flap-eared' (Flaccus).[1]

The eyes

139. The eyes lie beneath the brows and are the most invaluable part of the body; they distinguish life from death by the use they make of daylight.

141. Only men have eyes of different colours, whereas among all other creatures the eyes of each member of a species are alike. But some horses have grey eyes. In man the eyes are of widely differing types: larger than average; medium; small and protruding. Protruding eyes are thought to have poor vision, while deep-set eyes and those with the colour of goats' eyes have the best.

142. Furthermore, some people are long-sighted but others are myopic. In many people, sight depends on the brilliance of the sun, and they cannot see clearly on a cloudy day or after sunset; others see less clearly during the day, but have keener sight than the rest of their fellows at night.

144. The Emperor Claudius' eyes were frequently bloodshot. The Emperor Gaius had protruding eyes. Nero's eyesight was poor except when he screwed up his eyes to look at objects brought close to them. In the Emperor Caligula's school for gladiators, there were 20,000 in training, of whom there were only two who did not blink when they faced some threat of danger; they were invincible.

145. No other part of the body supplies more evidence of the state of mind. This is the same with all animals, but especially with man; that is, the eyes show signs of self-restraint, mercy, pity, hatred, love, sorrow, joy; in fact the eyes are the windows of the soul.

147. Nature has provided the eye with many thin membranes and hard outside coverings as a protection against cold and heat; she cleans the eyes with moisture from the tear-glands and makes them

1. The adjective *flaccus* literally means 'hanging down'.

slippery and able to dislodge foreign bodies that get into them.

148. Nature has provided the cornea, the horny centre of the eye, with a window in the form of the pupil. This has a narrow opening which prevents the gaze from wandering and, so to speak, focuses it so that it is not distracted by objects that it encounters in its field. The pupils are surrounded by rings which in some people are brown, in others grey, and in yet others blue, so that light from ambient brightness may be received in a suitable mixture and its strength, thus moderated, may not be troublesome.

149. Man alone is cured of blindness by the discharge of fluid from the eye. Many have regained their sight after being blind for twenty years. Some have been born blind without any defect in the eyes. Likewise, many have suddenly lost their sight without any previous injury. The leading experts state that the eyes are connected to the brain by a vein. I am inclined to believe that they are also connected to the stomach. For it is an established fact that if a man has an eye knocked out he is invariably sick.

150. It is a solemn ritual among Roman citizens to close the eyes of the dying and to open them again on the funeral pyre, since the custom has grown up that it is not right for the eyes to be seen by another human being at the moment of death, or for them not to be open to heaven.

The heart

181. The heart, in other animals, is in the middle of the chest, but in man alone it is below the left breast with its conical point projecting downwards. It is stated that the heart is the first organ formed, and the last to die. The heart is the warmest organ. It has a definite beat and a movement of its own as if it were a second living creature inside the body. The heart is covered by a very soft, firm membrane and protected by the rib-cage and chest so that it may provide the main impulse and origin of life.

182. The heart provides the main housing for the vital principle and the blood, in a winding recess which in large animals is threefold and in others, invariably double. This is the seat of the mind. From this source two large arteries run separately to the front and back of the body and, by means of a spreading network of branches, convey life-sustaining blood through other smaller arteries.

184. The Egyptians, whose custom it is to embalm bodies, believe that the human heart increases in size every year and that by the age of fifty reaches its maximum weight of about a quarter of an ounce,[1] and, that from that time forward it loses weight at a consistent rate so that because of heart failure man does not live beyond a hundred years.

Anthropoid apes

245. Some animals use their front feet as hands and sit while carrying food to their mouth with these, as, for example, squirrels.

246. Monkeys are exactly like human beings in their faces, nostrils, and the fact that they have ears and eyelashes on both the upper and lower lid – they are the only quadrupeds with these. They have nipples on their breasts, and arms and legs which can bend, like human beings, in opposite directions. Monkeys also have hands with fingers and nails, and a long middle finger. They differ a little in respect of their feet, for these are very long, like their hands, and make a footprint like the palm of a hand. Monkeys also have a thumb and knuckles, like a man, and besides a sexual member (in males only), also have internal organs following the human pattern.

Aristotle's belief that physiology provides clues to the course of one's life

273. I am indeed astonished that Aristotle not only believed but published his beliefs that our bodies contain clues to the course of our lives. But although I think his beliefs are without substance and only to be set down cautiously for fear that everyone may anxiously seek these prognostications in his own case, yet I will touch upon them because they were treated seriously by so great a man in the realm of science.

274. He considers few teeth, long fingers, a leaden complexion and several broken lines in the hand to be signs of a short life. He says, however, that those persons who have sloping shoulders, one or two long lines in their hands, more than thirty-two teeth and large ears enjoy a long life. Yet, as I see it, he does not observe all these signs in any one person, but in a number of individuals. These signs are trifling, to my way of thinking, but they are on everyone's lips. In a similar way, Trogus, one of the most critical authorities of our times, has added some external signs of character, which I will quote: 'When the forehead is large it denotes that the mind is

---

1. Clearly either Pliny or the text is in error: the adult male heart weighs between 10 and 12 ounces.

sluggish, while persons with a small forehead have an agile mind; those with a round forehead are irascible, as if the swelling denotes a swollen temper.' Trogus continues: 'When an individual's eyebrows are level, this shows he is gentle; when they are curved on the side of the nose, he is stern; when bent down at the temples, he is a scoffer; when entirely drooping, he is spiteful and ill-intentioned. If someone's eyes are narrow on both sides this indicates that he is malicious in character. Large ears are a sign of one who talks too much and is silly.' So much for Trogus.

Animals' bad breath

277. Lions' breath contains a virulent poison and bears' breath is unwholesome. No wild animal will touch things that have come into contact with a bear's breath, and things which bears have breathed upon putrefy more quickly. As for the other species, Nature has willed that only in man is the breath made bad in several different ways, namely by tainted food, decaying teeth, and most of all by old age.

BOTANY

Book XII

Trees and their products

1. The treasures within the earth were long hidden, and trees and forests were thought of as her ultimate gift to mankind. From trees first came food, and their leaves made men's caves more comfortable; their bark provided man with clothes. Even in our day and age some races live in this primitive manner.

2. So a growing feeling of amazement steals over us to think that, starting from these beginnings, man now quarries mountains for marble, seeks clothes from China, pearls from the depths of the sea, and emeralds from the depths of the earth. Because of this ear-piercing has been invented, for it was not enough to wear jewels on the hand, round the neck, or in the hair, without them actually piercing the body. Hence it is right to follow the natural order, to speak about trees before other things, and to mention the origins of our customs.

3. Trees were the temples of the gods, and, following old established ritual, country places even now dedicate an outstandingly tall tree to a god. Even images of shining gold and ivory are worshipped less by us than forests and their silence. Different types of trees are dedicated to their own deities and these relationships are kept for all time. For example, the Italian holm-oak is sacred to Jupiter, the laurel to Apollo, the olive to Minerva, the myrtle to Venus, and the poplar to Hercules. We also believe that the gods of the woods and Fauns and various goddesses are, as it were, assigned to forests by heaven.

4. With the passage of time, trees with juices more pleasant than corn mellowed human kind. Olive-oil from trees refreshed men's limbs and draughts of wine restored their strength. In short, all the many things that taste good, and come spontaneously thanks to the generosity of the year, are in demand. Similarly in demand are dishes served for the second course – in spite of the fact that wild beasts must be fought for them, and that fish feed on the bodies of shipwrecked men.

9. Among well-known plane-trees there is the one that grew in the walks of the Academy at Athens and, secondly, the celebrated plane in Lycia. A third plane-tree is connected with the Emperor Gaius Caligula, who on an estate at Velitrae was impressed by the 'flooring' of a single plane-tree and the horizontal branches serving as seats; he held a banquet in the tree – the leaves provided a partial awning – in a dining-room spacious enough to hold fifteen guests and the servants. Caligula called this dining-room his 'eyrie'.

11. There is a fourth plane-tree by a fountain at Gortyna on the island of Crete, celebrated in both Greek and Latin records for not shedding its leaves. A far-fetched Greek story about this tree survives, which asserts that Jupiter lay with Europa beneath it – as if there were no other trees of the same species on the island of Crete!

Pliny describes a number of exotic trees

22. The Indian banyan is remarkable for its fruit and is self-propagating. It spreads out its huge branches, and the bottom ones bend over towards the ground to such an extent that in the space of a year they root and produce new saplings for themselves in a circle round the parent tree after the manner of some ornamental garden. Inside this leafy shelter shepherds spend their summer, as it is both shaded and protected by the tree. The banyan is a pleasing sight with its domed appearance, when viewed from beneath or at a distance.

23. Its taller branches shoot up to a great height from the main body of the mother tree to form an extensive grove, in many examples enclosing a circle almost 100 feet in diameter; their shade has a spread of 400 yards. The broad leaves have the shape of an Amazon's shield and cover the fruit, so hindering its growth. The fruit is thus scarce and no larger than a bean. When the fruit is ripened by the rays of the sun shining through the leaves, it acquires a sweet taste and is worthy of this fantastic tree. The banyan grows mainly near the River Acesines.

24. The wise men of India live on another fruit, which is bigger and of a more agreeable flavour. The leaves resemble birds' wings and are just under a yard long and a yard wide. The tree puts out its fruit from the bark; the fruit has a wonderfully sweet juice and a hand provides enough fruit for four people. The tree is the plantain, its fruit, bananas. It grows most profusely in the land of the Sydraci, the furthest point reached by Alexander's expedition. There is another similar tree with sweeter fruit, but this upsets the bowels. Alexander gave orders that none of his troops should touch the fruit.

Pepper, ginger-trees and cane-sugar

26. Trees like our junipers but producing pepper occur everywhere, although some authorities say that they grow only on the south-facing slopes of Mount Caucasus. The seeds differ from those of the juniper, being in small pods like those we see in kidney beans. These pods, when picked before they have opened and dried in the sun, produce what is called 'long pepper', but if left to open gradually they reveal white pepper, and if this is then dried in the sun it changes colour and becomes wrinkled. Black pepper is sweeter but white pepper has a milder flavour than either black or long pepper.

28. The root of the pepper-tree is not, as some have thought, the substance called 'ginger', although its taste is similar. Ginger grows on farms in Arabia and in the country of the Ethiopian Cave-dwellers. Ginger is a small plant with a white root.

29. It is amazing that pepper is so popular. Some substances attract by their sweetness, others by their appearance, yet pepper has neither fruit nor berries to commend it. Its only attraction is its bitter flavour, and to think we travel to India for it! Who was the first man willing to try this on his food, who, because of his greedy appetite, was not content merely to be hungry? Both ginger and pepper grow wild in their own countries, yet they are purchased by weight as if they were gold and silver. Italy also has the pepper-tree; it is similar to the myrtle, but larger.

32. Although Arabia produces cane-sugar the Indian variety is more esteemed. The substance is a kind of 'honey' that collects in reeds. It is white like gum and brittle to the teeth: the largest pieces are the size of a hazelnut. Cane-sugar is used only as medicine.

Trees in Persia: the cotton-tree

37. Persia adjoins the land of the peoples mentioned above. On the shores of the Red Sea, at the place I have called the Persian Gulf, where the tides penetrate far inland, the nature of the trees is remarkable. For they are visible on the coast, when the tide is out, their naked roots embracing the sand like octopuses; they are eaten away by the salt and look like trunks that have been washed ashore and left high and dry. These trees remain where they are when beaten by the waves as the tide comes in. Indeed at high tide they are completely covered. It appears from the evidence that the trees are nourished by the salt water. Their size is amazing; in appearance they resemble strawberry-trees, but their fruit is like almonds on the outside and contains a twisted kernel.

38. Also in the Persian Gulf is the island of Tyros, which is covered with forests in the area facing east; the land here is also flooded by the sea at high tide. On the island's higher ground there are trees that bear 'wool', but in a different way from those of the Chinese, for the leaves of these trees have no such 'wool', and, were they not smaller, might be thought to be vine-leaves. The trees bear gourds the size of a quince. When these ripen and open up, they reveal balls of 'down' from which an expensive linen material for clothes is made.

39. Local people call this the 'cotton-tree'; it also grows, more profusely, on the smaller island of Tyros, which is 10 miles from the other island of the same name. Juba states that this shrub has a 'woolly' down around it and that the linen produced from this is of a higher quality than its Indian counterpart.

Resins: frankincense and myrrh

51. Next would have come cinnamon, if this were not an appropriate point to mention the riches of Arabia and the reasons that have given it the names 'Happy' and 'Blessed'. The principal products of Arabia are frankincense and myrrh; it shares myrrh with the country of the Cave-dwellers, but Arabia is the sole producer of frankincense – and even then, not the whole of Arabia.

54. It is said that not more than 3,000 families retain as a hereditary privilege the right to trade in frankincense; and so the members of these families are called sacred and not allowed to be defiled by meeting women or funeral parties when they are tapping the trees to obtain frankincense. In this way the price is inflated through religious scruples. Some authorities state that frankincense in the forests is available for all people without distinction, but others say it is shared out each year between different people.

55. There is no agreement about the appearance of the tree itself. We have conducted campaigns in Arabia, and Roman arms have penetrated a large part of the country – indeed, Gaius Caesar, son of Augustus, won renown there. Yet no Roman writer to my knowledge has so far described what this tree looks like. Greek descriptions of it vary.

58. When there was less opportunity for selling it, frankincense used to be collected only once a year. But present-day demand brings a second harvest. The early, natural harvest is about the time of the rising of Sirius, at the hottest part of the summer. They tap the tree where the bark appears full and stretched to its thinnest.

The bark is opened by a blow, and not stripped. A thick froth spurts out; this coagulates, thickens and becomes solid. This substance is caught on palm-leaves, where the nature of the ground requires this; elsewhere it is collected on a firm, trodden surface around the tree. The latter method gives a purer substance, the former a heavier yield. Any frankincense that remains on the tree is scraped off with an iron scraper, and so has fragments of bark in it.

59. The forest is divided up into sections, and because people are honest with one another, the allocations are safe. No one guards the trees that have been tapped, and no one is robbed by another. But at Alexandria, where the frankincense is processed, no amount of watchfulness is sufficient to guard the workshops! The workmen's aprons have a seal, they wear a mask or a close-meshed net over their heads, and have to take off all their clothes before they are allowed to leave. The workmen at Alexandria display much less honesty in handling the product than the growers in the forests.

63. Once collected, the frankincense is taken to Sabota on camels, and one of the gates of the city is opened to let the consignment in. The kings have made it a capital offence for the camels to turn off the main road. At Sabota a tithe, by volume not weight, is taken for the god they call Sabis before the incense can be put up for sale. This tithe is used to defray a public cost, for on a fixed number of days Sabis graciously entertains guests at a banquet. The frankincense can be exported only through the country of the Gebbanitae, and thus a tax is also paid on this to their king.

64–5. Their capital is Thoma, some 1,487 miles from the town of Gaza in Judaea on the Mediterranean coast. The journey is split up by camel-halts into sixty-five stages. Fixed amounts of the frankincense are given to the priests, the king's secretaries and in addition the guards; apart from these attendants, gatekeepers and servants all take a commission. All along the route they keep paying out – at one place for water, at another for fodder, as well as charges for lodging at halts and various dues.

Myrrh grows in many places in Arabia. The tree grows to a height of about six feet. It has thorns and a hard trunk, twisted and thicker than that of the frankincense-tree and even thicker at the root than the rest of it.

68. This tree is also tapped twice a year, at the same seasons as the frankincense-tree, but the incisions go from the root to those branches that are strong. Myrrh-trees spontaneously exude juice before they are tapped. This is called oil of myrrh, the most valued of all the varieties. Next is the cultivated kind.

Arabia's claim to the epithet 'Happy'

82. Arabia has neither cinnamon nor cassia, but nevertheless enjoys the epithet 'Happy' – a false and ungrateful title since she credits her happiness to the powers above, although she owes more thanks to the powers below. The extravagant habits of men, even in death when they burn over the departed products originally thought to have been created for the gods, have given rise to Arabia's epithet.

83. Leading authorities assert that Arabia does not produce as large a quantity of scent in a year as the Emperor Nero burned on the day his wife Poppaea died. Then add up all the very great number of funerals celebrated each year throughout the whole world and the scent piled up in heaps in honour of corpses – scent given to the gods a single drop at a time. Yet the gods were no less well disposed to worshippers offering salted spelt; indeed, as is evident, they were better disposed.

84. But the Arabian Sea still more deserves the epithet 'Happy', for it is from there that Arabia sends us pearls. And at the lowest reckoning, India, China and the Arabian peninsula take from our Empire 100 million sesterces every year: this is the amount our luxuries and our ladies cost us. For what fraction of these commodities, pray, finds its way to the gods above or to the gods of the lower world?

Cinnamon and cassia

85. An incredible story was told in far-off times – first of all by Herodotus[1] – that these two spices are obtained from birds' nests, especially from those of the phoenix in the region in which Bacchus was brought up; according to the story, the nests are dislodged from inaccessible rocks and trees by the weight of meat brought to them by the birds themselves, or by means of arrows weighted with lead. There is also a story of cassia growing round marshes and protected by a terrible species of bats that guard it with their claws, and by winged serpents. The locals inflate the price of the spices by concocting these tall stories.

86. There is an accompanying story that in the reflected rays of the midday sun an indescribably pervasive smell is given off from the whole peninsula: this is due to the blending of so many kinds of vapour. It is said that the first indication of Arabia's existence

1. III, 111.

received by Alexander the Great's fleets was carried by these odours out to sea. But all these stories are false if only because cinnamomum – which is the same as cinnamon – grows in Ethiopia, whose inhabitants are intermarried with the Cave-dwellers.

87. The Cave-dwellers buy cinnamon from their neighbours and carry it across vast seas in ships that are neither steered by rudders, nor driven by oars, nor drawn by sails, nor assisted by any device. Only man and his boldness stand in place of all these contrivances. Moreover, they choose the winter sea on about the shortest day for their journey since then it is mainly the east wind that blows.

88. This carries them on a straight course through the bays and, after they have rounded a cape, a west-north-west wind carries them to the harbour of the Gebbanitae, called Ocilia. Hence the Cave-dwellers make mostly for this port; they say that it is almost five years before their traders return and that many die *en route*. In return for their commodities they take home glass and copper ware, clothing, buckles, bracelets and necklaces; this trade, therefore, relies mainly on their confidence in women's demand for such items.

93. The right to sell cinnamon is granted by the king of the Gebbanitae, who opens trading by a proclamation.

Balsam

111. Balsam is superior to all other scents. The only country on which this plant has been bestowed is Judaea, where formerly it grew in two orchards only, both royal: the one was not more than 20 *iugera* in size, the other even smaller. The emperors Vespasian and Titus exhibited this tree in Rome and, remarkable to relate, from the time of Pompey the Great we have included trees in our triumphal processions.

112. The balsam-tree is now a subject of Rome and pays tribute along with the race to which it belongs. It differs completely in character from what Roman and foreign writers had recorded, being more like a vine than a myrtle. Quite recently it has been trained to grow from mallet-shaped slips tied up like a vine, and it covers whole hillsides like vineyards. A balsam-tree, without supports and carrying its own weight, is pruned in a similar manner when it puts forth its shoots; it thrives with raking, sprouts rapidly and bears fruit after two years.

113. Its leaf is very close to that of a kind of apple-tree and is evergreen. The Jews vented their anger on the balsam-tree, just as on their own bodies, but the Romans defended it and fought to

preserve a bush. The Treasury now plants the balsam-tree and it has never been more in evidence, but its height is less than 3 feet.

115. The tree is tapped by means of a piece of glass or stone, or with knives made of bone, for it dislikes having its vital parts harmed by iron and dies immediately – although it can stand having its superfluous branches pruned by an iron knife. The hand of the person making the incision has to be carefully balanced so as not to make a wound deeper than the bark.

116. The juice that oozes out is called balsam; it is extraordinarily sweet, but distils in tiny drops; this weeping liquid is collected by wool into small horns and is poured out of these and stored in a new earthenware pot. The balsam is rather like thick olive-oil and, while unfermented, is white; it then clears and turns red, hardening at the same time.

117. When Alexander the Great was campaigning in Judaea a shellful[1] was reckoned a fair day's production for a single tree in summer, and about 6 pints the production for the whole of a rather large orchard, and slightly over a pint for a smaller one, at a time when its price was twice its weight in silver. Now even a single tree produces more balsam. The balsam-tree is tapped three times every summer, and is then pruned.

118. Even the twigs are saleable; and within five years of the conquest of Judaea the cuttings and shoots realized 800,000 sesterces. These are called balsam-wood and are boiled to make perfumes; perfume-factories have used them instead of the actual juice of the shrub. Even the bark is valuable for drugs; but the 'tears' are most valued, then the seeds, then the bark, and finally the wood, which has the lowest value.

The adulteration of balsam and tests for its purity

121. Balsam is adulterated with attar of roses, with oil of Cyprus, of mastic, of behen-nut, of the turpentine-tree and with myrtle; but the worst adulteration is with gum, since then it both dries up on the back of the hand and sinks in water,[2] thus providing a double test.

122–3. Pure balsam ought to dry, but the kind mixed with gum turns brittle and forms a skin. Impure balsam can be detected by

1. Approximately a hundredth of a pint.
2. Pliny is inconsistent since later (paras. 122–3) he says that *pure* balsam sinks in water.

taste, or, when wax or resin has been added, by the use of hot coal, since then it burns with a blacker flame. Pure balsam thickens in warm water and settles to the bottom of the vessel, whereas adulterated balsam floats on the surface like oil. If it has been tainted by almond-oil, a white ring forms round it. The ultimate test is that pure balsam causes milk to curdle and does not leave stains on cloth. In no other field is there more flagrant fraud, since every pint purchased at a liquidator's sale for 300 denarii is sold again for 1,000 denarii; so much profit is there in increasing the volume of balsam by adulteration.

Book XIII

Perfumes

1. Forests are valuable in the realm of perfumes inasmuch as individual essences are not sufficiently remarkable in themselves, and luxury delights in compounding them to make a single perfume.

2. This led to the invention of perfume, but the name of the inventor is not recorded. Perfumes did not exist at the time of the Trojan War, nor was incense used in prayers to the gods. Men knew only the smell of cedar- and citrus-trees found in their own countries, or more accurately, the smell as it rose in columns of smoke at sacrifices, although attar of roses had been discovered.

3. Perfume ought to be considered the trademark of the Persians since they soak themselves in it and neutralize the odour from dirt by means of its excellent qualities. The first example of perfume that I have been able to discover was a chest of perfumes captured by Alexander the Great among the belongings of King Darius abandoned when his camp was captured. Afterwards our people also acknowledged the pleasure of perfume as among the most elegant and respectable good things in life, and it began to be a suitable offering to the dead. For these reasons I shall speak at greater length on this topic.

4. Perfumes are sometimes named after their country of origin, sometimes after the essences from which they are compounded, or after trees, or are named for a variety of reasons. The first thing one should realize is that fashions change and quite often their reputation fades. Once the most praised perfume was made on the island of Delos, then it was the perfume from Mendes in Egypt.

Athens, however, has consistently maintained the reputation of her 'Panathenaic' perfume.

The manufacture of perfume

7. The formula for making any perfume has two main ingredients: the essence and the solid base. The former generally contains various kinds of oil, and the latter is the substance providing the scent. The oils are called 'astringents', the scents 'aromatics', a third ingredient – which many people fail to mention – being the colourant. Cinnabar and anchusa should be added to provide the colour. A sprinkling of salt preserves the properties of the oil, but when anchusa is added, salt is left out. Resin, or gum, is added to preserve the scent in the solid base as it otherwise very quickly evaporates and disperses.

Types of perfume

9. I am inclined to believe that the perfumes most widely used are derived from the rose, which grows everywhere in profusion. So the simplest compound was, for a long time, attar of roses mixed severally with unripe olive or grape juice, rose and saffron blossoms, cinnabar, reed, honey, rush, flower of salt, anchusa or wine.

18. 'Royal' perfume is so-called because it is produced for the kings of Parthia; it is a blend of behen-nut juice, costus, amomum, Syrian cinnamon, cardamom, spikenard, cat-thyme, myrrh, cinnamon-bark, styrax-tree gum, ladanum, balsam, Syrian reed and rush, wild grape, cinnamon-leaf, serichatum, cypress, rosewood, panace, gladiolus, marjoram, lotus, honey and wine. Nine of the ingredients of Royal perfume are grown in Italy, the conqueror of the world; indeed no ingredients are grown elsewhere in the whole of Europe, except the iris in Illyria, and nard in Gaul. For wine, roses, myrtle-leaves and olive-oil may all be accepted as the common property of almost all countries.

19. The substances that we know as talcum powders are made of dried perfume. When perfumes are tested they are put on the back of the hand so that the warmth of the fleshy part of the hand may not spoil them.

20. Perfumes are the most pointless of all luxuries, for pearls and jewels are at least passed on to one's heir, and clothes last for a time, but perfumes lose their fragrance and perish as soon as they are used. Their highest recommendation is that, as a woman goes by, their use may attract even those who are otherwise occupied. They

cost more than 400 denarii for a pound. So much is paid for someone else's pleasure, since the wearer does not smell the perfume.

22. I have seen people put perfume on the soles of their feet, a trick that Marcus Otho taught the Emperor Nero. How, pray, could this be noticed, or how could there be any pleasure from that part of the body? Moreover, I have been told that someone with no imperial connections gave instructions for the walls of his bathroom to be sprinkled with perfume, and that the Emperor Caligula had his bath-tub perfumed; later one of Nero's slaves did the same, so that this would not seem to be a prerogative of emperors.

23. The most astonishing thing, however, is that this extravagant behaviour has found its way even into our military camps. At any rate the eagles and the standards, dusty and bristling with sharp points, are anointed on holidays. I wish I could say who first introduced this practice. The fact is, no doubt, that our eagles were bribed to conquer the world by such a reward! We seek their patronage for our vices so that we may have the right to use hair-oil under our helmets!

24. I could not easily say when perfume made its way to Rome. It is an established fact that in AUC 565[1] the censors Publius Licinius Crassus and Lucius Julius Caesar issued a proclamation forbidding anyone to sell 'foreign perfumes', as they are called.

25. But heavens alive! – at the present time some people actually add perfume to their drinks and consider the bitterness worth it so that their body can enjoy the strong scent inside and out.

Palm-trees

27. Palm-trees are not indigenous to Italy; they are found only in those parts of the world where it is warm.

28. Palms grow in light sandy soil – usually one containing alkaline salts. They enjoy well-watered locations and take up water all the year round, although they also like dry places. There are several kinds of palm, beginning with those not larger than a shrub.

29. The taller palms make forests, their pointed leaves shooting out from all round the tree like a comb; these trees, it must be understood, are wild types. The other kinds are slender and tall and have compact rows of knots[2] or circles in their bark which are easy

1. 189 BC.
2. Such knots are well illustrated by coin types of Carthage. See G. K. Jenkins and R. B. Lewis, *Carthaginian Gold and Electrum Coins* (RNS, London, 1963), pls. 5–6, 35.

for Eastern peoples to climb. They put a plaited noose round themselves and the tree, and this travels up with the climber at an amazingly rapid speed.

30. All the foliage is at the top of the tree, as is the fruit, which is not among the leaves as in other trees, but hangs in bunches from shoots of its own between the branches. The leaves are divided into two halves with razor-sharp edges, and gave the idea for folding writing-tablets. Today they are split up to make ropes, plaited wicker-work and sunshades.

36. Palms are propagated by layering – that is, by dividing up the trunk with incisions for a length of about three feet from the brain[1] of the tree and pegging this down. A slip, torn from the root, or from one of the youngest branches, makes a hardy young plant.

Figs

56. Egypt has many varieties of trees not found elsewhere – first and foremost the fig, which for this reason, is called the Egyptian fig. Its leaves resemble those of the mulberry in size and appearance. The fig produces its fruit not on the branches but on the trunk itself, and the Egyptian variety is exceptionally sweet and seedless. The tree's yield is extremely prolific, but only when iron hooks are used to make incisions in the fruit, which otherwise does not ripen.

57. When this is done the fruit is picked three days later, while another fig forms beneath it; the tree thus has seven crops of very juicy figs in a single summer. The wood derived from this fig-tree is of a peculiar kind but among the most useful of woods. As soon as it has been cut it is plunged into a marsh; it sinks to the bottom at first but afterwards begins to float on the surface; this indicates that the timber is seasoned.

The papyrus plant and the invention of paper

68. I have not yet touched upon marsh plants or shrubs that grow by rivers. But before I leave Egypt I shall also describe the nature of the papyrus plant since our civilization – or at any rate our written records – depends especially on the use of paper.

---

1. Pliny correctly explains the method of layering, but here, as elsewhere, uses terms of human physiology to describe the tree's parts.

69–70. According to Marcus Varro, paper was discovered as a result of the victory of Alexander the Great and of his foundation of Alexandria in Egypt. Prior to that, paper was not used, and people wrote first on palm-leaves and then on the bark of certain trees. Subsequently official records were inscribed on sheets of lead, and private documents were written on linen or on wax-tablets. We discover in Homer that the use of writing-tablets antedates Trojan times.[1] But in his day not even the land itself that is now known as Egypt existed, although at the present time papyrus grows in the Sebennytic and Saitic nomes of Egypt,[2] on land subsequently heaped up by the Nile. Homer had written that the island of Pharos, now joined to Alexandria by a bridge, took a day and a night to reach by sailing-ship from the land. Later, according to Varro also, when there was rivalry between King Ptolemy and King Eumenes about their libraries and the former stopped the export of paper, parchment[3] was invented at Pergamum. Afterwards the use of those materials, on which the immortality of human beings depends, spread indiscriminately.

71. Papyrus grows in the swamps of Egypt and in the sluggish waters of the Nile where they have overflowed and form stagnant pools not more than about three feet deep. The plant has a sloping root the thickness of a man's arm, and tapers upwards in a triangular shape, each plant not more than about fifteen feet tall, and tipped like a thyrsus. The roots of the papyrus plant are used by local people for timber, not only to serve as firewood but also for making various utensils and containers. Indeed they plait papyrus to make boats and they weave sails and matting from the bark and also cloth, blankets and ropes. They chew it when raw and when boiled, but only swallow the juice.

73. The papyrus plant also grows in Syria by the lake around which is found the scented reed already mentioned. King Antiochus would allow ropes made only from Syrian papyrus for his navy; the use of esparto grass for ropes had not yet become commonplace. It

---

1. *Iliad*, VI, 168. This is the only reference to writing in Homer's epics. The existence of Linear B encourages the belief that this was a memory surviving from the Mycenaean age. The Greeks borrowed the art of alphabetical writing from the Phoenicians in the eighth century BC.

2. Egypt was split into three governorships – the Thebaid, Lower Egypt and Central Egypt – each of which was further divided into nomes or administrative districts. There were forty-two nomes, each in the charge of a civil servant who was both judge and tax collector.

3. *Pergamena.*

has recently been realized that papyrus growing in the Euphrates near Babylon can also be used in the same way for paper. Yet until now the Parthians have preferred to embroider letters upon cloth.

74. Paper is manufactured from papyrus by splitting it with a needle into strips that are very thin but as long as possible. The quality of the papyrus is best at the centre of the plant and decreases progressively towards the outsides. The first quality used to be called 'hieratic' paper and in early times was devoted solely to books connected with religion, but, to flatter the emperor, was given the name 'Augustus'; the second quality was called 'Livia' after his wife, and so the term 'hieratic' was relegated to the third category.

75. The next quality had originally been called 'amphitheatre' after the place where it was made. This paper was taken over by the ingenious workshop of Fannius at Rome; they made its texture finer by careful dressing, and so upgraded ordinary paper to rank with the first quality – this paper was known by the name of the maker. But paper of this kind that did not have the additional processing remained in its own category as 'amphitheatre' paper.

76. Next is 'Saitic' paper, called after the town where it was produced in the greatest quantity, and made from low-grade scrapings; then comes 'Taeneotic' paper from a neighbouring place, made from fibre still closer to the outside covering of the papyrus. This paper is sold by weight, not quality. Finally there is what is called *emporitica*, 'packing' paper, which is no good for writing on but is used to cover documents and a wrapping for merchandise; for this reason it takes its name from the Greek word for a merchant.[1] After this is the outer layer – the actual papyrus – which is like rush and of no use even for ropes, except for those used in water.

77. All paper is 'woven' on a board dampened with water from the Nile; the muddy liquid acts as glue.[2] First an upright layer is smeared on the table – the whole length of the papyrus is used and both its ends are trimmed; then strips are laid across and complete a criss-cross pattern, which is then squeezed in presses. The sheets are dried in the sun and then joined together; each succeeding sheet decreases in quality down to the worst. There are never more than twenty sheets on a roll.

1. *emporos.*
2. This is incorrect: the river water is used to prevent the strips from drying out. The natural juice of the papyrus acts as the bonding agent.

Varieties of paper

78. There is a big difference in the width of the various types of paper. The best paper is 13 inches wide, and the hieratic 11 inches; Fannian is 10 inches, and amphitheatre 9 inches, while Saitic is not as wide as the mallet used in the manufacturing process. Other criteria of quality in paper are its fineness, thickness, whiteness and smoothness.

79. The Emperor Claudius imposed modifications on the best quality because the thinness of the paper in Augustus' time was not able to withstand the pressure of pens. In addition it allowed the writing to show through, and this brought fear of blots caused by writing on the back of the paper. Moreover, the excessive transparency of the paper looked unsightly in other ways. So the bottom layer of the paper was made from leaves of the second quality, and the cross-strips from papyrus of the first quality.

80. Claudius also increased the width of the sheet to a foot. There were also sheets 18 inches wide called *macrocola*;[1] examination of these revealed a defect: tearing off a single strip damaged several pages. Because of this Claudius paper is preferred to all other kinds, although Augustus paper still leads the field for letter-writing.

81. Surface roughness is smoothed out with a piece of ivory or a shell, but writing easily fades because the polished paper with its shinier surface does not absorb ink as well. The wetting process, when carelessly done, often causes initial resistance to writing and this flaw can be detected by a mallet-blow or even by the paper's musty smell. Another process is therefore added in the production of the paper.

The use of artificial adhesives in paper manufacture

82. A basic paste is made from the finest-quality flour mixed with boiling water and a sprinkling of vinegar; carpenter's paste and gum are too brittle an adhesive. A still better method of making paste is to mix breadcrumbs in boiling water. This results in the least amount of paste at the joins and produces paper softer even than linen. All pastes used should be just a day old, neither more nor less. When pasted, the paper is beaten thin with a mallet and its surface covered with paste; once more it has its creases removed by pressure and is flattened with the mallet.

1. cf. also Cicero, *Letters to Atticus*, XIII, 25, 3 and XVI, 3, 1.

83. This process may enable written records to last a long time. At the house of the poet and most distinguished citizen Pomponius Secundus, I have seen records in the hand of Tiberius and Gaius Gracchus[1] that were written nearly two centuries ago. Indeed I very often see autographs of Cicero, the late Emperor Augustus and Virgil.

The history of paper

84–85. There are important examples brought to light that go against Marcus Varro's views concerning the history of paper. Cassius Hemina, a historian of many years ago states, in his *Annals*, IV, that Gnaeus Terentius, a clerk, when digging his land on the Janiculum, unearthed a chest that held the body of Numa, king of Rome, and some books of his. This happened 535 years after Numa's reign.[2] Hemina further writes that the books were made of paper, which is all the more remarkable because they had remained intact, although buried.

86. Other people have wondered how those books could have lasted. Terentius gave the following explanation. In the middle of the chest there had been a square stone bound all round with cords covered in wax, and the three books had been placed on this stone. This was why, he thought, they had not decayed; moreover, the books had been soaked in citrus-oil, and for this reason had not been eaten by moths. The books in question contained the philosophical doctrines of Pythagoras. Hemina also says that the books had been burnt by the praetor Quintus Petilius because of their contents.

87. Piso the censor records the same story in his *Commentaries*, I, but he says there were seven volumes of pontifical law and the same number relating to Pythagorean philosophy. Tuditanus in his Book XIII states that there were twelve volumes of *Antiquities of Man*, and Antias says in his Book II that there were twelve volumes on *Pontifical Matters* in Latin, and the same number in Greek comprising *Doctrines of Philosophy*. Antias also mentions in his Book III a resolution of the Senate to the effect that these volumes should be burnt.

88. There is, however, universal agreement that the Sibyl took three volumes to Tarquin the Proud; two of these were burnt by

---

1. Tiberius (162–133 BC) and Gaius (153–121 BC) were sons of Titus Sempronius Gracchus, consul in 177 and 163 BC. They were tribunes who attempted to carry out agrarian and social reforms. They were murdered by the senatorial oligarchy.
2. 181 BC.

herself, while the third was lost when the Capitol was destroyed by fire during the Sullan crisis. Moreover, Mucianus, who was consul three times, asserts that during his recent governorship of Lycia he had read in a certain temple a letter of Sarpedon written on paper at Troy. I find this quite remarkable if Egypt did not exist when Homer was writing. And why, if paper was then already in use, is it known to have been customary to write on folding tablets made of lead or sheets of linen? Or why does Homer write that even in Lycia itself wooden tablets, not letters, were given to Bellerophon?[1]

89. Paper tends to be in short supply, and as early as the principate of Tiberius a shortage of paper led to the appointment of commissioners from the Senate to oversee its distribution; otherwise daily life would have been in chaos.

Citrus-wood

91. Adjoining Mount Atlas is Mauretania where there is a forest that produces a great number of citrus-trees; this is the origin of men's obsession for tables, which women use against them to counter the charge in respect of their own extravagance in pearls. There still exists a table once owned by Marcus Cicero for which, out of his meagre resources, he paid 500,000 sesterces.

Trees that grow beneath the sea

139. In the East it is a remarkable fact that immediately one leaves Coptos and passes through the desert the only vegetation one finds is the thorn known as 'dry thorn' – and even this quite rarely. On the shores of the Red Sea, however, there are flourishing forests mainly of bay- and olive-trees, both bearing berries and, when it rains, fungi that are changed into pumice by the sun's rays. These bushes are about 5 feet tall. The seas are full of sharks, so that a sailor can barely observe the trees in safety from his ship, as the sharks very often attack the actual oars.

140. Alexander's soldiers who sailed from India told of some trees growing in the sea, whose leaves were green under water but dried

1. *Iliad*, VI, 168. Bellerophon was the son of the Corinthian king Glaucus and Eurymede, and grandson of Sisyphus. He murdered Belerus, also from Corinth, and fled to Proteus, king of Argos, to be purified of the murder. He was eventually sent to Iobates, king of Lycia, with a letter in which Proteus asked him to put Bellerophon to death. Instead Iobates sent him to kill the chimaera thinking that he would perish in the attempt. With the aid of Pegasus, however, Bellerophon succeeded.

up in the sun as soon as they were taken out of the water and turned into salt. They also said that along the coasts there were bulrushes of stone exactly like real ones and that in deep water there existed certain shrubs the colour of cow-horn where they branched out, and red at the top. These were brittle when handled, like glass, but turned red-hot in fire, just like iron, their original colour returning when they had cooled down.

141. In the same region the tide submerges forests, although the trees are taller than the tallest planes and poplars. They have foliage like the bay-tree, and their flowers have the scent and colour of violets; their berries are like olives and have a pleasant smell. The berries form in the autumn and fall in the spring, but the tree is evergreen. The sea completely covers the smaller of these trees, but the tops of even the smallest stick up and ships are moored to them and to their roots when the tide goes out. I have been told by the same authorities that in this sea other trees have also been seen which always retain their leaves and have fruit like a lupin.

BOOK XIV

VINES AND VITICULTURE

Italian trees

1. The trees described so far have mostly been foreign species that cannot be trained to grow outside their original environment and do not travel well. I can now speak about trees shared by different countries; of all these Italy can be considered the special parent.

2. There is one matter concerning which my surprise exceeds my powers of expression, namely that our memory of some trees has perished, and similarly our memory even of the names listed by previous authors. For one would surely think, now that world-wide communications have been established thanks to the authority of the Roman Empire, that living standards have been improved by the interchange of goods and by partnership in the joy of peace and by the general availability of things previously concealed.[1]

The decay of science and the spread of avarice

3. Yet, in all conscience, people who know much of what has been published by earlier writers cannot be found. The research of men of former times was more productive, or their industry was more successful, a thousand years ago at the beginning of literature, when Hesiod began to expound his principles for farmers. His research was followed by several writers, and this has resulted in more work for us, since now we have to investigate not only subsequent discoveries but also those made by earlier authorities, because men's laziness has brought about a complete destruction of records.

4. What cause for this shortcoming could there be other than the state of world affairs generally? The thing is that other customs have crept in; men's minds are preoccupied with other matters and the only arts practised are those of greed. In earlier times peoples

---

1. Pliny's argument wanders. After the first sentence of this paragraph one would have expected him to say that the spread of Roman civilization has improved standards of *scholarship*, not of living. He is carried away, as often, by his obsession with the effects of avarice – a theme never far from the surface of his mind.

had their power limited to their own boundaries, and for that reason their talents were circumscribed; there was no scope for amassing a fortune, so they had to exercise the positive quality of respect for the arts. Accordingly they put the arts first, when displaying their resources, in the belief that the arts could bestow immortality. This was the reason why life's rewards and achievements were so plentiful.

5. The expansion of the world and the growing extent of our resources proved harmful to subsequent generations. Senators and judges began to be chosen by wealth, and wealth was the only embellishment of magistrates and commanders; lack of children began to exert the highest influence and power, and legacy-hunting was the most profitable occupation. In such a climate the only pleasure consisted in possession, whereas the true prizes of life went to rack and ruin and all the arts that were called 'liberal' – from liberty the greatest good – became quite the opposite. Obsequiousness began to be the sole means of advancement. Different men worshipped greed in different ways and different contexts, although every man's prayer had the same goal – namely, the acquisition of material possessions. Everywhere even distinguished people preferred to cultivate others' vices rather than their own good qualities. The result is, I declare, that pleasure has begun to live, while life itself has come to an end.

The vine (classed as a tree because of its size)

8. But where better to begin than with the vine? Pre-eminence in the vine is so much Italy's special distinction that even with this one asset she can be thought to have surpassed all the blessings of the world – even in respect of perfumes, since when the vine flowers, its pleasant scent is second to none.

9. The vine was rightly reckoned by men in times gone by to belong to the category of trees because of its size. In the city of Populonia we can see a statue of Jupiter that was carved from a single vine-stem and has remained sound for a great many years, and similarly a bowl at Massilia. The Temple of Juno at Metapontum is supported by columns made from vine-wood. Even now the way up to the roof of the Temple of Diana at Ephesus is by a staircase made from a single vine, grown, as the story has it, on Cyprus, since on that island vines grow to an exceptional height. No other timber has a longer life, but I am inclined to believe that the items mentioned above were constructed from the wood of the wild vine.

Viticulture

10. Our vines are kept in check by annual pruning and all their strength is drawn into the shoots, or driven downwards for layering. The only benefit obtained from vines is the juice which is recovered by a variety of methods, suited to the nature of the climate and the quality of the soil. In Campania the vines are tied to poplars; embracing their 'brides' and climbing with unruly arms in a knotted course among their branches, they rise level with their tops, so high up that those hired to harvest the grapes make a point of arranging for a pyre and a grave in their terms of employment. The vine indeed never stops growing.

11. I have even seen whole country houses, as well as other buildings encircled by the shoots and clinging tendrils of a single vine. And a thing that Valerianus Cornelius considered particularly noteworthy was that a single vine in the colonnades of the house of Livia at Rome shades the open walks with its trellises and at the same time produces 84 gallons of grape-juice annually.

12. There is an Italian tree, on the other side of the River Padus, called the maple, whose broad upper branches are covered by vines: these spread out with their bare snake-like growth to where the tree forks and then send out their tendrils along the upraised 'fingers' of the branches.

13. Vines, when supported by stakes about as tall as a man of medium height, make a shaggy growth and form a whole vineyard from a single cutting.

Varieties of vines and their wines

In some of the provinces the vine stands by itself without any prop, keeping its limbs close to itself and sustaining a luxuriant growth by virtue of its shortness.

14. In other regions the winds prevent this, for example, in Africa and in parts of the province of Gallia Narbonensis where vines are prevented from growing beyond their pruned stumps.

20. Democritus, who claimed to know all the different kinds of vines in Greece, alone thought it possible to count them, but all other writers have gone on record as saying that there is a countless number of types; the truth of this will become more apparent when we consider wines. I shall not, of course, mention all wines but only the most famous since there are almost as many wines as there are districts.

21. The vines of Aminaea are in the top category because of their body and their character, which clearly improves with age.

23. Next in order of quality are the vines of Nomentanum, the wood of which is red, and this leads some people to call these 'reddish vines'.

24. The Apian vine acquired its name because bees[1] are especially greedy for it. There are two kinds, both covered with down when young. The difference is that one ripens more quickly, although the other is not slow to ripen. These vines do not object to cold locations, but no others rot more quickly in the rain.

Foreign vines grown in Italy

25. So far the chief distinction has been given to vines indigenous to Italy. The rest have come from overseas. From Chios or Thasos there comes a light Greek wine of comparable quality to the Aminian vintages; the vine has a very tender grape and such small bunches that there is no profit in growing it except in a very rich soil. The Eugenia – its name indicates its high quality – was exported from the hills of Tauromenium to be grown only in the territory of Alba since, if it is transplanted elsewhere, it immediately loses its character. Indeed some vines have so great an attachment to certain localities that their reputation is indissolubly linked to them and they cannot be transplanted anywhere else without inferior results.

Cato on viticulture

44. The elder Cato was exceptionally famous for his triumph[2] and his censorship,[3] but even more for his literary fame and for the instruction he gave the Roman nation on all practical matters – in particular on agriculture. Although he was a man who, by the admission of his peers, was excellent and without rival as a farmer, Cato has only touched upon a few varieties of the vine, and the names even of some of these have passed into oblivion.

45. His views should be stated separately and given comprehensive treatment so that we are aware of the varieties that were the most celebrated in the whole catalogue of vines for the year

1. *apes*.
2. 194 BC.
3. 184 BC.

AUC 608[1] at about the time when Carthage and Corinth were captured and Cato died. Cato gives us an inkling of how far civilization has advanced in the subsequent 230 years and makes the following observations about vines and grapes.

46. 'Where the locality is said to be best for the vines and exposed to the sun, plant the small Aminian variety, the "double Eugenium" and small "Helvium". Where the soil is heavier, or the region is somewhat foggy, plant the larger Aminian, the Murgentine, Apician and Lucanian. All other vines, especially hybrids, are suited to any kind of soil. The small and large Aminian grape and the Apician are stored in jars, pips and all. They can also be preserved in new wine boiled down. The larger Aminian grapes with hard skins, which are hung up to dry to make raisins, should be kept in a blacksmith's forge.'

47. There are no older instructions in the Latin language on this topic – so close are we to its origins. In our own times there have been few examples of outstanding skill in viticulture; for this reason I must include the following details so that the rewards of success may be known, as this provides, in every sphere of activity, the greatest incentive.

Successful vineyards

48. Acilius Sthenelus, a plebeian and freedman, gained very great fame by his careful cultivation of a vineyard not much more than 40 acres in extent, in the region of Nomentanum, and which he sold for 400,000 sesterces. But the highest distinction, thanks to the work of the same Sthenelus, was enjoyed by Remmius Palaemon, famous in another field for his work on grammar. Within the last twenty years he bought a farm for 600,000 sesterces in that part of Nomentanum where the road branches, some 8 miles from Rome.

50. The low price of property in all directions on the outskirts of Rome is well-known, but especially in the area mentioned. Palaemon acquired farms that had been neglected through carelessness, on which the soil was no better than on the poorest estates. He set about cultivating the land not for any virtuous reason but at first out of vanity, which was an extremely well-known trait of his. Under Sthenelus' supervision, Palaemon had the vineyard dug and trenched, though he himself was merely playing at being a farmer. He improved it to a scarcely credible level and within eight years

1. 146 BC.

the vintage was sold to a buyer for 400,000 sesterces while the grapes were still hanging on the vines.

51. In the end Annaeus Seneca, the leading scholar of his time and a very powerful man, whose power finally became excessive and crashed around his ears – a person far from being an admirer of the trivial – became so passionately covetous of this estate that he was not ashamed to yield to someone he otherwise hated, and who was likely to brag about his coup. Accordingly he purchased the vineyards at four times the price Palaemon had paid – and this within less than ten years of its passing under the latter's control.

Famous wines of antiquity

53. The oldest wine to enjoy a reputation was that from Maronea grown in the coastal region of Thrace, as Homer attests.[1] However, I shall not follow up the legendary or variously reported stories about its origin, except the story that Aristaeus was the first from Maronea to mix honey with wine because of the especially pleasant natural flavour of each of these substances. Homer states that Maronean wine was mixed with water in the proportion of 1:20.[2]

54. Maronean wine even today keeps its strength and unconquerable body. Mucianus, who was consul on three occasions, discovered on a recent visit to Thrace that it is the practice to mix this wine with water in the proportion of 1:8, and that it is dark in colour, has a bouquet, and improves with age. Pramnian wine, praised by Homer, is also still famous; it is produced in the region of Smyrna, near the shrine of the mother of the gods.

55. Among the remaining wines no kind has been famous, but the year when Lucius Opimius was consul[3] and the tribune Gaius Gracchus was killed for stirring up the people to revolt, was known for the quality of its wines of all kinds. The weather was hot and sunny: men called it 'good ripening weather'. Wines from that year still survive today – some 200 years later – although they have now evaporated to the consistency of honey and have a rough taste. Such is the nature of wines when old.

56. If we assume that at historic prices these wines originally cost 100 sesterces for about half a gallon and that the interest on this sum has been multiplying at 6 per cent per annum, which is a legal and

1. See *Odyssey*, IX, 197 ff. Cf. Virgil, *Georgics*, II, 88 ff.
2. See *Iliad*, XI, 639 and *Odyssey*, X, 235.
3. 121 BC.

moderate figure, we can demonstrate that when Gaius Caesar, son of Germanicus, was emperor – 160 years after Opimius' consulship – they had appreciated in value by a factor of twelve. This is mentioned in my biography of Pomponius Secundus and the account of the banquet he gave to Gaius Caesar. So much money is tied up in our wine-cellars.

57. There is nothing else that experiences a sharper increase in value up until the twentieth year, or a greater decrease afterwards, if there is no inflation. Rarely to date, and only in the case of spendthrifts, has wine cost 1,000 sesterces a cask.

The physiological effects of wine

58. Drinking wine can heat the inner parts of the body, and when it is poured over the surface, it has a cooling effect. At this point it would not be out of place to recall what Androcydes, famous for his good sense, wrote to Alexander the Great in an effort to check his over-indulgence. 'When you are about to drink wine, O King, remember that you are drinking the earth's blood. Hemlock is poison to men and wine is poisonous to hemlock.' If Alexander had listened to his advice, assuredly he would not have killed his friends Clitus and Callisthenes in drunken stupors. So we may justifiably say that nothing else is more harmful to our pleasures when taken without moderation.

Italian wines

59. No one can doubt that some types of wine are more pleasing than others, or be unaware that of two wines from the same vintage one can be superior to the other. It can surpass its relation because of the jar or as a result of some chance accident.

60. Julia Augusta attributed her eighty-six years of life to the wine of Pucinum, which she drank to the exclusion of any other.

61. The late Emperor Augustus preferred wine from Setia to all others, and so did almost all his successors. Previously Caecuban had the distinction of being rated the best.

64. The third prize is gained, in varying degrees, by the wines of Alba in the neighbourhood of Rome. These are mostly very sweet but occasionally dry. The wines of Surrentum, which grow only in vineyards and are especially recommended for persons recovering from illness because of their thinness and health-giving properties, are also in this category. The Emperor Tiberius used to say that

doctors had conspired to make wine from Surrentum famous, but otherwise it was merely 'vintage vinegar'. Gaius, his successor, called it best-quality flat wine.

66. For public banquets the fourth position in the ratings has been held from the time of Julius Caesar onwards by Mamertine wines grown in the vicinity of Messana; Caesar was the first person to bestow on them a reputation, as witness his letters.

70. Wines from Pompeii are at their best within ten years and gain nothing from greater maturity. They are also observed to be injurious because of the hangover they cause, which persists until noon on the following day.

Some foreign wines

73. I shall now classify wines from overseas. The wines most highly regarded after those of the Homeric age were wines of Thasos and Chios, and of the Chian wines, the Ariusian in particular. In AUC 450[1] Erasistratus, a doctor of eminent standing, recommended in addition Lesbian wine. Currently the most popular of all is wine from Clazomenae, since the locals flavour it sparingly with sea-water; Lesbian wine, however, naturally tastes of the sea. The wine from Mount Tmolus is not popular for drinking undiluted because it is sweet; it is mixed with dry wines to add sweetness. Next in order are wines from Sicyon, Cyprus, Telmessus, Tripolis, Berytus, Tyre and Sebennys.

Early regulations relating to wine

88. Romulus poured libations of milk, not wine; proof of this lies in rites established by him that preserve this custom today. The Postumian law of King Numa says: 'Do not sprinkle wine on a funeral pyre.' Nor can anyone doubt that his reason for sanctioning this law was the scarcity of wine. By the same law he made it illegal to offer to gods libations of wine from an unpruned vine. He devised this scheme to compel people who were otherwise arable farmers, and careless about dangers to trees, not to neglect pruning. Marcus Varro tells us that Mezentius, king of Etruria, gave help to the Rutulians against the Latins, and that his price for this was all the wine then in Latium.

89. At Rome women were not allowed to drink wine. I find

1. 304 BC.

among various examples that the wife of Egnatius Maetennus was clubbed to death by her husband for drinking from a large jar and that he was acquitted of murder by Romulus. Fabius Pictor has written in his *Annals* of a matron who was starved to death by her family for having broken open the box containing the keys to the wine-store. Cato says that male relatives kiss women to find out whether they smell of drink.[1] The judge Gnaeus Domitius once delivered a verdict that a certain woman seemed to have drunk more wine than was needed for her health, without her husband's knowledge; he fined her a sum equivalent to her dowry. For a long time very strict economy was exercised in the use of wine.

91. The commander Lucius Papirius, just before his decisive battle against the Samnites, vowed a small goblet of wine to Jupiter, if he should be victorious.[2] Cato the Censor, on his voyage to Spain, whence he returned to a triumph, drank no wine other than that drunk by the crew – so little did he resemble those people who, even at banquets, serve their guests with different wines from those they themselves are drinking, or, as the meal progresses, surreptitiously substitute inferior wines.

Retsina

124. The method of flavouring wine is to sprinkle it in its raw state with pitch during the first fermentation – which takes nine days at the outside – so that the wine may acquire the smell of pitch and some hint of a sharp flavour. Some authorities think that a more effective way is to use unrefined flower of resin which enlivens the smoothness of the wine.

The storage of wine

127. The most highly thought of pitch for sealing vessels used to store wine comes from Bruttium; this is made from the resin of the pitch-pine.

132. Even when wine is ready for storage, differences in climate play an important part. In the Alpine region wine is put in wooden casks and these are covered with a tiled roof; in a cold winter fires

---

1. Pliny uses the word *temetum*, which denotes any alcoholic drink; the word for drunkenness (*temulentia*) is a derivative.

2. In 320 BC Lucius Papirius Cursor defeated the Samnites, thus reversing the disaster suffered by the Romans at the Caudine Forks (321 BC).

are lit to protect the wine from the effects of the cold. Although it is rarely recorded, the containers have occasionally been seen to burst in frost, leaving the wine standing in frozen blocks. This is little short of a miracle since wine does not normally freeze. Usually it is only chilled by the cold.

133. Districts with a mild climate store their wine in jars and bury them completely or partially in the ground, thus protecting them from the weather. In other regions people do this by roofing them over. They also lay down the following guidelines: one side of a wine-store, or at least one of its windows, ought to face north-east – or, at any rate, east. Manure-heaps and tree-roots should be a good distance away, and all things with a strong smell should be avoided as these easily affect wine. Certainly there should not be any cultivated or wild fig-trees in the vicinity.

134. Spaces must also be left between jars to prevent anything likely to affect the wine from passing from one to another, as the wine very soon becomes tainted. Moreover the shape of the jars is important, large-bellied, wide ones being less suitable. Immediately after the rising of Sirius, the jars should be coated with pitch, washed with sea-water, or water with salt in it, and then sprinkled with ashes or brushwood, or else with potter's clay, dried clean, and finally fumigated with myrrh. Likewise, the rooms in which the wine is stored should be frequently fumigated. Weak wines should be stored in jars sunk in the ground, whereas jars containing strong wine should be exposed to the air.

135. Jars ought never to be filled right up to the top, and the space above the surface of the wine should be covered with raisin-wine, or boiled grape-must mixed with saffron or sword-lily. The stopper of the jars should be sprinkled with the same mixture, with mastic or Bruttian pitch added. Jars must not be opened in mid-winter, except on a fine day when a south wind is blowing or when there is a full moon.

The effects of over-indulgence

137. Careful investigation reveals that no activity takes up more of a man's life than wine-making, as if Nature had not given us a perfectly healthy liquid to drink – namely, water – of which all other animals avail themselves! Yet we compel even our beasts of burden to drink wine! All this toil and effort is the price paid for something that alters men's minds and produces madness. Because of wine thousands of crimes have been committed, and drinking occasions so much pleasure that a huge section of mankind knows no other reward in life.

138. To allow us to drink more we reduce the strength of wine by straining it through a cloth bag, and other inducements for drinking are invented – even poisons: some men take hemlock beforehand so that fear of death may compel them to drink, while others take powdered pumice and substances to which I am ashamed to allude.

139. We see the most careful of drinkers cooking themselves in hot baths and being carried out lifeless, and others who cannot wait for dinner; some – without putting on a stitch of clothing, still naked and gasping – seize hold of a huge jar on the spot and, as if to demonstrate their strength, pour down the entire contents so as to sick it up again immediately and then drink another jar. This they repeat two or three times over, as if they were born to waste wine and as if the wine could be disposed only through the agency of the human body.

140. This is the *raison d'être* for the exercises that have been introduced from abroad – for rolling in mud, and throwing the neck back to show off the pectoral muscles. It is said that all these exercises are intended to raise a thirst. And what about all the drinking contests and vessels engraved with scenes of adultery, as though drinking on its own were not enough to teach debauchery? So wines are drunk as a result of debauchery, and drunkenness is even promoted by the offer of a prize and – heaven help us! – is actually purchased. One man get a prize if he eats as much as he drinks. Another drinks as many cups as a throw of the dice demands.

141. Then lecherous eyes calculate their chances of seducing a married woman, while intense looks betray this to the husband. Then the secrets of the heart are revealed. Some men proclaim the content of their wills; others openly reveal facts of fatal consequence and do not keep to themselves words that will return to them through a cut throat. How many men have met their end in this manner! As the proverb says, '*In vino veritas*'.

'*The morning after*'

142. Meanwhile, even in the most favourable circumstances, the intoxicated never see the sunrise and so shorten their lives. This is the reason for pale faces, hanging jowls, sore eyes and trembling hands that spill the contents of full vessels; this the reason for swift retribution consisting of horrendous nightmares and for restless lust and pleasure in excess. The morning after, the breath reeks of the wine-jar and everything is forgotten – the memory is dead. This is what people call 'enjoying life'; but while other men daily lose their yesterdays, these people also lose their tomorrows.

143. Some forty years ago when Tiberius was emperor, it was the fashion to drink on an empty stomach and to have wine as an aperitif – again a result of foreign habits and the determination of doctors to advertise themselves by some gimmick.

Famous drinkers

146. Torquatus had the rare distinction – for even this 'science' has its own rules – of never having slurred his speech or relieved himself by being sick or by any other bodily function while drinking. Indeed he was always present for duty with the morning guard without any mishap, even though he had consumed a larger amount at one draught than anyone else. He also added to his record with smaller draughts, without taking breath or spitting while drinking.

147. Tergilla reproaches Marcus Cicero with the charge that his son used to down 9 or 10 pints at a time and that when drunk he threw a goblet at Marcus Agrippa. These are indeed acts associated with intoxication. But no doubt young Cicero wished to deprive Mark Antony, his father's murderer, of fame in this sphere.

148. For Antony had very early won the prize for drinking by bringing out a book about his own habits. Indeed by daring to champion his claim in this work, he clearly demonstrates, in my view, the extent of the evils he had inflicted on the world through his drink problem.

Beer

149. Western people also have their own liquor made from grain soaked in water. Alas, what wonderful ingenuity vice possesses! We have even discovered how to make water intoxicating!

BOOK XV

THE OLIVE AND OTHER FRUIT-TREES

The history of the olive-tree and the production of olive-oil

1. Theophrastus, one of the most famous of Greek authors, whose literary career was at its height in AUC 440,[1] says that the olive grows only within 40 miles of the sea. Fenestella states that in AUC 173[2] in the reign of Tarquinius Priscus there were no olive-trees at all in Italy, Spain or Africa. Now the olive has spread even across the Alps and into the heart of the provinces of Gaul and Spain.

2. In AUC 505[3] during the consulship of Appius Claudius, the grandson of Appius Claudius Caecus and Lucius Junius, 12 pounds of olive-oil cost 10 *asses*, and later, in AUC 680,[4] Marcus Seius, son of Lucius, throughout his year as curule aedile, provided the Roman people with oil at 1 *as* for slightly less than 10 pounds.

3. Anyone who knows that twenty-two years later, when Gnaeus Pompey was consul for the third time, Italy exported olive-oil to the provinces, will find these facts less surprising. Hesiod, also, who considered that instruction in agriculture was one of the most important things in life, says that no one who plants an olive-tree ever lives long enough to enjoy its fruit – so slow was cultivation in his day. Now, however, olive-trees bear fruit even in nurseries, and when they are transplanted, they provide a harvest the year after.

4. Fabianus says that the olive will not grow in regions that experience extremes of temperature. Virgil identifies three varieties of olive: the *orchites*, *radius* and *posia*;[5] he adds that olive-trees do not need hoeing, pruning or any special attention. Undoubtedly, however, soil and climate are singularly important. The trees are pruned at the same time as vines, and like the ground between them to be hoed.

5. Olive-gathering follows the grape harvest, and even more

---

1. 314 BC.
2. 581 BC.
3. 249 BC.
4. 74 BC.
5. *Georgics*, II, 85 ff.

expertise is needed to produce olive-oil than wine, because the same tree produces different kinds of oil. The green olive, which has not yet begun to ripen, gives the first oil and this has an outstanding taste. Moreover, the first oil from the press is the richest, and the quality diminishes with each successive pressing, whether the olives are crushed in wicker sieves or by enclosing the branch in narrow-meshed strainers – a recently invented technique.

6. The riper the berry the more greasy and less pleasant is the flavour of the oil. The best time for gathering olives, striking a balance between quality and quantity, is when the berries begin to turn dark – locals call these *druppae*, the Greeks, *drypetides*. Apart from this, it makes a difference whether the berries mature in the presses or on the branches, and whether the tree has been watered or the berry has only been moistened by its own juice, or has taken in nothing but dew from the sky.

7. Age imparts an unpleasant taste to oil, which is not the case with wine, and after a year it is old. Nature shows forethought in this, if one chooses to interpret in this way, since it is not necessary to use up wine which is produced for getting merry; indeed the pleasant overripeness that comes with maturity encourages us to keep it. She did not, however, wish to be niggardly with oil, and has made its use widespread, even among the masses, because of the need to use it up quickly.

8. Italy has won first place throughout the world, especially in the region of Venafrum and the area that produces Licinian oil, thanks to which the Licinian olive-tree has acquired particular fame. Africa does not produce olive-oil, as its soil is more suited to grain.

Mistaken ideas about olive-trees

9. Olives consist of a stone, oil, flesh and lees, which is a bitter liquid formed from water, so that there is very little of this in dry areas and a great deal in wet places. The oil is a juice peculiar to the olive.

10. Those who think that the onset of decay is the beginning of the ripening process are wide of the mark. It is also a mistake to believe that the amount of oil increases with the growth of the flesh, for the juice becomes solid material and the stone of the olive gets larger. Watering, whether artificial or by means of repeated rainfall, uses up the oil unless good weather follows to impede the growth of the solid part of the berry. For the cause of oil, as of other things, is warmth, as Theophrastus concludes, and this is the

reason why our presses and store-rooms are warmed by large fires.

11. A third mistake results from false economy: because of the cost of picking, some people wait for the olives to fall. Those who adopt a middle course shake down the fruit with long poles, but even this harms the trees and causes a loss of fruit in the following year. In fact there was a very old rule for olive-gatherers: 'Don't strip the trees, or beat them.' Those who are careful use a reed and strike the branches lightly and from the side. But even this makes the tree bear fruit in alternate years since the buds get knocked off – and this is no less the case if people wait for the olives to fall: by remaining attached to the branches beyond their proper time, they use up the food for the coming crop and usurp its place. This is proved by the fact that if they are not gathered before the west wind blows, they obtain fresh strength and fall off less easily.

Olive-oil and its uses

19. The natural properties of olive-oil provide the body with warmth and protect it against the cold; it cools the head when hot. The Greeks, progenitors of all vices, have diverted the use of olive-oil to serve the ends of luxury by making it available in gymnasia. (It is known that those in charge of gymnasia have sold oil that has been scraped off bodies for 80,000 sesterces.) The majesty of Rome has accorded the olive-tree great honour by crowning our cavalry squadrons with wreaths of olive on the Ides of July[1] and also when celebrating minor triumphs. Athens also crowns its victors at Olympia with wreaths of wild olive.

Cato's instructions for olive-growing and the production of olive-oil

20. I shall now repeat Cato's instructions about olives.[2] He recommends that the larger *radius*, the Sallentine, the *orchites*, the *posia*, the Sergian, the Caminian and the wax-white olive should be planted in a warm rich soil, and adds, with remarkably good sense, that whichever are thought to be the best species for particular localities should be planted in those localities. He recommends planting the Licinian olive in a cold, poorish soil, since a rich or warm soil ruins

---

1. The fifteenth. On this day the knights were reviewed by the censors. Under the principate this function was taken over by the emperor.
2. *De Re Rustica*, VI, 1 f.

its oil and the tree becomes exhausted by its very productiveness; moreover, it is attacked by moss and red rust.

21. Cato expresses the view that olive-groves should be in a place exposed to the sun and facing west; he does not advise any other location. The *orchites* and *posia* are best kept in brine when they are green, or crushed and stored in mastic-oil. The best olive-oil is made from the most bitter olive one can obtain. As for the rest, the olives should be collected from the ground as soon as possible, washed off if they are dirty, and dried for three days, which is sufficient. If the weather is cold and frosty, the olives must be pressed on the fourth day and then sprinkled with salt. If the olives are kept on a wooden floor their oil diminishes and deteriorates in quality, and the same occurs if the oil is left with the lees and residue.

22. The oil should therefore be ladled several times a day with a shell and put into cauldrons made of lead, since it is spoilt by copper. All these processes must be carried out with presses that have been heated and tightly closed, in the presence of as little air as possible. No wood ought to be cut in the area where the oil is being produced and so the most suitable fire is fuelled by olive-stones themselves. The oil must be decanted from the cauldrons into vats so as to leave the residue and the lees. For this reason the vessels must be changed often and the osier filters wiped with a sponge to ensure the greatest degree of cleanliness.

23. A later discovery, writes Cato, was to wash the olives in boiling water and immediately put them whole into the press – for in this way the lees are squeezed out – and then to crush them in oil-mills and press them a second time. This is known as 'a pressing', and the first liquid to be squeezed out after the initial crushing is called the 'flower'. A reasonable output for a team of four men in a twenty-four-hour period, using a double container, is three pressings.

Artificial oil

24. Originally there was no artificial oil, and I assume that this is the reason why there is no mention of it in Cato's account. Nowadays there are several kinds and first I shall discuss those that are produced from trees, and chief among these is oil derived from the wild olive. This is a thin oil which has a more bitter taste than that obtained from the cultivated olive; it is only useful for medicines. Very similar is the oil from the ground-olive, a rock-shrub not

more than 3 inches high, with leaves and fruit like those of the wild olive.

25. Then there is the kind of oil obtained from the castor-oil plant, which grows abundantly in Egypt; some authorities call it *croton*, others, *sibi*, yet others, wild *sesamon*. It is boiled in water, and the oil, which floats on the surface, is skimmed off. In Egypt, however, where it is prolific, salt is sprinkled on the pod and the oil is squeezed out, without the use of fire or water. This oil is disgusting as food and thin for lamps.

Cato on the uses of the lees of olive-oil

33. The lees have attracted particular praise from Cato. He explains how vats and casks for holding oil are impregnated with lees to prevent them from soaking up the pure olive-oil; how threshing-floors are smeared with lees to keep off ants and to prevent cracks from forming. Moreover, the clay of walls and the plaster and flooring of granaries – even clothes cupboards – are sprinkled with lees, and seed-corn is dipped in them as a protection against wood-borers and harmful insects. The lees are a remedy for diseases of animals and trees, and, in human beings, for mouth ulcers.

34. Cato also states that reins, all leather articles, shoes and the axles of wheels are greased with boiled lees – as are copper vessels, to prevent verdigris and to give them a more attractive colour. All wooden utensils are similarly smeared, and so are attractive jars for storing dried figs, or sprays of myrtle with leaves and berries attached, or indeed anything else of a similar kind. Finally, logs of wood steeped in olive-lees produce smokeless fuel.

Apples and pears

47. There are many varieties of apple. I have spoken about citrons when describing the tree of that name.[1] The Greeks, however, call these fruits 'Medic' apples after their place of origin.[2] Equally foreign are the jujube-tree and tuber apple; these have only recently come into Italy, the former from Africa, the latter from Syria. Sextus Papinius, who was consul in our time,[3] was the first person to bring them to Italy, in the last years of the principate of the Emperor

1. XIII, 91 ff.
2. Media.
3. AD 23.

Augustus, having grown them in his camp. The fruit is more like berries than apples, but is especially decorative for terraces since nowadays we have whole forests climbing even over the roofs of our houses.

53. An equally early kind of pear is called the 'proud' pear. This is small but ripens very quickly. The best of all varieties is the Crustumnian. Then there are the Falernian pears used to make drinks since they contain such a large volume of juice.

54. But pears that have drawn attention to their sponsors, by names accepted at Rome, include the Decimian and its offshoot called the pseudo-Decimian, the one known as the Dolabellian that has the longest stalk, the variety of Pomponian called 'breast-shaped', the Lateran, and the Anician, which is ready when autumn has passed; the last named has a pleasantly tart flavour.

Grafts

57. Grafting has long since been perfected, since people have tried every possibility – Virgil mentions the grafting of nuts on to an arbutus, apples on to a plane-tree, and cherries on to an elm. Nothing more can be devised – at any rate, it is now a long time since any new variety of fruit has been discovered.

Picking and storage of fruit

59. The general rule for keeping fruit is that storage-rooms should be constructed in a cool, dry place with boarded floors, north-facing windows which are opened on a fine day, and glass windows to keep out the south winds. The draught from a north-east wind also spoils the appearance of pears by shrivelling them. Windfalls should be kept widely spaced. Fruit should have a bed of closely packed straw or chaff and be placed far apart so that the gaps between the rows admit a uniform draught. It is said that the Amerian apple is the best and the honey-apple the worst for keeping.

60. Quinces should be stored in a secure place from which all draughts are excluded; otherwise, they should be boiled or soaked in honey. Pomegranates should be hardened by hot sea-water, then dried in the sun for three days and hung up in such a way as to be protected from the dew at night. When wanted for use they should be thoroughly washed in fresh water. Marcus Varro recommends keeping pomegranates in large jars of sand, and also, while they are unripe, covering them with earth in pots with the bottom broken

out, but with all air excluded, and with their stalks smeared with pitch, since, kept in this way, they grow to an even larger size than they do on the tree. Varro also says that all other fruit of the apple variety should be wrapped up separately in fig-leaves (but not fig-leaves that have fallen) and stored in wicker baskets or else smeared with potter's clay.

61. He adds that pears should be stored in earthenware jars covered with pitch and placed upside-down in a hole in the ground, with earth heaped on top.

62. Some of the most recent authorities, who have researched more deeply into the problem, recommend that fruit and grapes should be picked early for storing, when the moon is on the wane and after about 9 p.m., in fine weather with a dry wind blowing. They also add that the fruit ought to be chosen from dry places and before it is completely ripe.

66. Columella writes that grapes should be stored in earthenware containers carefully coated with pitch, and that these should be kept under water in wells or in cisterns. In the coastal region of Liguria, close up against the Alps, local people dry their grapes in the sun and wrap the raisins in bundles of rushes; they store the raisins in casks sealed with plaster. The Greeks employ the same method but use leaves from plane-trees or from the vine itself, or fig-leaves that have been dried for a day in a shady place; they put layers of grape-skins in the cask between the grapes. This method is used on Cos and in Berytus, where the grapes are second to none in sweetness.

67. Some people prefer to keep the grapes in sawdust, or in shavings of pine, poplar, or ash. There are also people who recommend that grapes should be hung in a granary as soon as they are picked – but not near any apples – because they claim that the dust from the corn dries them best. The hanging bunches can be protected from wasps by sprinkling them with oil sprayed from the mouth.

Figs and well-known stories about figs

68. Of the remaining varieties of fruit the fig is the largest, and some figs rival even pears in size.

71. The earliest variety is the purple fig, which has a very long stalk, while the latest is the swallow fig.

74. The variety called 'African' by Cato reminds me that he used this fig for an important demonstration. For, burning with a deadly

hatred of Carthage and troubled with anxiety about the safety of his descendants, Cato used to shout at every meeting of the Senate: 'Carthage must be destroyed!' Now one day he brought into the Senate House an early ripe fig from Africa, showed it to his fellow senators and said: 'I ask you, when do you think this fig was plucked from the tree?'

75. All agreed that it was fresh, so he said: 'Know this, it was picked two days ago in Carthage; that's how near the enemy are to our walls!' Immediately they began the Third Punic War, in which Carthage was destroyed.[1] Cato, however, had died the year after this story was told. What should we wonder at most in this episode? Ingenuity or coincidence? The speed of the journey or the fervour of the man?

76. The outstanding feature – which I consider more remarkable than anything else – is that so great a city as Carthage, which for 120 years had competed to control the world, was overthrown by the evidence of a single fruit – a result that neither Trebia, nor Trasumenus, nor Cannae, which saw Rome's reputation buried, nor the Carthaginian camp pitched at the third milestone from Rome and Hannibal riding up to the Colline Gate in person, were able to achieve! So much nearer did Cato bring Carthage by means of a single fig!

77. A fig-tree that grows in the Forum itself, the meeting-place of Rome, is sacred because things struck by lightning are buried there, and all the more so for being a reminder of the fig-tree under which the nurse of Romulus and Remus first tended the founders of our empire, on the Lupercal. This tree is known as Ruminalis because the she-wolf was discovered beneath it giving her teats[2] to the infant boys. A bronze statue was dedicated near by to celebrate this marvel. And the present fig-tree prophesies some future happening whenever it shrivels up; then it is replanted thanks to the priests. There was also a fig-tree in front of the Temple of Saturn which was removed in AUC 260[3] after a sacrifice had been made by the Vestal Virgins, because its roots were threatening to overturn a statue of Silvanus.

78. Another fig-tree, self-seeded, grows in the middle of the Forum at the spot where, when the foundations of the Empire were crumbling – a portent of disaster – Curtius had filled up the hole

1. 146 BC.
2. *rumis.*
3. 494 BC.

with the greatest treasures – namely, virtue, a sense of duty, and his own glorious death.[1]

Cherry-trees

102. Prior to Lucius Lucullus' victory against Mithridates, that is, until AUC 680,[2] there were no cherry-trees in Italy. Lucullus was the first to bring them back from Pontus, and in the span of 120 years they have crossed the ocean and have spread as far as Britain. But cherries fail to grow in Egypt, whatever the amount of attention. Apronian cherries are the reddest and Lutatian the blackest, while Caecilian are completely round.

103. The Junian cherry has a pleasant taste, but only if eaten under its tree, since it is so tender that it cannot stand being transported. Pride of place belongs to the hard cherry, which the people of Campania call the Plinian cherry; in Belgic Gaul and likewise on the banks of the Rhine the best variety is the Lusitanian.

104. There are also Macedonian cherries which grow on a small tree, rarely exceeding 4½ feet in height, and ground-cherries which grow on a still smaller bush. The cherry is among the first fruit to pay its yearly thanks to the farmer. It likes a north-facing position and cold conditions. Cherries can also be dried in the sun and stored in casks just like olives.

The myrtle and its uses

119. The myrtle retains its Greek name, which shows it to be of foreign origin. When Rome was founded myrtles grew on the present site of the city, since the story goes that the Romans and Sabines were intent on battle because of the carrying-off of the young women, but laid down their arms and purified themselves with sprigs of myrtle at the place that now houses the statues of Venus Cluacina (*cluere* was an archaic word meaning 'to cleanse').

120. This tree contains a kind of incense for fumigation, which was chosen for this purpose on the occasion mentioned because Venus, the guardian spirit of the tree, also presides over marriages. I

1. In 362 BC a chasm opened in the Forum and soothsayers said that this could only be filled by throwing into it Rome's greatest treasure. So Marcus Curtius mounted his horse and leapt into the hole; the earth closed over him. The spot was marked by a circular pavement and known as the Lacus Curtius. See further Livy, I, 19 and VII, 6.

2. 74 BC.

am inclined to think that the myrtle was the first of all trees to be planted in public places in Rome – a tree with a remarkable power of prophecy and augury. The shrine of Quirinus – that is, Romulus himself – is considered to be one of the most ancient temples. In this there were two sacred myrtles, which grew in front of the building for a long time: one was called the patricians' myrtle, the other the plebeians' myrtle.

121. For many years the former was the more flourishing of the two and was vigorous and happy. As long as the Senate flourished, this was a huge tree, while the plebeians' tree was shrivelled and in a poor condition. But when the latter grew strong, as the patricians' myrtle turned yellow – namely, from the time of the Marsian War – the influence of the Senate grew weak and gradually its authority withered away to nothing.

125. The myrtle has also become involved in warfare. Publius Postumius Tubertus, during his consulship, celebrated a triumph over the Sabines and was the first of all men to enter the city with an ovation;[1] because he had won easily, without bloodshed, he entered crowned with a wreath of the myrtle of Venus Victrix. As a result, that tree was sought after even by our enemies. Subsequently, those celebrating a triumph wore a myrtle wreath, except Marcus Crassus, who wore a laurel wreath on entering Rome after his victory over Spartacus and the slaves.[2]

127. The laurel is especially reserved for triumphs and is certainly very much favoured in houses; it is the guardian of the doorways of emperors and high priests, where it hangs alone adorning their homes and keeping a vigil before the threshold. Visitors to Delphi are crowned with laurel, as are generals celebrating a triumph at Rome.

133. The laurel itself is a messenger of peace, inasmuch as holding out a branch of laurel, even between enemy armies, is a sign of a truce. For the Romans in particular the laurel is a messenger of

---

1. An ovation was a minor form of triumph: the general entered Rome on foot or horseback instead of in a chariot, dressed in a *toga praetexta* without a sceptre, and wearing a wreath of myrtle instead of laurel.

2. A band of gladiators, led by a Thracian named Spartacus, who had gained military experience in the auxiliary forces, broke out of their barracks in Capua and called the rural slaves to liberty. They won initial victories (73–72 BC) but, after a campaign in south Italy, Marcus Crassus prevented the slaves from escaping by ship at the Straits of Messana. Spartacus and most of his followers died in 71 BC. Six thousand of the surviving slaves, whose masters could not be found, were displayed on crosses set up along the whole length of the Appian Way.

rejoicing and victory; it accompanies dispatches and decorates soldiers' spears and javelins and generals' *fasces*.

134. A branch of laurel is placed in the lap of Jupiter Optimus Maximus whenever a new victory brings rejoicing, and this is not because the laurel is evergreen, nor because it brings peace – since the olive is preferable for both these things – but because it flourished on picturesque Mount Parnassus and is thus thought to be pleasing to Apollo. For even the Kings of Rome in their time used to send gifts to Apollo's shrine and ask him for oracles in return; the case of Brutus bears witness to this. My point here is that Brutus, guided by the oracle's response, won freedom for the people by kissing the famous patch of earth that bore the laurel.[1] Another possible reason is that the laurel, alone of the shrubs used as an indoor plant by man, is never struck by lightning.

135. Personally, I am inclined to believe that it is for these reasons that the laurel has the place of honour in triumphs rather than because it was used, as Masurius states, for the purpose of fumigation and purification from the blood of the enemy. It is forbidden to pollute the laurel and the olive in unholy uses, so that they must not be employed even for making a fire at altars and shrines when divinities are to be propitiated. The laurel clearly shows its objection to fire by crackling and making a sort of solemn protest. It is said that the Emperor Tiberius used to put a laurel wreath on his head when there was thunder, to protect himself from the danger of lightning.

136. There are also noteworthy occurrences involving laurel connected with the late Emperor Augustus. Livia Drusilla, who on marriage received the name of Augusta, had just been betrothed to Caesar and had taken her seat when an eagle dropped into her lap from the sky a hen of unusual whiteness, without doing it any harm. As she looked at it in amazement, but with a certain sangfroid, she experienced a further cause for wonder in that the eagle was holding a laurel branch with berries in its beak. The augurs gave instructions that the eagle and any young it produced should be protected and the branch planted in the ground and guarded with due religious observance.

---

1. Lucius Junius Brutus, a nephew of Tarquin the Proud, accompanied Tarquin's sons to Delphi to consult the oracle about the portent of a snake that had appeared in the king's palace. The two princes took the opportunity of asking the oracle who would succeed to the throne. The answer was: 'The one who first kisses his mother.' Brutus pretended to stumble and kissed the earth – the Mother of all men.

137. This was carried out at the country house of the Caesars on the banks of the Tiber, about 9 miles from Rome along the Via Flaminia. The house is called 'The Roost', and the laurel-grove that sprang from the original branch has flourished marvellously. Subsequently the emperor, when celebrating a triumph, held a laurel branch from the original tree and wore a wreath on his head; all emperors adopted this practice. Furthermore, the custom has grown up of planting branches that they have held; and laurel-groves, distinguished by the emperors' individual names, still survive.

138. The laurel is the only tree whose leaves have a special name – that is, bay-leaves. Incidentally, it should be stressed that the laurel can be grown from a cutting, since Democritus and Theophrastus have expressed doubts about this.

BOOK XVI
FOREST TREES

1. Next I would have given an account of acorn-bearing trees, which first provided men with food and were foster-mothers to them in their helpless and savage state, but I was compelled by wonder, arising from personal experience, to consider first the kind of life enjoyed by people who exist without any trees or shrubs.

2. I have pointed out that in the east, on the shores of the ocean, many peoples suffer this deprivation. Indeed, the peoples known as the Greater and Lesser Chauci, whom I have visited in the north, are in this situation. There, twice in every twenty-four hours, the ocean's vast tide sweeps in a flood over a large stretch of land and hides Nature's everlasting controversy about whether this region belongs to the land or to the sea.

3. There these wretched peoples occupy high ground, or man-made platforms constructed above the level of the highest tide they experience; the Chauci live in huts built on the site so chosen and are like sailors in ships when the waters cover the surrounding land, but when the tide has receded they are like shipwrecked victims. Around their huts they catch fish as they try to escape with the ebbing tide. It does not fall to their lot to keep herds and live on milk, like neighbouring tribes, nor even to fight with wild animals, since all undergrowth has been pushed far back.

4. The Chauci make ropes of sedge and rushes from the marshes to weave into nets for catching fish. They scoop up mud in their hands and dry it in the wind rather than the sun. Using peat turves as fuel, the Chauci cook their food and warm their bodies that are frozen by the north wind; their only drink comes from rain-water stored in tanks in the forecourts of their houses. These are the peoples who, now they have been conquered by the Roman nation, claim that they are reduced to slavery![1] This is indeed the situation. Fortune often spares men to punish them.

5. Another wonder results from the existence of forests. These fill the whole of the rest of Germany and accentuate the cold climate

---

1. This was certainly the attitude of the Britons to Roman conquest and rule. See Tacitus, *Agricola*, 15.

with their shadows; the tallest grow not far from the Chauci, especially round two lakes.[1] The shores themselves are occupied by oaks which have a vigorous growth rate, and these trees, when undermined by the waves or driven by blasts of wind, carry away vast islands of soil trapped in their roots. Thus balanced, the oak-trees float along in an upright position, with the result that our fleets have often been terrified by the 'wide rigging' of their huge branches when they have been driven by the waves – almost deliberately it would seem – against the bows of ships riding at anchor for the night; consequently our ships have had no option but to fight a naval battle against trees!

6. In the same northern region is the huge expanse of the Hercynian oak forest,[2] which, impervious to the passage of time, is coeval with the world and exceeds all marvels with its almost limitless age. Leaving aside reports that would strain credibility, it is accepted that hills are raised up as roots collide, or where the ground has fallen away from them. In their struggle with one another their arches rise as high as branches and curve in the manner of open gateways so that squadrons of cavalry can pass through. These oak-trees are largely of the acorn-bearing variety, always honoured by the Romans.

7. From these trees comes the Civic Wreath,[3] most famous token of military prowess, and likewise long regarded as an emblem of the emperor's clemency: because of the unholy nature of the Civil Wars, to refrain from killing a citizen began to be considered worthy of reward. Next come mural, rampart and gold wreaths[4] and below them are beaked wreaths[5] especially famous in two examples; these are the cases of Marcus Varro and Marcus Agrippa, awarded this honour for campaigns against pirates under Pompey the Great and the Emperor Augustus respectively.[6]

8. Before this, the 'beaks' or rams of ships placed in front of the speakers' platform were the pride of the Forum, like a wreath for

1. The IJsselmeer, formerly Zuyder Zee.
2. The Black Forest and beyond.
3. A Civic Wreath was voted by the Senate to Julius Caesar. Thereafter such a wreath hung on the door of the palace of Augustus and all his successors.
4. A gold wreath, decorated with turrets, was given to the first man to scale the walls of a besieged city; similarly, one with a palisade to the first man to cross a palisaded trench. A further variety of triumphal wreath was called *corona aurea*.
5. A beaked wreath (*corona navalis*, or *rostrata*) was given to the first sailor to board an enemy ship, and, later, to a commander who won an important naval victory.
6. Varro in 67 BC, and Agrippa in 36 BC.

the Roman nation.[1] But later they began to be trampled on and defiled by seditious tribunes, and power began to pass from the state to private individuals. After this the beaks passed from beneath the feet of the speakers to the heads of citizens. Augustus gave Agrippa a Beaked Wreath, but from all mankind he received the Civic Wreath.

The history of the award of wreaths as a mark of honour

9. In far-off times Civic Wreaths were given only to gods, and it is for this reason that Homer bestowed a wreath only on heaven[2] and on a whole battle,[3] not on any individual – not even in the context of single combat. It is said that Bacchus was the first to put a wreath of ivy on his head. Then people wore wreaths when making sacrifices in honour of the gods, and at the same time put wreaths on the sacrificial victims.

10. Wreaths were also used until very recently in sacred athletic contests; nowadays, however, they are not awarded to the victor but bestowed on his native city in virtue of his achievements. From this has arisen the practice of giving wreaths to generals who are about to celebrate a triumph, so that they can be dedicated in temples, and also of giving wreaths at games. However, it would be a lengthy task to discuss who was the first Roman to receive each kind of wreath, and not relevant to the purpose of this work.

11. Romulus crowned Hostus Hostilius, the grandfather of King Tullus Hostilius, with a wreath of leaves because he was the first to enter Fidenae. The army saved by Publius Decius, their military tribune in the First Samnite War,[4] when Cornelius Cossus was in command, bestowed on him a wreath of leaves. The Civic Wreath was originally made from the leaves of the holm-oak, but subsequently the leaves of the winter oak were preferred; the latter tree is sacred to Jupiter.

12. Rigorous conditions were imposed, comparable to those attached to the foremost wreath of the Greeks, given under the sponsorship of Zeus himself. These were as follows: the recipient had to have saved the life of a Roman citizen or to have killed an

---

1. These 'beaks' (*rostra*) were taken from ships belonging to Antium after its unsuccessful revolt (338 BC).
2. *Iliad*, XVIII, 485.
3. *Iliad*, XIII, 736.
4. 343–341 BC.

enemy; the place in which the deed was done should not have been left in enemy hands at the end of the day; the person saved had to admit this – otherwise witnesses were of no avail; finally the recipient himself had to be a Roman citizen.

13. Auxiliary forces do not award this honour even if it is a king who is saved. Nor does the honour admit of degrees, even if the person saved is a general, because the founders wished the honour to be the highest for any and every citizen. The recipient is allowed to wear the wreath for ever more. As he makes his entrance at the games, it is customary for even the Senate to stand up on every occasion, and he has the privilege of sitting next to the senators. He, his father and his father's father are exempt from all public duties.

14. Siccius Dentatus won fourteen Civic Wreaths, and Capitolinus, six. How worthy of eternal fame is the Roman character which rewards such noble deeds with honour alone! Although it enhanced the value of other wreaths by means of gold, yet it refused to allow a price to be set on the safety of a citizen, so proclaiming that it is wrong to save the life of a fellow citizen for the sake of mere gain.

Acorns are ground to make substitute flour

15. Nowadays for many peoples acorns are their wealth, even in peacetime. Furthermore, when corn is scarce, acorns are dried and ground into flour; the flour is then kneaded to make bread.

The cork-tree

34. The cork-tree is very small and its acorn-shaped fruit rare and of poor quality. Its only useful product is the bark, which is very thick and when cut grows again; when flattened it can form a sheet 10 feet square. This bark is mostly used for drag-ropes for ships' anchors, floats for fishermen's nets, and stoppers for wine jars; it is also employed in making women's winter shoes. For this reason the Greeks call this, not inappropriately, the 'bark' tree. Some call it the female holm-oak. Where the holm-oak does not grow, as around Elis and Sparta, they use wood from the cork-tree as a substitute, especially in chariot- or carriage-makers' workshops. The cork-tree is not found in Italy or Gaul.

35. The bark of the beech, lime, fir and pitch-pine is also much utilized by country people. From it they make containers – including baskets, and large, flat trugs for carrying corn and grapes at harvest-time; they also use it for the eaves of cottage roofs. A scout writes dispatches for his superiors by cutting letters on fresh bark.

Pitch-pine and resin-producing trees

38. In Europe six related kinds of tree produce pitch. Of these the pine and its wild variety have a long and narrow leaf like hair, ending in a needle-point; this yields the smallest amount of resin, scarcely enough to warrant inclusion here.

40. The pitch-pine likes mountains and cold locations; it is a funeral tree placed at doors as a sign of bereavement and grown on graves. Now, however, it has been taken into our homes because it can easily be clipped. The pitch-pine yields a very large volume of resin in the form of white drops, and this is so like incense that when the two substances are mixed, the eye cannot distinguish them.

41. Similarly the fir, greatly sought after for ship-building is found high on mountains as if it had fled from the sea; its shape is the same as that of the pitch-pine.

42. Wood from the fir-tree is excellent for beams and several everyday items.

43. The larch – the fifth kind – grows in the same region and has the same appearance; its wood is superior, not affected by age, resistant to damp, and red in colour, with a penetrating smell.

44. The 'torch' pine – the sixth kind – so named with good reason, exudes more resin than the rest, and is good for fires and torchlight at religious ceremonies. These trees – certainly the male ones – also produce a liquid with a very strong smell which the Greeks call *syce*.

48. All these varieties of resinous tree produce a great quantity of soot when alight and spit out charcoal, shooting it a long way with a crackling sound – all trees, that is, except the larch.

49. The biggest of the whole group is the fir.

50. The yew, with its forbidding appearance, has no sap, and of all the trees of its type is the only one that has berries. The male berries are poisonous. Sextius states that the Greeks call the yew *milax*, and that the poison of the Arcadian variety has such an instantaneous effect that it is fatal to sleep under it or eat one's food beneath it. Some authorities claim that this is the reason why poisons were called 'taxic' – 'toxic'[1] meaning 'that in which arrows are

---

1. *toxon* (Greek) = bow; *taxus* (Latin) = yew.

dipped'. I find it recorded that a yew-tree is rendered harmless if a copper nail is hammered into the tree.

The production of pitch from the pitch-pine

52–3. In Europe liquid pitch is extracted from the pitch-pine by heating and is used for protecting ships' gear and for many other purposes. The wood is chopped up and put into furnaces heated from the outside by a surrounding fire. The first liquid produced flows like water through a pipe. This is known as 'cedar-juice' in Syria and is so strong that the Egyptians pour the liquid over dead bodies to embalm them. The liquid that follows is thicker and now produces pitch; this in turn is collected in copper vats and thickened by vinegar which makes it coagulate.

The tapping of pitch

57. An incision is made in a pitch-pine on the side facing the sun; this is not a narrow cut, but a wound produced by stripping the bark for a length of at most 24 inches and at least 20 inches from the ground. The distillation from the whole tree flows into the wound. A similar procedure is used for the torch-pine. When the flow stops, a second opening is made elsewhere in a similar way, and then in a third place. Finally the whole tree is felled and burned for pitch.

Tree-management, including propagation and transplantation

131. It is not unusual for toppled trees to be set upright again and to revive when the earth has made a sort of scab over the wound. This is most commonly experienced in the case of plane-trees, which trap a large volume of wind because of the density of their branches; when these are cut off and the weight of the tree is lessened, it is replanted in its own hole. This has been achieved with walnut-trees, olives and others.

132. There are examples of trees that toppled over without being laid low by a storm or any cause other than some supernatural one, and that rose up again of their own accord. This omen was witnessed by Roman citizens during the wars against the Cimbrians, when an elm in the grove of Juno at Nuceria, after its top had been cut off because it was leaning right on to the altar, recovered spontaneously to such an extent that it immediately flowered. From that time

forward the power of Rome recovered after being ravaged by disasters.

134. Trees, which are a gift of Nature, have three modes of growth: spontaneously, from seed, or from roots. As shown, not all trees will grow in all locations or survive if transplanted. In some cases this is because of a dislike of the region, in others because of contrariness but more often it is because of the weakness of the trees transplanted or the fact that the climate is unfavourable or the soil incompatible.

Ivy

144. Ivy is said to grow in Asia Minor at the present time. Theophrastus had said[1] that it did not grow there, or in India, except on Mount Meros, and that Harpalus had made every effort to cultivate it in Media without success, while Alexander the Great, in imitation of Bacchus, had returned victorious from India with his army wearing wreaths of ivy because of its rarity. Among the Thracians it adorns the wands of the wine god at sacred festivals, and likewise the helmets and shields of his worshippers, although it is very harmful to all trees and plants and destroys tombs and walls. Ivy is very popular with cold-blooded snakes. It is surprising, therefore, that any honour has been accorded to ivy.

Aquatic shrubs and reeds

156. Among plants that enjoy cold situations it may be fitting to mention aquatic shrubs. Reeds will hold the first place among these, necessary as they are for warlike and peaceful purposes; they are also welcome for entertainment. Northern peoples thatch their houses with reeds, and thatched roofs are durable. In other parts of the world reeds provide very light ceilings for rooms. They serve as pens for writing on paper – especially Egyptian reeds, because of their affinity, as it were, with papyrus. The reeds from Cnidus and those that grow around the Anaetic Lake in Asia Minor are more esteemed. Reeds taper gradually towards the top and carry a rather thick tuft of hair which is not without value; this either serves as a substitute for feathers to stuff innkeepers' palliasses, or, in places where it grows very hard and woody, as in Belgic Gaul, it is pounded and inserted between the joints of ships to caulk the seams;

I. *c.* 314 BC.

this material holds fast better than glue and is more reliable than pitch for filling cracks.

159. In the East people use reeds to make war. By means of reeds with a feather attached, they hasten the approach of death.

Bamboo and reeds

162. The Indian bamboo is the size of a tree, as we frequently see in our temples. A single length between knots – if this can be believed – can serve as a boat. The bamboo grows in exceptional profusion in the vicinity of the River Acesines.

164. There are several kinds of reed. One is rather compact with its joints close together; another has its joints further apart, and is thinner. Yet another reed is hollow throughout its whole length, and the Greeks call this the 'flute' reed, since the absence of pith and flesh render it very useful for making flutes. The reed from Orchomenus has a passage running even through the knots, and the Greeks refer to this as the 'pipe-reed'.

173. In Italy the reed is used mainly for propping up vines.

Rushes

178. Thatch and mats are made from rushes. Without their outer covering reeds are used for candles and funeral torches. In some places rushes are stronger and more rigid, and are employed to carry sails not only by the boatmen of the River Padus but also at sea by African fishermen, who hang their sails in an extraordinary manner between masts. The Moors use reeds for roofing their huts, and close investigation suggests that rushes are a substitute for papyrus in the inner areas of the world.

The structure of trees

181. In general the bodies of trees, like those of other living things, possess skin, blood, flesh, sinews, veins, bones and marrow.

182. The bark acts as skin; next to this most trees have layers of 'fat' called *alburnum*[1] because of its white colour. This is soft, and is the worst part of the wood; it rots easily, even in hard oak, and is prey to beetles and should always be cut away. Under the 'fat' is the flesh, and then the bones – the best part of the wood.

1. Sap-wood.

184. The flesh of some trees contains fibres and veins, which are easily distinguishable, the veins being broader and whiter. Veins occur in wood which is easily split, so that if you put your ear to one end of a beam of such wood, however great its length, you can hear even the tapping of a pen at the other end since the sound travels along passages that run straight through the wood. In this way you can also discover whether the wood is twisted or interrupted by knots.

186. A plank of wood floats horizontally, depth of submergence depending on proximity to the roots before the timber was felled.

Unusually large trees

200. What is thought to be the largest tree so far seen in Rome was the one that the Emperor Tiberius had displayed as a curiosity on a deck that had been rigged up for a mock naval battle. This had been brought to Rome with the rest of the wood used and lasted until the construction of Nero's amphitheatre.[1] It was a log of larchwood 120 feet long and with a uniform thickness of 2 feet; from this could be estimated the almost incredible height of the rest of the tree, by calculating the distance to the top.

201–2. A particularly remarkable fir-tree served as the mast of the ship that conveyed from Egypt the obelisk set up in the Vatican Circus and brought four shafts of the same stone for its base, on the orders of the Emperor Gaius. Certainly nothing more impressive than this ship has ever been seen afloat. It was ballasted with 150 cubic feet of lentils and its length occupied a large part of the left side of the harbour at Ostia. Under the Emperor Claudius it had been sunk there and used as a base for three moles that rose as high as the ship's towers erected on it; these moles were constructed of cement from Puteoli, especially dug and taken there for this purpose. It took three men linking arms to encircle the tree.

203. German pirates sail in boats made of a single tree that has been hollowed out; some of these boats carry thirty people.

Wood-borers

220. There are four types of pest that attack wood. Borers[2] have a very large head in proportion to their body and gnaw the wood

1. AD 59.
2. Such worms are commonly found in ships' timbers.

with their teeth. These are seen only in the sea and are the only ones that can properly be called by this name. The land-based pests are known as 'moths', and those like gnats as 'thrips'. The fourth type belongs to the maggot class, some pests of this type being produced by the wood itself when its sap decays, and others by the worm known as the 'horned-worm' – as in trees. When the horned-worm has gnawed away enough to be able to turn round, it gives birth to another.

221. In some trees, as in the cypress, the bitter taste, in others, as in the box-tree, the hardness of the wood, prevents the birth of these insects. Men claim that the fir does not rot if it is stripped at the time of budding – that is, between the twentieth and thirtieth of the month. Those who accompanied Alexander the Great observed that on the island of Tylos in the Red Sea there were trees used for ship-building whose timber had been found free from rot after 200 years, even though submerged in water.[1]

225. Fir used upright is strongest and is very suitable for door-panels and any kind of inlaid work one cares to mention, whether in the Greek, Campanian or Sicilian style. When planed it makes wood-shavings that twist in a spiral like the tendril of a vine.

Veneers

231. The principal woods that are cut into thin layers and used as veneers for other woods are the citrus, the turpentine-tree, types of maples, the box, the palm, the holly, the holm-oak, the root of the elder and the poplar.

232. This was the origin of extravagance involving trees: one wood was covered by another and an outer finish for a cheaper wood was made from a more expensive variety. Veneers were invented to make a single tree go as far as possible. Nor was this sufficient! The horns of animals began to be dyed, tusks cut up and wood inlaid with ivory, then, later on, veneered. Next men decided to search for material in the sea; tortoise-shell was cut up for inlay. During the principate of Nero people discovered with extraordinary ingenuity how to make tortoise-shell lose its natural colour by painting it in imitation of wood. A short while ago wood was not considered a luxury item, but now it is even made from tortoise-shell!

1. Pliny probably means teak, although any wood that remains totally immersed does not rot.

Mistletoe

245. There are three types of mistletoe; the one that grows on the fir and larch is called *stelis* in Euboea and *hyphear* in Arcadia; the type found on the oak, the hard-oak, the holm-oak, the wild pear, the turpentine-tree and most other trees is called mistletoe. There is a very prolific variety on the oak called *dryos hyphear*. With the exception of those that grow on the holm-oak and the oak, the smell and poison of the berry and the unpleasant odour of the leaf differ; berry and leaf are both bitter and sticky.

247. Invariably when mistletoe is sown it does not grow unless it has been passed through bird-droppings – especially pigeons' and thrushes'. Such is its nature that it will not sprout unless matured in the stomach of a bird. Its maximum height is 18 inches and it is always green and bushy. The male mistletoe is fertile, the female barren, except that even a fertile plant sometimes does not bear berries.

249. While on this topic I must not omit the veneration shown to mistletoe by the provinces of Gaul. The Druids (the name they give to their priestly caste) hold nothing more sacred than mistletoe and the tree on which it grows, provided it is a hard-oak. They also choose groves of hard-oak for its own qualities, nor do they perform any sacred rites without leaves from these trees, so that from this practice they are called Druids after the Greek word for oak.[1] For they believe that anything growing on oak-trees is sent by heaven and is a sign that the tree has been chosen by God himself.

250. Mistletoe, however, is rarely found on hard-oaks, but when it is discovered, it is collected with great respect on the sixth day of the moon. Then, greeting the moon with the phrase that in their own language means 'healing all things', the Druids with due religious observance prepare a sacrifice and banquet beneath a tree, and bring two white bulls whose horns are bound for the first time.

251. A priest in a white robe climbs the tree and with a golden sickle cuts the mistletoe, which is caught in a white cloak. Then they sacrifice the victims praying that God may make his gift propitious for those to whom he has given it. They think that mistletoe given in a drink renders any barren animal fertile and is an antidote for all poisons. So great is the power of superstition among most peoples in regard to relatively unimportant matters.

1. *drys.*

BOOK XVIII

AGRICULTURE

Early farming[1]

1. My next topic is the nature of various types of grain, gardens and flowers, and what, besides trees and shrubs, results from Earth's benevolence.

6–7. In the earliest days of Rome Romulus instituted the Priests of the Fields and named himself the twelfth of their number. At that time just over an acre each was enough land for the Roman people and no one had more. Of those who were slaves of the Emperor Nero a short while ago, how many would have been satisfied with an ornamental garden of that size? Now they want fish-ponds larger than that, and we are lucky if they do not want kitchens covering a larger area.

9. A parcel of land that one pair of oxen could plough in a day used to be called a *iugerum*.[2] The distance over which oxen could draw a plough in a single haul of reasonable length was called an *actus*.[3] The most generous gift given to generals and brave citizens was the largest area of land that a man could plough in a day.

11. Cato tells us that to praise a man by saying he is a good farmer and cultivator of the soil is the highest accolade. This is the origin of the word for 'wealthy', *locuples*, which means 'full of room', that is, land. Our word for money, *pecunia*, was derived from the word for cattle, *pecus*. Even today in the censor's accounts all national revenue is called 'pastures', because rent from pasture-land was for a long time the only source of public income. Furthermore, fines were reckoned only in terms of payment of sheep and oxen.

12. King Servius was the first to stamp bronze coinage with the likeness of sheep and oxen.

---

1. See generally K. D. White, *Roman Farming* (London, 1970).
2. Approximately two-thirds of an acre.
3. 120 feet.

Treatises on agriculture

22. To write treatises on agriculture was an occupation of the highest merit, even among foreign peoples, in that kings like Hiero, Attalus, Philometor and Archelaus did so, and also generals like Xenophon and the Carthaginian Mago, whom our Senate held in very great honour after the capture of Carthage. So much so that when it gave away that city's libraries to African princes, it decreed in Mago's case alone that his twenty-eight volumes should be translated into Latin, although Marcus Cato had already written his treatise. They also passed a decree that this task should be assigned to those skilled in the Carthaginian language, in which expertise Decimus Silanus, a man of most distinguished background, surpassed everyone else.

Cato's advice on buying a farm

26. The bravest men and most active soldiers and those least given to dishonesty are produced by the farming community. When buying a farm do not appear eager. In matters pertaining to the country, spare no trouble – least of all when buying land. A poor purchase always gives rise to regret. Cato adds that those about to buy land should before all things pay attention to the water-supply, the road and the neighbour.

27. As regards neighbouring farmers, he recommends that one should consider how prosperous they look; 'for, in a good district,' he writes, 'the people have an air of prosperity.'

28. Land which occasions its owner continuous struggle is bad. High on the list of priorities, Cato sets the following: the land should be fundamentally sound, there should be a plentiful supply of labour near to hand, a flourishing town and routes for carrying away produce by water or road, and furthermore that the farm should have good buildings and its land should have been well-farmed. 'It is better,' says Cato, 'to purchase land from a good owner.'

29. In Cato's view the most lucrative farm activity is viticulture. Understandably , since, in general, he is cautious about the outlay of money. Next in profitability he puts kitchen gardens with a good water-supply, and this is especially true if they are near a town.

Farmhouses

32. The basic requirement is that the farmhouse should not be too small for the farm, nor the farm for the farmhouse. The proper

arrangement needs a degree of skill. An occasion springs to mind when Gaius Marius, seven times consul, built a country house in the district of Misenum, making use of the skill he had employed in the layout of a camp.[1] So much so that even Sulla the Fortunate said that all other men had been blind compared with Marius.

33. It is generally agreed that a country house ought not to be located near a marsh or facing a river. Homer's words are indeed very true, that there are always unhealthy mists rising from a river before dawn. In hot regions the house should face north, in cold regions south, and in temperate situations due east.

35. Old authorities thought that the most important consideration was not to take on too large a farm, since they reckoned it better to sow less land but cultivate it more thoroughly; this opinion, I see, was held by Virgil.[2] And truth to tell, large estates have ruined Italy and indeed are now also ruining the provinces.

On choosing a farm-manager

36. Suffice it to say that a farm-manager ought to be as near as possible to the owner in intelligence, but not regard himself as such. Farming carried out by criminals is abominable, as is everything else done by desperate men. It may seem rash to quote a saying of writers of old whose truth is only apparent when it is inwardly digested and considered – namely that 'nothing less than the highest standard of farming is worthwhile'.

The secrets of good farming

41. Gaius Furius Chresimus, a freed slave, obtained much greater returns from a smallish farm than his neighbours derived from vast estates. As a result he was very unpopular, as if he had been spiriting away other people's crops by magic.

42. He was indicted by the curule aedile, Spurius Albinus. Afraid that he would be found guilty when the tribes had to vote, he brought all his agricultural implements – including his splendidly made iron tools, heavy mattocks and ponderous ploughs – into court. He also produced his farm labourers – strong men, and, according to Piso's description, well looked after and clothed – and, finally, his well-fed oxen.

---

1. Gaius Marius (157–86 BC) became consul for the first time in 107 BC and was appointed commander against Jugurtha, king of Numidia.
2. *Georgics*, II, 412.

43. Then he said: 'These are my magic spells, citizens, and I am not able to exhibit or summon as witnesses my midnight labours, early risings and sweat and toil.' This ensured that he was acquitted by a unanimous vote. Indeed labour is essential to farming, and this is the reason behind our forefathers' saying that 'on a farm the best fertilizer is the owner's eye'.

Grain and pulses

48. I shall now describe the nature of the various types of grain. There are two main varieties: cereals, such as wheat and barley; and pulses, such as beans and chick-peas. The differences between them are too well known to warrant description.

Wheat

63. There are many types of wheat produced by various peoples. Speaking personally, I would not compare them with Italian wheat for whiteness and weight. Foreign wheat can be compared only with that from the mountain regions of Italy. Boeotian is the best foreign wheat, then Sicilian, and finally African.

65. The dramatist Sophocles in his *Triptolemus* praised Italian corn in preference to all other kinds: translated, he says: 'That happy land of Italy glows white with bright, shining wheat.'

Barley and porridge

71. The first cereal crop to be sown in the season is barley. India has both cultivated and wild barley and the Indians use it to make their best bread and also porridge.

72. Barley is the oldest food, as proved by the Athenian ceremony, recorded by Menander, and also by the name once given to gladiators, who used to be called 'barley men'. The Greeks use this for porridge in preference to any other grain.

Methods of milling

97. The milling of all types of grain is easy. Etruria crushes the ears of emmer-wheat – after it has been roasted – with an iron-tipped pestle or in a hand-mill that is serrated and denticulated inside with grooves radiating from its centre. Most of Italy uses a bare pestle and a millstone driven by water-wheel.

Bread and bakers at Rome

105. It seems pointless to describe the different kinds of bread. In some places bread is called after dishes eaten with it, such as oyster-bread, in others from its taste, such as cake-bread; it may also derive its name from the method of baking – for example, oven-bread, tin loaf, or pan-bread.

107–8. There were no bakers at Rome until after AUC 580[1] and the war with King Perseus. The citizens used to make their own bread and this was the special task of the women, as it is even now in most nations. Plautus had already mentioned bakers; he uses the Greek word in his *Aulularia*,[2] which has caused much discussion among scholars as to whether the line involved is genuine. The expression that occurs in Ateius Capito proves that it was usual in his day for cooks to bake bread for the well-to-do, and only those who ground spelt were called millers; nor did people have cooks on their permanent staff, but instead they hired them from the provision market. The Gallic provinces invented a kind of sieve of horsehair, while Spain made sieves and sifters of flax, and Egypt, of papyrus and rush.

Methods of harvesting and threshing corn

296. There are various methods of harvesting. On large estates in the provinces of Gaul huge frames with toothed edges are mounted on two wheels and driven through the corn by a team of oxen pushing from behind; the ears of corn are ripped off and fall into the frame.[3] Elsewhere the stalks are cut through with a sickle and the ears are stripped off between two pitchforks. In some places the stalks are cut off at the root, in others the roots are pulled up as well.

297. Where people thatch their houses with straw they store it as long as possible, but where there is a shortage of hay they require chaff for spreading as litter.

298. The ear itself, when reaped, is beaten out in some places by threshing-sledges on a threshing-floor, in others by being trodden by mares, and in yet others by being threshed with flails. Wheat is

---

1. 174 BC.
2. Line 400.
3. This is the earliest reference to such a machine, the existence of which is attested by representations on four monuments from north-east Gaul.

found to give a greater yield if it is reaped late, but is of finer quality and stronger the earlier it is reaped. The most obvious rule is to reap grain before it hardens and when it has begun to get its colour. There is indeed an oracular saying: 'Better to be two days too soon than two days too late with your reaping.'

The storage of grain

301. Some authorities recommend building elaborate granaries with brick walls a yard thick and filling them from above and making them draught-proof and windowless. But in other places they build wooden granaries supported on pillars, preferring to allow the air to circulate on all sides and even from below.

303. To us the storage of corn at the proper time will seem most important, since if it is gathered in when insufficiently ripened and firm, or stored while hot, pests are certain to breed in it.

MATERIA MEDICA

BOOK XX

DRUGS OBTAINED FROM GARDEN PLANTS

Man's food and plants

1–2. From here onwards I shall begin to outline an extremely important role of Nature and will explain to man his proper foods, compelling him to admit that he does not realize how his life is sustained. Let no one be deceived by the mundane terminology – consider this a slight and unimportant assignment. I shall describe Nature at peace or in conflict with herself, and the likes and dislikes of things that are voiceless and even inanimate. Moreover – to increase our wonder – all these exist for man's benefit. In matters relating to the fundamental principles of things the Greeks have employed the terms 'sympathy' and 'antipathy', as, for example, in the cases of water putting out fire, the sun swallowing up water, or the moon generating water. The sun and moon are eclipsed, each through the transgression of the other. To turn from heavenly things, magnetite attracts iron while another type of stone repels it. The diamond, that rare pleasure afforded by wealth, unbreakable by and resistant to all other forces, is destroyed by goat's blood. I shall mention other things, equally or even more remarkable, in their appropriate contexts. Let me only crave forgiveness for beginning my account with very small but health-giving objects. First I shall describe plants from the vegetable garden.

Wild and cultivated cucumbers

3. There is a variety of cucumber that grows wild and is much smaller in size than the cultivated one. Its seed is crushed and put in rain-water, where it sinks to the bottom. Then the action of the sun coagulates it, and this substance is made into tablets for the benefit of man. These are beneficial for the treatment of poor eyesight, eye-diseases and styes on the lids. Cucumber-roots, dried and compounded with resin, heal impetigo, scabies and what is commonly called psoriasis and ringworm.

Onions

39. Onions do not grow wild. Cultivated ones provide a cure for poor vision through the tears caused by their very smell. Even more effective is the application of some onion-juice to the eyes. Onions are said to be a soporific and, eaten with bread, to be capable of healing mouth-sores.

42. The school of Asclepiades claims that eating onions promotes a healthy complexion, and that if they are eaten every day on an empty stomach they maintain good health, are beneficial for the stomach, and ease the bowels by moving gas along; when used as a suppository they disperse haemorrhoids. Finally, added to that extracted from fennel, onion-juice is marvellously efficacious when used in the early stages of dropsy.

Garlic and lettuce

50. Garlic is very potent and beneficial against ailments caused by changes of water and of location. It drives away snakes and scorpions by its smell and, according to some authorities, every kind of wild beast.

64. Lettuces have special properties, in addition to those already mentioned; they are soporific, and can check sexual appetite, cool a feverish body, purge the stomach and increase the volume of blood. Lettuces disperse flatulence, suppress belches and aid digestion. No other food is more effective in stimulating or diminishing the appetite. In either case the amount taken is critical: too much loosens the bowels, a moderate amount causes constipation.

About cabbage

78. It would be a lengthy task to list the good points of the cabbage since Chrysippus the doctor devotes a volume exclusively to this vegetable, with specific headings according to its effects on different parts of the body. Likewise, Dieuches and, above all, Pythagoras and Cato have no less wholeheartedly sung the praises of the cabbage. We should examine these opinions more closely so as to learn what medicine the Roman people used for 600 years.

80. Cato gives his highest recommendation to the curly variety, and then to the smooth with its large leaves and big stem. He records that when the cabbage is taken raw in the morning together with oxymel, coriander, rue, mint and silphium root, in doses of 5 fluid ounces, it is good for headaches, impaired vision, spots before

the eyes, for the spleen and stomach, and for hypochondria. He claims that the power of the ingredients is so great that the person who pounds them together feels himself become stronger as he pounds the mixture.

81. According to Cato's treatise, fresh or old wounds, and even malignant lesions that cannot be healed by any other treatment, should be fomented with hot water and have an application of pounded cabbage twice a day. He advises the same treatment for fistulas and sprains, and similarly for tumours that must be drawn out or dispersed. Cato states that boiled cabbage prevents dreams and insomnia, if sufferers eat it with oil and salt while fasting as much as possible.

84. Having dealt with the views expressed by Cato, I should now record those of the Greeks. I shall limit myself to what Cato omitted. For example, as cabbage is an enemy of the vine, men say it counteracts wine and consequently, if taken in food beforehand, prevents intoxication. If taken after drinking it neutralizes the effects.

85. The Greeks assert that cabbage greatly improves the eyesight and that the benefit is considerable even if the juice of a raw cabbage mixed with Attic honey merely touches the corner of the eyes. The followers of Erisistratus claim that nothing is better for the stomach and sinews and so prescribe cabbage for paralysis as well as for the coughing-up of blood.

Pennyroyal

152. Pennyroyal and mint are allies in their ability to revive people who have fainted; both are put in glass bottles full of vinegar. For this reason Varro states that a garland of pennyroyal, rather than one of roses, is more suitable for bedrooms, since an application of this is said to relieve a headache. Moreover its very smell protects the head, so it is said, from injury, cold, heat and thirst. People in the sun do not suffer from heat exhaustion if they carry two sprigs of pennyroyal behind their ears.

153. Pennyroyal, boiled with honey and soda, cures intestinal disorders. Taken in wine it acts as a diuretic and, if the wine is Aminian, it disperses kidney stones and all internal pains.

156. The wild variety is even more effective for the same purposes.

Poppies

198–200. The calyx of the cultivated white poppy is taken in wine to induce sleep. The seed cures leprosy. A sleep-inducing drug is also obtained from the dark poppy, by making incisions in the stalk when the buds are forming, as Diagoras advises, or when the tree is dropping its blossoms, according to Iollas. Both recommend that the incision should be made beneath the head and calyx, and neither in this or any other variety is the incision made in the head itself. Poppy-juice is plentiful and becomes thick naturally and, fashioned into tablets, is dried in the shade. It is not only sleep-inducing but, if too much is swallowed, brings about a fatal coma. Men call the juice opium. I am told that the father of Publius Licinius Caecina, a man of praetorian rank, died of opium poisoning at Bavilum, in Spain, when an unbearable illness made his life not worth living; it was the same with several others. For this reason a great dispute has arisen. Diagoras and Erisistratus have completely rejected opium as a fatal drug, forbidding it to be administered because it damages the sight. Andreas adds that the only reason it does not cause immediate loss of sight is the fact that it is adulterated at Alexandria. But since then in the form of that well-known drug called codeine, its use has not been condemned. Poppy-seed, when pounded into tablets and added to milk, is used to induce sleep; it is also used with rose-oil for headaches and, as drops, for earache.

Tests of the purity of opium

202. The best poppies grow in dry soils and where the rainfall is light. The first test for opium is its characteristic smell; indeed the smell of it, in its pure state, is unbearable. The second test is to put it in the flame of a lamp in which it should burn with a bright, clear flame and only give off a smell when the lamp has been extinguished. This does not happen if the opium is adulterated, in which case it is harder to ignite and continually goes out. Another test for pure opium is by means of water on which it floats as a miniature cloud, while the impure substance gathers in blisters. Most amazing, however, is the fact that pure opium is detected by the summer sun, for it sweats and melts until it becomes like fresh juice.

A famous specific used by Antiochus the Great

264. Finally here is a very famous remedy compounded from garden plants to counteract the poison of venomous animals. The prescription is contained in a lapidary inscription in the Temple of Aesculapius on Cos: 2 denarii of wild thyme, and 2 denarii of opopanax and spignel, 1 denarius of trefoil seed, of aniseed, of fennel-seeds, of

ami and of parsley, 6 denarii of vetch and 12 denarii of flour made from vetch that has been ground and sieved then kneaded with the best wine into tablets each of 1 victoriatus in weight. The dose to be one of these tablets in 5 fluid ounces of wine. King Antiochus the Great is said to have taken this prescription as an antidote for the poison of all venomous creatures except the asp.

FLOWERS AND HERBS

Thyme

56. There are two kinds of thyme: the pale and the dark. Thyme flowers about the period of the summer solstice, when the bees collect from it. It offers a rough guide to the yield of honey, for beekeepers hope for an abundance of honey if the thyme flowers profusely. Thyme is damaged by rain and sheds its flowers. The seed is invisible to the eye; but the seed of wild marjoram, although very small, is large enough to be seen. But what does it matter that Nature has hidden the seed of thyme?

57. We know that it is inside the flower itself and that a plant grows from the flower when sown. Is there anything with which men have not experimented? Attic honey has a greater reputation than that of any other kind in the whole world. And so thyme has been imported from Attica and, as I am informed, grown with difficulty from the flower. But another characteristic of Attic thyme proved a hindrance: survival depends on a sea breeze. The same view was held in olden times about all kinds of thyme, and people believed that this was the reason why thyme did not grow in Arcadia. They also thought that the olive was found only within 35 miles of the sea, whereas we know today that thyme covers even the stony plains of the province of Gallia Narbonensis, and this is almost the only source of revenue for the inhabitants. Thousands of sheep gather there from distant places to graze on the thyme.

Bees and honey

70. Bees and apiaries are particularly associated with gardens and flowers. In favourable conditions, beekeeping offers large returns for minimal outlay. So, for the sake of the bees, you should plant thyme, wild parsley, roses, violets, lilacs and many other flowers.

73. What I have discovered about bees' food is amazing and worth recording. Hostilia is a village on the River Padus. When their food-supply fails in this region, the local people put the hives on boats and carry them 5 miles up river by night. At dawn the

bees come out, feed and return every day to the boats, whose position alters until such time as they have settled low in the water under the very weight of honey – an indication that the hives are full. They are then taken back to Hostilia and the honey is extracted.

74. In Spain the locals transport the hives about on mules for a similar reason. The food that the bees eat is of such great importance that even their honey may become poisonous. At Heraclea, in Pontus, the honey is extremely harmful in certain years, even though it comes from the same bees. The authorities for this have not said from what flowers this honey is obtained but I will record the findings. There is a plant called 'goat's bane' from its fatal effect on cattle, especially goats. When the flower of this plant withers in a rainy spring, the bees take from it a harmful poison. Consequently the ill-effects are not experienced in all years. The signs of poisonous honey are that it fails to thicken, causes sneezing and is heavier than pure honey. Cattle that have eaten poisonous honey throw themselves to the ground, seeking to cool their bodies which are running with sweat.

Helenium

159. Helenium, which had its origin in the tears of Helen,[1] is believed to preserve physical charm and to keep the fresh complexion of our womenfolk unimpaired – whether of their face or the rest of their body. Moreover, women think that they acquire a kind of aura of attractiveness and sex appeal by its use. People attribute an exhilarating effect to this plant when taken in wine – the power that nepenthes had of driving away sorrow; Homer sang the praises of the latter plant.[2] Helenium also has a very sweet juice. The root, taken in water while fasting, is good for asthma. It is also taken in wine for snake-bites.

Greek weights and measures

185. As I have frequently had recourse to Greek terms in specifying weights and measures, I will add a gloss on them at this juncture. The Attic drachma – for the Attic standard is generally adopted by

1. Daughter of Zeus and Leda, wife of Menelaus, whose abduction by Paris led to the Trojan War.
2. *Odyssey*, IV, 221.

doctors – is the equivalent in weight of a silver denarius, and 6 obols make a drachma; there are 10 chalci to an obol. As a unit of dry measure the cyathus weighs 10 drachmae. When an acetabulum[1] is indicated it means one quarter of a hemina,[2] that is, 15 drachmae. The mna, which the Romans call 'mina', weighs 100 Attic drachmae.

1. Approximately 2½ fluid ounces.
2. Approximately 10 fluid ounces.

BOOK XXIII

THE VINE AND THE WALNUT

Fruits: their products and uses

1. Pomona, not content merely to protect and nourish with the shade of her trees the plants that I have recorded, has given healing qualities to hanging fruit.

2. Nature has bestowed healing powers on the vine especially, not satisfied with having supplied it with an abundance of delightful flowers, scent and unguents in its unripe juice and its varieties of wild grape. 'Mankind,' says Nature, 'thanks to me, enjoys a very great deal of pleasure. For I create the juice of the grape, the oil from the olive, and dates and fruit in so many varieties. I am unlike Mother Earth, all of whose gifts must be won by hard work – ploughing with oxen, beating on threshing-floors, and grinding between millstones, all to produce food with uncertain delay and with any amount of toil. My gifts are ready-prepared and demand no laborious effort. They offer themselves spontaneously, and even fall of their own accord if it is troublesome to reach them.' Nature has striven to outdo herself, creating more for our benefit even than for our pleasure.

The power of vinegar

54. Wine still provides remedies even when sour. It's most noted strength lies in its cooling properties, although it is no less effective in causing things to disintegrate. Thus earth fizzes when vinegar is poured on it. I have many times claimed, and will continue to make the claim, that vinegar is often beneficial in combination with other substances. It checks nausea and hiccoughs. Inhaling vinegar fumes stops sneezes, and holding vinegar in the mouth wards off the effects of too hot a bath. In the case of many convalescing people vinegar swallowed with water is good for the stomach. A gargle of vinegar and water helps in heat exhaustion, and the same mixture is a very good restorative for the eyes when used as a fomentation.

56. Vinegar checks chronic coughs, catarrh of the throat, asthma and shrinkage of the gums. It is harmful, however, to the bladder

and to weak tendons. Doctors were unaware of its power as an antidote to asp-bites. Recently, however, a man bitten by an asp he had trodden on while carrying a skin of vinegar felt the wound every time he put down the container, but otherwise it was as if he had never been bitten. He deduced that vinegar was an antidote, and he was helped by drinking a draught.

57. Similarly those who suck out poison rinse their mouths with vinegar. Its power is extensive; it is not limited to foods but applies to many other things. Vinegar poured on rocks splits them when fire has been of no avail. No other sauce seasons food so well or increases its flavour so much.

Walnuts

147. Walnuts[1] have acquired their name in Greek from the heaviness of head that they bring about. The trees themselves and their leaves give out a poison that affects the brain. If the kernels are eaten they have the same effect, although the pain is less severe. Walnuts are more pleasant when fresh; when dried, they are oily, harmful to the stomach and difficult to digest; they cause headaches and are also bad for coughs.

148. The more walnuts one eats, the easier it is to expel tapeworms. Very old walnuts cure gangrene, carbuncles and bruises. The bark of walnut-trees cures ringworm and dysentery. The pounded leaves mixed with vinegar cure earache.

149. When the mighty King Mithridates had been defeated, Gnaeus Pompeius found in a notebook of his, written in his own hand, a prescription for an antidote: 2 dried walnuts, 2 figs and 20 leaves of rue were to be pounded together with the addition of a pinch of salt. Anyone taking this on an empty stomach would be immune to all poison for the whole day.

1. *kara* = head; *karyon* = nut.

BOOK XXIV

DRUGS OBTAINED FROM FOREST TREES

Trees and remedies derived from them

1. Even woods and the wilder aspects of Nature furnish medicines, for there is no place where Nature has not provided remedies for mankind – so that the desert itself has become a chemist's shop, although at every point there occur amazing examples of incompatibility. Thus the oak-tree and the olive are at odds as a result of a long-lasting hatred, so that if the one is planted in the hole from which the other has been dug out, it dies. The oak also dies if planted near a walnut. The hatred between cabbage and vine is fatal. Yet the vegetable that puts the wine to flight withers when planted opposite cyclamen or wild marjoram.

3. Inanimate objects also – even the smallest – have their own particular poisons. Cooks remove excessive salt from food by means of linden bark and fine flour. Salt corrects our aversion when we find something over-sweet. Nitrous or bitter water is sweetened by the addition of pearl barley and rendered drinkable within two hours, and this is why pearl barley is put into linen wine-strainers. Fuller's earth from Rhodes and potter's clay in our own country have similar properties. Gum is more easily removed by vinegar, and ink by water.

Medicine

4–5. This was the origin of medicine. For Nature had decreed that such things alone should be our remedies – universally available, easy to discover, and free, things that support our daily life. Since then, men's deceitfulness and ingenuity have discovered bogus dispensaries in which each individual has been promised a new lease of life – at a price. Immediately, complicated mixtures that defy explanation began to multiply. Arabia and India are considered to be the home of these remedies. Medicine is imported from the Red Sea for a tiny sore, although even the poorest people consume perfectly good remedies for their dinner each day. But if remedies were sought in the vegetable garden, or obtained from a plant or

shrub, no art would be less expensive than medicine. And so greatness has led the Roman people to abandon their traditional practices,
and as a result of our conquests we ourselves have been conquered.
In one art we are subject to foreigners and they rule their rulers. I
will amplify this assertion elsewhere.

Remedies from trees and plants

7. Ground acorns compounded with salted axle-grease cure callouses
that the Greeks call 'malignant'. The acorns of the holm-oak are
more efficacious, and in all acorns the shell and the immediately
underlying pith are the most potent part. When boiled these help
stomach disorders. In dysentery even the acorn itself is used. The
same mixture is used for snake-bites, catarrh, and suppurations.

Cork- and juniper-trees

17. The large juniper, which Greek botanists call *cedrelate*, provides
pitch called *cedria* and is very useful for toothache. For the berry
breaks the teeth and makes them fall out, thus relieving the pain.
The juice would be of great use for book-rolls,[1] were it not for the
headache caused by inhaling it. Oil of juniper preserves dead bodies
against corruption by time, but causes living ones to decay – a
strange anomaly to rob the living of their life and to give a kind of
life to the dead!

18. The juice also makes clothes disintegrate and kills animals.
For this reason I do not think that it should be taken orally as a
remedy for quinsy or indigestion, as some authorities have recommended. I should also be apprehensive about rinsing teeth with it in
vinegar when they ache, or dropping it into the ears for deafness or
worms. There is an extraordinarily tall story to the effect that if one
covers the penis with it before intercourse, it acts as a contraceptive.
I should not hesitate to use it as an ointment for phthiriasis or for
dandruff.

Tree-moss

27. Tree-moss grows in Gaul as I have indicated. It is useful in the
bath for infections of the female sexual parts and, beaten up with
cress and salt-water, also helps the knees and swellings on the thighs.

---

1. The oil would presumably be applied to papyrus to soften it prior to rolling.

Taken with wine and resin it very quickly acts as a diuretic. Combined with wine and juniper-berries, it drains off the water in dropsy.

Resin

35. Resin is dissolved in oil for treating wounds; its medical properties are to close up wounds, to act as a cleansing agent and to disperse gatherings. Similarly, the resin of terebinth cures chest ailments. When warmed it is used as an embrocation to relieve pains in the limbs. Slave-dealers are particularly fond of this ointment for rubbing all over their slaves' bodies in order to correct thinness. They loosen the skin of every limb by walks and thus enable the slaves' bodies to absorb more food.

Agnus castus *and* genista

59. The *agnus castus* is not very different from the willow, in either its use for wickerwork or the appearance of its leaves, except that its smell is more pleasant. There are two species: the larger grows into a tree, like the willow; the smaller is many-branched and has paler, downy leaves. Both grow on marshy plains.

60–61. The seed, taken in drink, tastes not unlike wine and is said to alleviate fever, to stimulate perspiration when mixed with oil and applied as an embrocation, and to dispel tiredness. The trees provide drugs that are diuretic and promote the menstrual flow.

62. These trees also furnish drugs that check violent sexual desire, and thus act as antidotes to penis-stimulating bites from poisonous spiders.

65. The *genista* is used for cords, and bees are very fond of its flower. I wonder whether this is the plant that Greek writers called *sparton*? Homer had this in mind when he said: 'The ships' cords were loosed.' It is certain that Spanish or African esparto grass was not yet in use at that time and, although ships were built with caulked seams, they employed flax for this purpose, never esparto.

Ivy and some of its many uses in medicine

76. Clusters of ivy-berries, either taken internally or applied locally, cure disorders of the spleen; for liver trouble, however, they must be applied externally.

77. Ivy-juice dropped into the nostrils clears the head and is even

more effective if soda is added. It is taken as drops in suppurating or painful ears; it also removes ugly scar tissue. The juice of white ivy-berries, warmed with hot iron, is beneficial for complaints of the spleen. Six berries in 3 fluid ounces of wine are sufficient. Furthermore, these berries, taken three at a time in oxymel, expel tapeworms. For toothache Erisistratus recommends five berries of the golden-berried ivy pounded in rose-oil and warmed in the skin of a pomegranate. The drops should be put in the ear next to the pain.

Holly and brambles

116. A mixture of crushed holly-leaves and salt is good for diseases of the joints, while holly-berries are beneficial for menstruation pains, stomach disorders, dysentery and cholera. Taken in wine the berries arrest diarrhoea. If the boiled root is applied to the skin it extracts embedded objects, and it is extremely useful for dislocations and swellings. A holly-tree planted in a town or country house wards off magic. Pythagoras writes that water is solidified by its flowers and that a holly-stick thrown at any animal, even if it falls short because the thrower lacks strength, rolls nearer the target of its own accord – such is the natural power of the holly-tree.

117–18. Nature did not intend even brambles to have only harmful purposes and so she provided them with blackberries which are food even for men. They have a drying and astringent property, very good for gums, tonsils and sexual parts.

Magical plants

156. In keeping with my promise to discuss plants that are out of the ordinary, it occurs to me to say a few words about those species with magical properties. Surely no plants give rise to greater wonder. Pythagoras and Democritus, who both follow the accounts of the Magi, first brought these to the notice of our part of the world. Pythagoras writes that water is congealed by *coracesia* and *calicia*, but I find no mention of these in other authorities, nor anything else about them in Pythagoras.

157. Pythagoras also gives the name *minyas*, or *corinthia*, to a plant whose juice, if used as a fomentation, immediately heals snake-bites. He adds that if this juice is poured on the grass and someone happens to walk on it – or if by chance it is sprinkled on the body – the person inevitably dies. The poison from this plant is fiendish, and yet it counteracts other poisons.

158. Pythagoras also refers to a plant called *aproxis*, whose roots catch fire at a distance like naphtha. He tells us the symptoms of diseases that have attacked the human body when the cabbage is in flower. Persons who have suffered from these, even when cured experience phantom symptoms of the diseases every time the cabbage flowers.

Ophiusa and thalassaegle

163. The snake-plant, says Democritus, grows at Elephantine in Ethiopia. It is leaden in colour and disturbing to look at. If one takes a draught, it causes such terrible visions of threatening snakes that men commit suicide from terror. Therefore persons who commit sacrilege are forced to drink it; its antidote is palm-wine. The thalassaegle is found along the banks of the Indus and is thus called *potamaugis*. A draught of this rouses men to frenzy and plagues them with hallucinations.

176. The Greeks jokingly call one plant the 'friendly' plant[1] because it sticks to clothes. A band made of this and placed on the head relieves headaches.

1. Goose grass.

BOOK XXV

THE NATURE OF WILD PLANTS

Plants used in medicine

1–2. The renown of the plants about which I intend to speak (sometimes Mother Earth produces these solely for use in medicine) makes one admire the care and application of men in olden times. They left no stone unturned and kept nothing secret – that is, if they wished to benefit posterity – whereas we wish to hide and suppress their discoveries and to cheat human life even of the good things that have been won by others. Indeed, those who have acquired a little knowledge keep this secret; by not sharing their learning they increase its prestige. So averse have we become to new research and improving life. For a long time the main object of our great intellects has been to confine to individual memory the achievements of the ancients so that they fall into oblivion. But, goodness knows, some have been added to the roll of the gods for a single discovery, men whose earthly life, at any rate, has been more widely celebrated because plants have been named after them.

3. They have searched pathless mountain heights, unexplored deserts and all the bowels of the earth, discovering the power of every root and the uses to which vegetation may be put.

4. This topic was treated less than appropriately by those of our own writers who have a voracious appetite for all things useful and good. The first, and for a long time the only, authority was Marcus Cato, the master of all useful crafts; he dealt with veterinary science, but, none the less, merely touched upon the subject of herbal medicine. After him only one of our distinguished scientists, Gaius Valgius,[1] attempted the subject. He was well known for his scholarship and left an unfinished work dedicated to the late Emperor Augustus. He began this with a reverent preface praying that His Imperial Majesty should always, and above all others, be the healer of every human ill.

---

1. A Roman poet; contemporary of Virgil and Horace.

Mithridates' interest in medicine

5. Mithridates, the greatest king of his time, defeated by Pompey the Great, was, as we know from direct evidence and by report, a more attentive researcher of life than any man born before him.

6. Alone and unaided, he devised a plan to drink poison every day after first taking remedies, in order that by accustoming himself to the poison he might become immune to it. He was the first to discover the different antidotes, one of which bears his name. Mithridates also mixed the blood of Pontic ducks with these antidotes because they lived on poison. Still extant are treatises addressed to him, written by the famous doctor Asclepiades, who when invited to come from Rome sent written instructions instead. It is well attested that Mithridates was the only person to speak twenty-two languages, and that he never addressed any of his subject peoples through an interpreter during all fifty-six years of his reign.

7. Mithridates, with his breadth of intellect, was especially interested in medicine and amassed detailed knowledge from all his subjects, who covered a substantial part of the world. He left among his private possessions a bookcase of these treatises, together with specimens and descriptions of their individual properties. Pompey the Great, when all the royal booty had fallen into his hands, ordered his freedman Lenaeus, a man of learning, to translate these works into Latin. As a result, this great victory was of no less benefit to everyday life than to the state.

Greek writers on herbal medicine

9. Other authorities have for the most part been content to record just the names since they thought these sufficient as a guide to the properties and nature of a plant for those willing to search for this information. This knowledge is not difficult to obtain. For my part I have had the good fortune to examine all except a very few plants, thanks to the research of Antonius Castor, the supreme botanical authority of our times. I used to visit his garden, with which he was obsessed; in this he would grow a large quantity of specimens, even after he had reached the age of 100. In spite of his great age he suffered no loss of memory or physical strength. Nothing else aroused greater wonder in antiquity than the science of botany.

10. Stories abound everywhere about Medea[1] of Colchis and other practitioners of magic, especially Circe of Italy who has been numbered with the gods.

11. This is why, in my opinion, Aeschylus, one of the earliest dramatists, writes that Italy is full of herbs,[2] and may have said the same about Circeii where Circe[3] lived. There is strong evidence for this even today, for people agree that the Marsi, who are descended from Circe's son, are snake-charmers. Homer, however, the father of ancient learning, while a great admirer of Circe in several passages, gives pride of place for herbs to Egypt even though at that time the irrigated part of Egypt, familiar to us now, did not exist; this region was later formed by the alluvial mud of the Nile.

12. He writes that Egyptian herbs were freely given by the king's wife to Helen, and these included nepenthes which brings oblivion and forgetfulness of sorrow; these Helen was to pass on to all mankind. But the first of all those traditionally associated with more precise utterances on botany was Orpheus.

16. Such was the state of medicine in the old days when the whole science found its way into the dialects of Greece. The reason why more herbs are not known is because only illiterate country folk try them out, for they indeed are the only ones who live among them. Nobody cares to search for herbs when confronted by hordes of doctors.

Moly

26. The most famous plant, according to Homer's testimony, is the one he thinks the gods call moly; he assigns its discovery to Mercury, and likewise the revelation of its power over the most potent sorcery. People say that at the present time it grows around Pheneus in Arcadia, and on Cyllene. It is said to resemble the description in Homer;[4] that is, it has a round black root the size of an onion and the leaves of a squill; it is not difficult to dig up.

1. Daughter of Aeetes, king of Colchis; celebrated for her skill in magic. When Jason went to Colchis in search of the Golden Fleece, she fell in love with him and assisted him in his quest. Medea fled with Jason to Greece but he deserted her for the youthful daughter of King Creon. Medea took vengeance on him by murdering their two children and destroying his new wife.

2. Quoted by Theophrastus, *Historia Plantarum*, IX, 15, 1.

3. Daughter of Helios by Perse and sister of Aeetes. Like Medea, she was distinguished for her magic arts. She lived on the island of Aeaea and entertained Odysseus' companions, whom she turned into pigs. Odysseus received the magic root moly from Hermes and compelled Circe to restore his men to human form.

4. *Odyssey*, X, 302–5.

27. Greek authorities have painted its flowers yellow, although Homer describes it as white. I know a doctor, one of those specializing in herbs, who says that moly also grows in Italy and that a specimen can be brought for me from Campania within a few days, dug up there in spite of the difficulties occasioned by the rocky soil; it has a root 30 feet long, and indeed this was not even its whole length since it had been broken off.

Mandrake and its use as an anaesthetic

150. The juice from the mandrake's leaves, when touched by dew, is deadly. Even when kept in brine the leaves retain their harmful properties. The mere smell brings a heavy feeling to the head, and, although it is eaten in certain countries, those who, through ignorance, inhale too much of the smell are rendered speechless, while too large a draught causes even death. When the mandrake is used as a sleeping-draught, the quantity administered to the patient should be proportionate to his strength – a medium dose being about 2 fluid ounces. It is also taken in drink as an antidote for snake-bites and before surgery and injections to produce anaesthesia. Some find the smell sufficient to induce sleep.

Hemlock

151. Hemlock is poisonous and has a bad reputation because the Athenian state employed it for inflicting capital punishment.[1] However, its use for many other purposes must be included. It has a poisonous seed, but the stem is eaten as a salad by many people. The seed and leaves have the power to chill and so cause death. Persons who have drunk hemlock begin to grow cold at their extremities.

152–3. The remedy lies in the warming nature of wine, if taken before the hemlock reaches the vital parts. Hemlock kills by thickening the blood – its other powerful property – and so spots are seen on the bodies of people killed by it. A poultice is made from hemlock-juice for cooling the stomach. The main use, however, is for checking running eyes in summer and for alleviating pains in them. Hemlock is an ingredient of eye-salves and checks all kinds of catarrh. The leaves also relieve all types of swelling, pain or discharges.

1. Socrates' death sentence was carried out by giving him a draught of hemlock. See Plato, *Phaedo*, 117 ff.

154. Anaxilaus is the authority who states that if young girls' breasts are rubbed with hemlock they will always be firm. If it is rubbed on mens' testicles at puberty it suppresses sexual desire. The most potent hemlock grows at Susa, in Parthia, the next most potent grows in Laconia, Crete and Asia; in Greece the most potent is found around Megara, and the next most potent in Attica.

Erigeron and toothache

167. If one draws a line round erigeron with an iron tool and digs it up and then touches a painful tooth with the plant three times, spitting after each touch, and finally replants it in its original spot so as to keep it alive, it is said that the tooth will never ache again.

BOOK XXVI

DISEASES AND THEIR REMEDIES

New skin diseases

1. Men's faces have also become prone to new diseases, unknown in former times in Italy, unknown indeed to almost the whole of Europe. Even so, these have not spread all over Italy, or through Illyricum, or the provinces of Gaul and Spain generally. In fact they are limited to Rome and its environs. These diseases are painless and do not endanger life but cause so much disfigurement that any kind of death would be preferable.

2. Men gave the most severe of these diseases the Greek name *lichen*, while in Latin it was known as *mentagra* because it usually began on the chin;[1] at first this was by way of a joke – for men are quick to mock others' afflictions – but shortly the term was adopted generally. The disease in many instances affected the whole of the face and only the eyes escaped; it also spread to the neck, chest and hands, covering the skin with a disfiguring scale.

3. Our fathers and forefathers did not experience this plague. It first spread into Italy in the middle of Tiberius Claudius' principate when a Roman knight from Perusia, a quaestor's secretary, introduced the disease from Asia Minor where he had served. Women were immune to it, likewise slaves and the lower middle classes. Nobles, however, were particularly susceptible through the quick contact of kissing. The scar, which remained on many who had undergone treatment, was more disfiguring than the actual disease itself. The disease was cured by the application of caustic, but flared up again unless the flesh was cauterized through to the bone.

4. There arrived from Egypt, where the disease originated, doctors who specialized solely in this complaint to the exclusion of others. They made a fortune, since it is a well-established fact that Manilius Cornutus, a man of praetorian rank and governor of the province of Aquitania, paid 200,000 sesterces for treatment. On the other hand it has been more usual for new kinds of disease to be of epidemic proportions at their outset. But surely nothing could be

1. *mentum.*

more remarkable than that some diseases should break out suddenly in a particular part of the world and attack particular limbs or particular age groups or even people of a special position. It is as if plagues choose their victims: one attacks children, another adults; in one case the nobility is especially susceptible, in another the lowest rung of society.

Leprosy

7. I have said that leprosy did not occur in Italy before the time of Pompey the Great, and that although the affliction usually begins on the face, as a kind of spot on the end of the nose, the skin soon dries up all over the body and is covered with spots of varying colours. The skin is uneven in places, being sometimes thick, sometimes hard as with a rough scab. Finally it goes black and presses the flesh against the bones, while the toes and fingers swell.

8. Leprosy is endemic in Egypt. When kings contracted the disease it had deadly consequences for the people because the tubs in the baths were prepared with warm human blood for its treatment. Leprosy quickly died out in Italy.

9. It is a strange phenomenon that some diseases disappear in our society, yet others, like colic, persist.

Medicine in former times

10. The remedies that I record were used by people of old – when, in a way, Nature herself made medicine – and they endured a long time. At any rate, the works of Hippocrates[1] are full of references to herbs. He was the first to draw up rules for medicine, with great distinction. So too are the works of Diocles of Carystus, who is next after Hippocrates in date and reputation. Similarly those of Praxagoras and Chrysippus and then of Erisistratus of Ceos. Gradually, however, experience, the most effective teacher of all things – especially in the field of medicine – gave way to words and hot air. For people find it more agreeable to sit listening in lecture theatres than to go out into lonely places searching for different plants at the appropriate season.

---

1. For Hippocrates and other medical writers, see pp. 261 ff.

Asclepiades

12–13. The ancient system of medicine remained firm until, in the time of Pompey the Great, Asclepiades, a teacher of rhetoric who found too little profit in that profession but had sufficient wisdom and intelligence to succeed in spheres other than the forum, suddenly changed careers. Since he had not practised medicine and knew nothing about remedies, he was forced to reduce the whole science to the discovery of causes which he based on guesswork. In particular he recognized five generally applicable principles. These were: fasting from food, abstinence from wine, massage, walking, and outings in the fresh air. Since each individual realized that he could provide these things for himself and since all fell in with his opinions as if that which was very easy was also true, Asclepiades brought almost the whole of mankind round to his point of view just as if he had been sent down from heaven.

14. Asclepiades used to win people over by the empty device of promising the sick now wine, which he gave as the opportunity presented itself, now cold water. According to Marcus Varro, he preferred to be known by the nickname 'cold-water-giver'. He also invented other popular treatments, for example, suspending beds to relieve diseases, or by rocking his patients to induce sleep. He introduced hydrotherapy because of man's extreme obsession with baths, and also a number of other things that it gives satisfaction and pleasure to enumerate; all these enhanced his reputation. His fame was no less when he encountered a stranger's funeral procession, had the body removed from the pyre and saved it.

16–17. The many distressing and crude practices of ancient medicine helped Asclepiades. It was the custom to bury patients under a cover and to promote sweating by every means, to roast the body before a fire or continually to make patients seek sunshine in our rainy city – or, rather, throughout rain-drenched imperial Italy. Then, for the first time, baths with floors suspended over a hypocaust were used, and this innovation was much to people's taste. Moreover, he abolished the extremely painful treatments prescribed for certain diseases, as in the case of quinsy which doctors cured by thrusting an instrument into the throat.

18. Above all, untrustworthy magic practices helped Asclepiades. These were current to such an extent that they were able to destroy confidence in all herbs. It was believed that the plant known as the Ethiopian sage dried up rivers and pools; that by the touch of onothuris all things which were shut, were opened; that if achaemenis were thrown into the ranks of an enemy, the troops would panic and flee; and that the Persian king used to give latace to his envoys so that wherever they went they would lack nothing.

19. What use were these plants when the Cimbri and Teutones shouted their blood-curdling war-cries, or when Lucullus, with a few legions, laid low so many Magian kings? Or why did Roman generals always make supplies a priority in their campaigns? Why did Caesar's troops go hungry at Pharsalus, if an abundance of food could have been supplied through the virtues of a single plant? Would it not have been better for Scipio Aemilianus to have opened the gates of Carthage by means of a plant rather than to have shaken the defences with battering-rams for so many years? Let the Pomptine Marshes be dried up today by the Ethiopian sage, and much land round Rome could be restored to Italy.

20. It would certainly be extraordinary if the ancients' credulity, albeit grounded on solid fact, could go so far, were the human mind not incapable of moderation and were I not about to show, in the appropriate context,[1] that Asclepiades' system went far beyond even that of the Magi. In everything, it is the nature of the human mind to begin with necessity and end in excess.

1. XXIX, 6 ff.

BOOK XXVII

MORE PLANTS USED IN MEDICINE

1. Without doubt, merely dealing with this topic has increased my admiration for men of the past. The greater the number of plants that remain to be described, the more one respects the careful research of older authorities and their generosity in handing down their results. Doubtless the munificence of Nature herself would seem to have been surpassed by men in this respect, had the discoveries been the result of human effort.

2–3. But it is obvious that this munificence was derived from the gods – or at any rate, even when a man actually made the discovery, was divinely inspired – and that the same Mother of all things both produced the plants and revealed them to us. Nor is there any greater miracle of life, if we are willing to admit the truth.

Aconite

4. Who could show sufficient respect for the diligent research of men of former times? It is agreed that aconite takes effect more quickly than all other poisons. If the sexual parts of a female are touched by aconite, death ensues on the same day. Aconite was the poison that Marcus Caelius accused Calpurnius Bestia[1] of using to kill his wives in their sleep. Hence that famous, bitter peroration of the prosecution denouncing the defendant's finger. According to legend, aconite came from the saliva of the dog Cerberus when Hercules dragged him from the underworld, and this is why it grows in the vicinity of Heraclea, in Pontus, where people point out the entrance to Hades used by Hercules.

5. But men have turned this plant to the advantage of their health, having found by experiment that when given in warm wine it counteracts scorpion-stings. Its nature is to kill a human being unless it finds something else in him to destroy.

6. Scorpions touched by aconite grow numb and pale and are

---

1. The defeat of a Roman force of 140,000 men, under Spurius Postumius Albinus, by Jugurtha (110 BC) led to judicial proceedings against the nobles. Calpurnius Bestia was one of those sent into exile.

stupefied, admitting defeat. They find help in white hellebore since its touch dispels the numbness.

7. Men kill panthers with a mere taste of it; otherwise these animals would overrun the regions in which they occur. But it has been shown that panthers are saved from death by tasting human faeces. This remedy was surely found by chance, and on every occasion – even today – it is surely a new discovery, since wild animals possess neither reason nor memory of experience to enable them to pass on results among themselves?

8. Therefore it is Chance, that great deity, which has made most of the discoveries that enrich life.

Wormwood

45. There are several types of wormwood: the Santonic comes from the territory of the Santoni in Gaul; the Pontic from Pontus, where cattle fatten on it and so are found without gall. There is no more outstanding wormwood than the Pontic. Italian wormwood is far more bitter, whereas the pith of the Pontic variety is sweet. Wormwood is singularly honoured in the religious rites of the Roman people; at the Latin Festival there is a race for four-horse chariots on the Capitoline Hill in which the winner drinks wormwood because, as I believe, they think that health is a splendid prize to give.

48. With sil, nard from Gaul and a little vinegar, it removes bile, is a diuretic, soothes the bowels, expels worms from the stomach, and counteracts nausea and flatulence.

52. Wormwood taken on voyages averts sea-sickness; worn under a cummerbund it prevents swelling of the groin. It promotes sleep if inhaled or secretly placed under the patient's head. But amongst clothes it keeps off moths. If mixed with oil, wormwood drives away gnats when the body is rubbed all over with it; similarly, when ignited, its smoke repels gnats. Ink mixed with wormwood protects writings from mice, and, compounded with ointment and rose-oil, dyes the hair black.

Factors affecting the potency of drugs

143–4. This is the sum total of the worthwhile information that I have received or discovered about plants. In conclusion I think it not inappropriate to warn that their efficacy varies with age. Nor is there any doubt that the powers and effectiveness of all roots are lessened if the fruit ripens before they are dug, and it is the same with seeds if the root has been previously tapped for its juice. The properties of all plants are weakened by habitual use and, if they have been taken daily, cease to be beneficial when needed; it is the same with harmful plants. All plants grown in cold regions with north-east winds, and likewise those grown in dry places, have greater potency and effectiveness.

Different diseases are endemic among different races

145. There are quite significant differences between races. For example I have heard that tapeworms and intestinal roundworms infest peoples of Egypt, Arabia, Syria and Cilicia, while these infestations are never found among the Thracians and Phrygians. This, however, is less remarkable than their being found among the Thebans, but not among the Athenians.

BOOK XXVIII

DRUGS OBTAINED FROM MAN; MAGIC AND SUPERSTITION

1. My account of the nature of all things extracted between heaven and earth – that is, leaving aside what is dug out of the ground itself – would have been completed by now, except that by dealing with remedies derived from plants and shrubs, I have digressed to the wider subject of medicines obtained from the very living creatures that are themselves restored to health by them. I have also described plants and the appearances of flowers and many rare things that are difficult to find. So I can hardly remain silent about the benefits to man that are to be found in man himself, and about other kinds of remedies that are found among us, especially since life itself becomes a punishment for those wracked with pain and disease.

2. I shall devote all my attention to this subject, although I realize the risk of causing disgust, as I am determined to be less concerned about courting popularity than about benefiting human life. Indeed I also intend to investigate foreign things and unusual customs. Belief can appeal only to authority, although when collecting data I have tried to draw on views that have gained almost universal acceptance. My criterion is quality of research rather than quantity of material.

Remedies obtained from man himself

4. I shall begin with man's earnest search for help for himself from himself, in which we are immediately confronted with a major problem. People think it extremely effective to suck warm, living blood from a man and, putting their lips to the wound, to drain his very life – but men are not in the habit of putting their mouths to the wounds of wild animals! Our own reaction is one of revulsion when we see wild beasts drinking blood. Others seek to obtain bone-marrow from legs, and brains from infants.

5. Indeed there is considerable frustration if these remedies are not effective. To inspect human entrails is considered sinful. What can it be to eat them?

6. Who was the first, Osthanes,[1] to invent such a practice as yours? For the responsibility lies at your door, you the destroyer of human rights and practitioner of monstrous deeds; you initiated these things, I suppose, to preserve the memory of your existence. Who was the first to envisage eating human limbs one by one? What soothsayer egged him on? What origin could your 'medicine' have had? Who made magic potions more innocent than their remedies?

7. Apollonius writes that scraping sore gums with the tooth of a man who has met a violent death is extremely effective, and Meletus says that a human gall-bladder cures cataracts. Antaeus made pills from the skull of a hanged man to cure bites from rabid dogs.

9. I do not think that life should be so sought after that it is prolonged by every possible means. Whoever you are who holds this view, death is no less certain for you, even though you may have protracted your existence through foul practices or sin. Therefore let every man remember that there is no greater comfort for the soul than this, namely that of all the bounty bestowed on man by Nature, nothing is better than a timely death; and the greatest blessing is that each man can accomplish this for himself.

The power of words and incantations

10. The first of the human-derived remedies raises a question of paramount importance and one that admits of no definite answer. Do words and incantations have any power? If they do, it would be right and fitting for man to be given the credit for this. Individually, all our wisest men reject the belief, while people *en masse* unthinkingly trust in their power at all times. Moreover, the slaughter of victims is thought to be ineffectual without a prayer; and without a prayer the gods are considered not to have been properly consulted.

11. Again, there are formulae for obtaining favourable omens, others for warding off evil, and yet others for commendation. We observe that our chief magistrates pray according to fixed formulae; to prevent any word being omitted or misplaced, someone reads out in advance from a written text; another person is appointed as a monitor to keep watch, and yet another is charged with ensuring silence, while a flautist plays to prevent anything but the prayer being heard.

13. Even today we believe that our Vestal Virgins can root runaway slaves to the spot by a spell, provided that the slaves have not

---

1. A Persian magician (fifth century BC) who wrote works on oriental magic.

left Rome. Yet if we support the view that the gods hear certain prayers or are moved by any form of words, we must answer 'yes' to the whole question of whether words and incantations have power.

14. Lucius Piso,[1] in his *Annals*, I, records that King Tullus Hostilius used the same sacrificial ritual as Numa, which he discovered in Numa's books: he attempted to summon Jupiter down from the sky, but was struck by lightning because he failed to follow the ritual to the letter. Many claim that the destinies and portents associated with great events are changed by words.

15. Workmen, excavating the foundations for a shrine on the Tarpeian Hill, discovered a human head. As a result, envoys were sent to Olenus of Cales, the most distinguished soothsayer in Etruria, who saw that this signified fame and success and tried by cross-examination to transfer this good fortune to his own people. First of all he traced the outline of a temple with his stick on the ground in front of him and asked: 'Is this then what you mean, Romans? Will the Temple of Jupiter Optimus Maximus be here where we found the head?' The *Annals* insist that the destiny of Rome would have passed to Etruria, if the Roman envoys had not been forewarned by Olenus' son and replied: 'Assuredly not, but we are saying that the head was discovered in Rome!'

18. Again, in the laws of the Twelve Tables, there are these words: 'Whoever shall have bewitched the crops', and elsewhere, 'Whoever shall have cast an evil spell'. Verrius cites reliable authorities to show that it was the practice before laying siege to a town for Roman priests to call forth the place's patron deity and promise him the same or more lavish worship under the Romans. This sacred office remains the responsibility of the pontiffs and it is agreed that this is why the patron deity of Rome has been kept secret, for fear that any enemy should act in a similar manner.

19. Indeed everyone fears being jinxed. This is what makes us break the shells of eggs or snails immediately we have eaten them, or pierce them with the spoon we have just used. Hence Theocritus among the Greeks, and Catullus and more recently Virgil among our authors have copied love-spells. Hence also walls are covered with prayers to ward off fires.

21. Homer says that Odysseus checked the bleeding from his wounded thigh by a magic spell. Theophrastus claims there is a magic formula to cure sciatica. Cato published one formula for

---

1. Lucius Piso (consul 133 BC) was an opponent of the Gracchi.

setting dislocated limbs and Marcus Varro another for gout. The dictator Caesar, after a dangerous accident to his carriage, is said always, as soon as he was seated, to have repeated a prayer three times for a safe journey – a practice we know most people follow today.

Superstitions

22. I would like to supplement this part of my account by referring to individuals' personal experience. Why do we wish one another a 'Happy and Prosperous New Year' on the first day of the year? Why do we choose people with lucky names to lead the victims on days of general purification? Why do we meet the evil eye with a special attitude of prayer, some calling on the Greek Nemesis? – which is why there is an image of the goddess on the Capitol in Rome, although she does not have a name in Latin.

23. Why, at any mention of the dead, do we protest that we do not attack their memory? Why do we believe that in everything odd numbers are more powerful? Why do we say 'Bless you' when people sneeze? Even Tiberius Caesar, the most gloomy of men, is said to have uttered this when in his carriage; some think that it is more pleasing to the gods if one adds the person's name to the wish.

24. Moreover, it is accepted that absent people can tell that others are talking about them by a tingling of the ears. Attalus assures us that if we say 'Two' when we see a scorpion, it stops in its tracks and does not sting. Talking of scorpions reminds me that in Africa no one makes a decision without first saying 'Africa';[1] among all other peoples a man calls on the gods first for their approval.

26. If fires are mentioned at a banquet, we counter the omen by pouring water under the table. It is considered very unlucky to sweep the floor when a guest is leaving a banquet, or for a table or a serving-trolley to be removed while a guest is still drinking.

27. It has also been observed that guests at a banquet suddenly fall silent all at once only when the number of those present is even and that this spells danger to the reputation of each of them.

28. To cut nails on market days at Rome in silence, beginning with the first finger, is a superstition shared by many. Similarly, to cut hair on the seventeenth and twenty-ninth of the month is believed to prevent hair-loss and headaches.

1. In the time of Hadrian (AD 138–161) Africa was personified as a woman with a scorpion in her hand or on her head.

29. Marcus Servilius Nonianus, a leading citizen of Rome who not so long ago was afraid of conjunctivitis, was in the habit, before referring to the disease himself or letting anyone else speak to him about it, of tying a sheet of paper with thread; on it were written the two Greek letters P and A. Similarly, Mucianus, who had held three consulships, fastened a white linen bag with a live fly inside.

30. The whole bodies of certain men confer benefits, for example the bodies of those families that frighten snakes. These, by a touch or by sucking out the venom, relieve victims who have been bitten. In this category are the Psylli, the Marsi and the Ophiogenes, as they are called, who live in Cyprus. An envoy from this family, called Evagon, was thrown into a large open container full of snakes by the consuls as an experiment; to their astonishment the snakes licked him all over.

Magic powers attributed to human saliva

36. I have pointed out that the saliva of someone fasting is the best protection against snakes, but daily life may teach us other effective uses. A surprising fact, but one easily tested, is the following: if one regrets inflicting a blow, whether with the hand or by a weapon, and immediately spits into the palm of the hand responsible, the resentment of the person struck is lessened.

38. It acts as a charm if a man spits on his urine; similarly if he spits in his right shoe before putting it on or when passing a place where he has encountered some danger.

Some remedies suggested by the Magi

47. The Magi tell the following lies: when a whetstone on which iron tools have often been sharpened is placed without his knowledge under the pillows of a man dying from poisoning, it causes him to reveal what has been administered, when and where, but not the name of the poisoner. It is universally accepted that if a man struck by lightning is turned over on to his wounded side, he immediately speaks.

48. To prevent a wound from being painful the Magi bid people wear as an amulet a nail attached by thread or by anything else that has been trodden on. Warts can be removed by those who, after the twentieth of the month, lie on their backs on a path, gazing upwards with their hands stretched over their heads, and rub the wart with whatever they have grasped.

49. People say that if one pares a corn when a star is falling, it is very quickly cured, and one can relieve a pain in the head by applying a poultice from vinegar poured over door-hinges. Similarly, a rope used by someone who has hanged himself relieves headaches if tied round the temples. A fish-bone lodged in the throat falls out if the feet are immersed in cold water. If it is another kind of bone, however, bits of bone from the same container should be applied to the head. If a piece of bread sticks, pieces from the same loaf should be placed in both ears.

Remedies dependant on will-power

53–4. It would not be proper for me to omit remedies that depend on man's will-power. Fasting from all food and drink, sometimes abstaining from baths, when health demands one of these courses of action, are all considered among the most efficacious remedies. Among other remedies are physical exertion, voice exercises, anointing and massage, if done skilfully: for violent massage hardens the body, while gentle massage softens it; and too much massage reduces the body, while a moderate amount builds it up. Of special benefit are walking, carriage-rides of various kinds, and horse-riding, which is good for the stomach and hips. A sea voyage is good for tuberculosis; for chronic diseases, a change of locality and self-treatment by sleep, bed rest and occasional emetics are all beneficial. Lying on the back is good for the eyes, face and coughs; and lying on either side, for catarrh.

55. Sunshine is the best of the self-administered remedies, as is the vigorous use of towels and strigils.

Sexual intercourse

58. Democritus dismisses love-making as being merely the means by which one human springs from another, and says that the less it is indulged in, the better. Athletes when sluggish are revitalized by love-making, and the voice is restored from being gruff and husky. Sexual intercourse cures pain in the lower regions, impaired vision, unsoundness of mind and depression.

Remedies from elephants and lions

88. Elephant's blood, especially that of the male, cures all excess mucus which is referred to as catarrh. Shavings of ivory and Attic

honey are said to remove dark spots on the face and ivory-dust removes whitlows.

89. Lion fat and rose-oil keep the skin free from blemishes and protect the complexion. The falsehoods of the Magi claim that those rubbed with this fat enjoy more popularity with nations and kings, especially when the fat is from between the brows – where there cannot be any fat!

The Magi discredited

104. The tricks of the Magi are ineffective, since they are not able to call down the gods, or speak with them, whether they try with lamp, bowl, water, glove, or any other thing.

Milk and butter

133. Butter is produced from milk; it is the choicest food among barbarian tribes and one that distinguishes the well-to-do from the lower class. The milk comes mostly from cows – hence the name 'cow's cheese'; the richest is that made from sheep's milk, and butter is similarly made from goat's milk. In winter the milk is warmed. In summer the butter is extracted by vigorously shaking the milk in tall containers; a small hole just under the mouth admits air, and apart from that the containers are completely sealed. A little water is added to turn the milk sour.

134. The part that curdles most, floating on the top, is butter, a fatty substance. The stronger the taste the more the butter is esteemed. When matured it is used in several recipes. It is naturally astringent, emollient, fattening and cleansing.

Fat, suet, marrow, gall and blood

135. Of remedies common to animals, the next most highly reputed is fat, especially pig's fat, which was sacred to men even in olden times. At any rate, even today brides ritually touch the doorposts with it on entering their homes. Lard is matured in two ways: either with salt or on its own; it is much more beneficial when matured.

147. Fresh blood from a bull is considered poisonous except at Aegira. There the priestess of Earth, when about to prophesy, drinks the bull's blood before she goes down into the caves.

148. Drusus, while tribune of the people, is reported to have

drunk goat's blood because he wished, by his paleness, to accuse his enemy Quintus Caepio of having poisoned him and so to arouse hatred against him. So great is the power of he-goat's blood that its use in tempering gives a finer edge than any other method, and a rough surface is smoothed more thoroughly by it than by a file.

Poisons, dandruff and baldness

158. Those who have swallowed quicksilver find a remedy in lard. Poisons – especially henbane, mistletoe, hemlock, sea-hare, and other substances – are neutralized by drinking ass's milk.

163. Bear-grease, mixed with ladanum and maidenhair, prevents baldness and cures mange and thin eyebrows, if compounded with lamp-black from lamp-wicks and the soot that collects in the curved nozzles. Mixed with wine, bear-grease is a cure for dandruff.

Beauty treatment

183. Women think that ass's milk removes wrinkles from the face and makes the skin white and soft, and some are known to bathe their cheeks in it seven times a day and keep scrupulously to this routine. Poppaea, wife of the Emperor Nero, began this custom, even filling her bath-tubs with the milk; for this purpose she was always accompanied by a string of asses.

Soporifics and aphrodisiacs

260. Cato thought that eating hare induced sleep, and the man in the street believes that it gives a person charm for nine days – a frivolous pun[1] – but there must be a reason for such a strong belief. According to the Magi the gall of a nanny-goat brings on sleep if it is placed on the eyes or put under the pillow.

262. As an aphrodisiac, Salpe bids one plunge an ass's penis seven times in hot oil and then rub the appropriate part with this. Dalion prescribes the ash from it to be taken in drink.

Some strange beliefs about animals

263. If a horse casts his shoe, as often happens, and someone retrieves it and puts it aside, it is a cure for hiccoughs for those who remember

1. *lepus*: 'hare'; *lepos*: 'charm'.

where they have put it! Horses burst if, when being ridden, they follow the tracks of wolves. Wolves will not enter a field, if a wolf has been caught, his legs broken, a knife driven into his body, the blood sprinkled a little at a time round the edges of the field and the body itself buried in the place from which it was first dragged.

BOOK XXIX

MEDICINE, DOCTORS AND MEDICAL PRACTICE

The early history of medicine

1. The great number of remedies already discussed, or remaining to be discussed, and their nature, compel me to enlarge on the art of medicine itself. I am well aware, of course, that to date no one has written about this subject in Latin,[1] and that one can make only tentative suggestions on new fields of study especially where these are devoid of intrinsic appeal and where their exposition is fraught with such difficulty.

2. However, since this question is likely to occur to all those who are familiar with medicine – namely, why available and efficacious remedies are no longer prescribed in medical practice – one is instantly reminded of the malign influence of fashion on medicine, more than on any other science. Even nowadays it is subject to fads, although no science is actually more profitable.

3. Medicine bestowed upon its earliest practitioners a seat among the gods and a place in heaven. What is more, even today, medical advice is sought from oracles in a variety of ways. Medicine increased its standing through an offence against the gods, for, according to legend, Aesculapius[2] was struck by lightning for bringing Tyndareus[3] back to life. And medicine continued to advertise the fact that others had been restored to life with its help. Medicine was famous in Trojan times, when its reputation was on a surer footing, but only to the extent of providing treatment for wounds.

1. Pliny disregards Scribonius Largus' *Compositiones Medicamentorum* and the medical writings of Aulus Cornelius Celsus.

2. Aesculapius was a hero and the god of healing. On his cult see E. D. Philips, *Greek Medicine* (London, 1978), Appendix, pp. 197–201. He had two sons who fought at Troy and also acted as 'doctors' – namely, Machaon and Podalirius (see Homer, *Iliad*, II, 73).

3. The son of Perieres and Gorgophone, husband of Leda. One night Leda had intercourse with both Zeus and Tyndareus: the result was the birth of Pollux and Helen (by Zeus) and Castor and Clytemnestra (by Tyndareus). Tyndareus invited Menelaus to come to Sparta and surrendered his kingdom to him.

Hippocrates and his successors

4. The subsequent history of medicine, strange to relate, is cloaked in total darkness until the Peloponnesian War, when it was brought back to light once more by Hippocrates; he was born on Cos, an especially famous and powerful island dedicated to Aesculapius. It had been the custom for patients cured of their diseases to write on the walls of the Temple of Aesculapius details of the help they had been given, so that afterwards other people might benefit from similar remedies. Hippocrates, so the story goes, recorded these writings and, as Varro[1] believes, after the temple had been burnt down, founded the practice we refer to as bedside examination. Afterwards there was no limit to the profit to be derived from medicine, since Prodicus, a native of Selymbria and one of Hippocrates' students, founded the art of healing with ointments and discovered a source of income for apothecaries and unqualified practitioners.

5. Chrysippus[2] reconstructed the scientific principles of medicine by extensive argument, and, following Chrysippus, Erasistratus[3] (a grandson of Aristotle) made very significant changes. As a first illustration of prizes obtained from the practice of medicine – Erasistratus was given 100 talents by King Ptolemy I for curing his father Antiochus.

6. Another group of doctors set up practice in Sicily, calling themselves 'empiricists' because of their reliance on experience. There Acron from Agrigentum enjoyed the backing of Empedocles,[4] a natural scientist. Such schools disagreed amongst themselves and were all rejected by Herophilus, who described the venous pulse in terms of metrical feet, relating it to the age of each patient. The empiricists were disbanded because members were obliged to have high academic qualifications. As I have recounted, the sect that Asclepiades subsequently founded also came to adopt a different

1. Marcus Terentius Varro (116–27 BC) was a prolific writer on grammar, agriculture and other subjects.
2. Chrysippus of Cnidus (c. 270 BC), a physician associated with the tradition of the Cnidian centre of medical instruction, wrote a treatise on vegetables, noting especially the health-giving properties of cabbage!
3. Erasistratus of Ceos (first half of the third century BC) was a physician and researcher who founded a medical school at Alexandria.
4. Empedocles of Chalcedon (second century BC) was pupil of Praxagoras and peripatetic before studying medicine at Cos. He was the founder of sphygmology, the systematic study of the pulse.

approach. Themison,[1] a pupil of Asclepiades and at first one of his followers, likewise altered his scientific views when he was older. Antonius Musa,[2] another student of Asclepiades, introduced even further changes with the backing of the late Emperor Augustus, whom he had saved, when critically ill, by giving him the opposite treatment to what he was receiving.

Some doctors and their treatments

7–8. I pass over many famous medical men – doctors like Cassius, Calpetanus, Arruntius and Rubrius. Under the Empire their annual emolument was 250,000 sesterces. Quintus Stertinius[3] denounced the emperors because, although he had been satisfied with 500,000 sesterces a year, by counting up the houses in the city he showed that he had in fact earned 600,000 sesterces; an equal emolument was showered on his brother by the Emperor Claudius, and his estate, although drained by the cost of works for the beautification of Neapolis, provided his heir with 30 million sesterces, an amount equalled during the same period only by Arruntius. Then Vettius Valens came on the scene, well known because of his affair with Messalina, wife of the Emperor Claudius, and equally renowned for his eloquence. He gained followers and power and founded a new sect.

9. In the principate of Nero, all that generation rushed to join Thessalus,[4] who was then undermining all scientific doctrines and haranguing doctors of every age with a sort of mad frenzy. The nature of Thessalus' intelligence and ability can be easily judged from a single piece of evidence: on his monument on the Appian Way he wrote that he was 'the conqueror of doctors'. No actor or charioteer had a larger escort than the writer Crinas[5] from Massilia

1. Themison of Laodicea (*fl.* mid first century BC) was a pupil of Asclepiades, and reputed to have been the founder of the 'Methodists' school of medicine. He taught the use of leeches for bleeding and experimented with drugs.

2. Antonius Musa, a physician of Augustus, was the first to introduce hydropathy at Rome. He also wrote a treatise, in several books, on the properties of drugs.

3. Quintus Stertinius (a contemporary of Augustus) translated Stoic tenets into Latin verse.

4. Thessalus of Cos (*fl. c.* 420 BC) was the more famous of Hippocrates' two sons. Galen considered him to be the author of Books VI and VII and part-author of Books II and V of the Hippocratic work *Epidemics*.

5. An annalist who treated the history of Rome from its beginnings to the Second Punic War, writing at some time before 149 BC. His interests included etymology, religious and social matters and antiquities. His work reflects the influence of Cato.

when he was out and about. A rather cautious and superstitious man, he added another dimension to medical practice. Keeping a close eye on the time factor, he fed his patients according to the movements of the stars, in line with astrological almanacs. Crinas outstripped even Thessalus in authority, and recently left 10 million sesterces having spent not much less than that sum on the walls and fortifications of his native city.

10. These men ruled the roost when suddenly Charmis, also from Massilia, descended on us and condemned not only former doctors but also hot baths, and persuaded people to take cold baths even during the winter cold spells. He immersed his patients in tanks, and we were treated to the sight of old men of consular rank making themselves stiff with cold just to show off. We have Annaeus Seneca's confirmation of this.

11. There is no doubt that all these doctors sought fame by means of some innovation, and irresponsibility trafficked with our lives. This accounts for those wretched arguments at the sick-bed when no two doctors gave the same opinion for fear that a colleague's diagnosis might appear to carry more weight. It also accounts for the sad inscription occurring on some monuments which says: 'A gang of doctors killed me.' The art of medicine changes daily and is constantly given a new look: we are swept along by the empty words of Greek intellectuals. It is well known that those who are successful speakers have the power of life and death over us, just as if thousands of people do not exist without doctors or medicine. The Romans did so for more than 600 years, although they are not slow to accept advances – and indeed were even avid for medicine until they put it to the test and rejected it!

12. It is now appropriate to review the distinguished achievements of men of former times in this profession. Cassius Hemina,[1] one of the oldest authorities, states that the first doctor to come to Rome from the Peloponnese was Archagathus, son of Lysianas, in AUC 535[2] in the consulship of Lucius Aemilius and Marcus Livius. He also says that Archagathus was accorded the rights of a Roman citizen and was given a surgery at the crossroads of Acilius, bought for that purpose at public expense.

13. They say that he was a surgeon and that on his arrival he was at first very welcome, but soon, as a result of unlimited recourse to surgery and cautery, he acquired the nickname 'butcher', and his

---

1. Said to have been the first person to practise medicine scientifically at Rome.
2. 219 BC.

profession, together with all doctors, became loathed. This can be seen most clearly from the evidence of Marcus Cato, whose triumphant term of office as censor added little to his influence. More important were his personal qualities. I intend, therefore, to quote him verbatim.

Cato's views on doctors

14. I shall speak about Greek doctors at the appropriate point, Marcus, my son. I shall tell you what I discovered at Athens and convince you that it is good to skim through their literature, but not to study it in depth. They are a worthless and intractable lot – in this area accept me as a prophet. For when the Greeks give us their literature it will undermine our whole way of life, and even more so if they send us their doctors. They have sworn among themselves to kill all foreigners with their medicine, but they do so for a fee, to win credit and destroy us easily. They also persistently call us foreigners and treat us with less consideration than others and belittle us as country bumpkins. I have forbidden you to have anything to do with doctors!

15. Cato died in AUC 605[1] aged eighty-five, so he undoubtedly had ample time for public life and, quite apart from his terms in office, lived long enough to gain experience. What conclusions should we draw then?

Pliny's assessment of medicine

Are we to believe that Cato condemned a highly beneficial science? Most certainly not! For he appends details of the treatment by which he prolonged his own and his wife's life to old age by the very remedies that we are now discussing, and he professed to have a notebook of remedies by which he cured his son, his slaves and household: these remedies I have arranged according to their different uses.

16. Previous generations rejected not medicine but rather the medical profession, mainly because they refused to ransom their lives with doctors' fees. This is why they are said to have built the Temple of Aesculapius outside the city, even when he was acknowledged as a god in Rome, and at another time they raised a temple to him on an island. Long after Cato's time they expelled the Greeks from Italy and took care to make specific mention of doctors

1. 149 BC.

– I shall add my voice to the praises of their foresight. Of the Greek sciences it is only medicine that the Romans have not followed, thanks to their good sense.

17. Very few Roman citizens have practised medicine, although it is a highly profitable profession: the few who have, have not hesitated to adopt Greek practices. Indeed no authority has any weight unless he researches and publishes on medical matters in Greek – even in the eyes of laymen and those with no knowledge of Greek. Where their health is concerned, people have less faith if they understand what is written. So, here indeed is a phenomenon unique among the sciences; anyone claiming to be a doctor is immediately trusted, although in no other profession is an untruth more dangerous.

18. But this we choose to ignore, so enticing is the pleasure each one of us derives from wishful thinking. Moreover, culpable ignorance is not punishable by law; there is no precedent for retribution. Doctors learn by exposing us to risks, and conduct experiments at the expense of our lives. Only a doctor can kill a man with impunity. Indeed the blame is transferred to the deceased, who is criticized for want of moderation, and it is thus the dead who are censured. Custom demands that juries should be vetted by the emperors' censorial scrutiny; the walls of our homes are no barrier to this examination. That judgement may be passed over a trifling sum of money a man is summoned from Gades and the Pillars of Hercules; a sentence of exile cannot be given unless forty-five men have been empanelled to vote on this.

19. But as for the juryman himself, what kind of men conspire to kill him without a second thought? All this is deserved so long as we all choose to remain ignorant of what is good for our health. We walk with another's feet, recognize people with another's eyes, make our greeting thanks to another's memory and live thanks to another's care. Nature's precious gifts and the fundamentals of life have disappeared. We have nothing left but our luxuries.

20. I will not leave Cato exposed to the hatred of such a grasping profession, or the Senate that shared his opinion, but failed (contrary to what one might have expected) to seize any opportunities to bring accusations against medicine. For there has been no richer source of poisonings or of conspiracies to benefit from wills – not to mention adultery committed in emperors' palaces, as in the case of Eudemus with Livia, the wife of Drusus Caesar, and likewise Valens with a certain royal personage to whom his name has been linked.

An indictment of doctors and their practices

21. The blame lies not with medicine but with the rest of us. Cato, as I see it, was no more anxious on Rome's behalf in this connection than he was about the presence of royal ladies in Rome. Let us not indict the greed of doctors, their financial rapacity in dealing with patients whose life is in the balance, the fees charged for curing ailments, the payments on account that lead to death, and the secret instructions to ease aside a cataract rather than remove it completely. In the final analysis, the large number of sharks using medical practice to prey on people seems to be the best thing about the situation, in that competition between rivals – rather than any sense of decency – reduces fees.

22. It is common knowledge that a sick man, a provincial, was handed over as a patient to Alcon who specializes in the treatment of wounds, and that Charmis received an introduction fee of 200,000 sesterces. Charmis was convicted and fined one million sesterces by the Emperor Claudius. As an exile in Gaul, and subsequently on his return, Charmis made good this sum within a few years. The blame for such incidents should be laid at the door of individuals.

23. I will not cast blame on even the dregs of that gang of doctors or on its ignorance, or on the doctors' lack of restraint and their newly fashionable use of hot water in illnesses, and their regime of fasting for sick patients, who then, when they faint, are fed too frequently. Nor will I cast blame on the thousand ways in which doctors backtrack on their treatments or their requests to the kitchen, or their mixtures and ointments. Indeed none of life's attractions has escaped their attentions.

Bizarre remedies

24. I should like to believe that our forefathers frowned on foreign imports and price-fixing – of course not even Cato foresaw all this when he condemned medicine. A complicated mixture called *theriace* is compounded of innumerable ingredients, although Nature has provided many remedies each of which on its own would suffice. The 'Mithridatic antidote' contains fifty-four ingredients, no two substances having the same weight. The prescription for some is a sixtieth of a denarius – which god, in the name of Truth, is responsible for this?

25. No human intellect could have been so subtle! It is sheer medical exhibitionism and an outrageous boast on the part of science

– not even the doctors themselves know these things. I have learned that through ignorance of names, red lead is commonly added to medicines instead of Indian cinnabar. This is a poison, as I shall make clear when dealing with pigments.

26. These matters concern the health of individuals; but it was what Cato feared and foresaw – though much less deleterious and of less account, as the leading lights of medicine themselves admit – that destroyed the moral fibre of the Empire. I refer to treatments we undergo when in good health, namely, wrestlers' ointment, although this was invented to promote well-being; hot baths, which they have persuaded us cook the food inside our bodies, but which leaves everyone enfeebled, with the most susceptible being carried out feet first; the potions taken during fasts, followed by vomitings and yet larger potions; the effeminate removal of hair through the use of resins, and, likewise, women's pudenda exposed for all to see thanks to depilatories.

27. Assuredly, there is no greater reason for the decay of morals than medicine; and in this we have daily proof that Cato was a prophet and oracle when he said that one needed only to skim through the literature of Greek intellectuals rather than make a detailed study of it.

28. On behalf of the Senate and the 600 years of the Roman Republic, I feel I must speak out like this against the profession of medicine. In moments of extreme crisis, good men put themselves in the hands of the worst. At the same time I must counter the misguided notions of those laymen who consider nothing beneficial unless it costs the earth!

BOOK XXX
MAGIC

The origins of magic

1. Previously in my work I have often shown the lies of the Magi for what they are, whenever the argument or occasion demanded, and I shall continue to expose their untruths even now. It is hardly surprising that the influence of magic has been very great since, alone of the arts, it embraces three others that exert the greatest power over men's minds, and these it has made subject to itself alone.

2. No one will doubt that the origin of magic lay in medicine, and that it crept in surreptitiously under the pretence of furthering health, as if it were a loftier and holier form of the healing art. In this way it acquired the enticing and welcome promises of religion which even now remains very much a closed book to the human race; and with this success it also took control of astrology, because there is no one who is not eager to learn his destiny or who does not believe that the most accurate method of so doing is to observe the sky. So magic, with its triple bond on men's emotions, has reached such a peak that even today it has power over a great part of the world and in the East commands kings of kings.

3. Undoubtedly magic began in Persia with Zoroaster, as authorities are agreed. But there is insufficient agreement about whether he was the only man by that name, or whether there was another and later Zoroaster. Eudoxus, who wished magic to be recognized as the most noble and useful of the schools of philosophy, asserts that this Zoroaster lived 6,000 years before the death of Plato, and Aristotle confirms this.

4. Hermippus, who wrote most painstakingly about the whole art of magic and interpreted 2 million verses written by Zoroaster, also added lists of contents to his books and handed down the name of Agonaces as the teacher who had instructed him, placing him 5,000 years before the Trojan War. What is particularly surprising is that the tradition and craft should have endured for so long; no original writings survive, nor are they preserved by any well-known or continuous line of subsequent authorities.

5–6. For few people know anything by reputation of those who survive only in name and lack any memorials, as, for example, Apusorus and Zaratus of Media, Marmarus and Arabantiphocus of Babylon, or Tarmoendas of Assyria. The most surprising thing, however, is that there is absolutely no reference to magic in the *Iliad*, although so much of the *Odyssey* is taken up with magic that it forms a major theme – unless people put another interpretation on the story of Proteus,[1] the songs of the Sirens,[2] Circe and the summoning of the dead from Hades. Nor has anyone in later times said how magic came to Telmessus, a city in which superstition is rife, or when it was taken up by the old women of Thessaly who for a long time were a byword with us, although magic was unfamiliar to the Thessalians of the time of the Trojan War, who were content with Chiron's[3] medicine and with Mars as the only thunderer.

7. I am indeed surprised that this reputation stuck with Achilles' people for so long that Menander, who was endowed with an unrivalled sensitivity and taste for literature, named a comedy *Thessala*, which has as its theme the tricks of women for drawing down the moon. I would have said that Orpheus was the first to import magic to his native land from abroad and that superstition evolved from medicine, if the whole of Thrace had not been free of magic.

8. The first person, at any rate, as far as I can ascertain, to write a book on magic – a book that still survives – was Osthanes, who accompanied Xerxes on his expedition to Greece and nurtured the seeds, as it were, of this monstrous art, spreading the disease to all corners of the world on his way. However, some very thorough researchers place another Zoroaster, who came from Proconnesus, somewhat before Osthanes' time. One thing is certain. Osthanes was chiefly responsible for stirring up among the Greeks not merely an appetite but a mad obsession for this art. I observe that from times past the highest literary fame and distinction have almost invariably been sought through magic.

9–10. Pythagoras, Empedocles, Democritus and Plato went

1. A sea-god who had the power of changing shape at will.
2. Sea-nymphs whose songs could enchant any who heard them. Odysseus escaped by putting wax in his ears and tying himself to the mast of his ship.
3. Chiron was the son of Cronus and Philyra, the wisest of all the Centaurs. Instructed by Apollo and Artemis, he was renowned for his skill in medicine, music, prophecy, hunting and gymnastics. Many heroes of the ancient world, including Jason, Castor and Pollux, Peleus and Achilles, are described as having been his pupils in these arts.

overseas to learn magic, going – to put it more accurately – into exile rather than on a journey. On their return they taught this art and considered it among their special secrets. Democritus popularized Apollobex the Copt and Dardanus the Phoenician, entering the latter's tomb to seek out his works and basing his own on their doctrines. It is a marvel without equal in life that these works were accepted by anyone and transmitted by memory, for they are so lacking in credibility and propriety that those who admire Democritus' other writings deny his authorship in this case. To no purpose, however, for there is general agreement that it was Democritus in particular who beguiled men's minds with the charm of magic. Another remarkable consideration is that both arts, medicine and magic, flourished side by side. For Hippocrates expounded the principles of medicine at the same period as Democritus explained magic, that is, about the time of the Peloponnesian War in Greece, which began in AUC 300.[1]

11. There is another type of magic derived from Moses, Iannes, Lotapes[2] and the Jews, but dating from many thousands of years after Zoroaster: Cypriote magic is much more recent. At the time of Alexander the Great another Osthanes made a not inconsiderable addition to the influence of the profession. He was famous because he accompanied Alexander and must have travelled all over the world.

12. The Twelve Tables still retain traces of magic among Italian tribes. It was only in AUC 657[3] that the Senate passed a decree forbidding human sacrifice.

13. Magic continued to be practised in the two Gallic provinces within living memory. The principate of Tiberius saw the removal of the Druids[4] and of the whole pack of soothsayers and doctors. But these remarks are of little interest when one considers that magic has crossed the ocean and reached Nature's empty wastes. Today even Britain, in awe, practises magic with such impressive rites that one might think that she had given the Persians the art of magic. So much agreement is there worldwide on the subject of magic, although nations are otherwise at loggerheads or are ignorant

1. 454 BC.
2. Pliny here means Iotape, that is, Yahweh.
3. 97 BC.
4. The ruling classes in Gaul were the Druids and the nobles. Druidism taught the transmigration of souls and claimed to possess a secret lore that was jealously guarded from the profane world. It was the unifying principle that gave the Gauls a sense of community.

of one another's existence. An incalculable debt is owed to the Romans who destroyed these monstrous practices, in which human sacrifice was considered an act pleasing to the gods and eating the victim was thought to be beneficial to one's health.

14. As Osthanes stated, there are many kinds of magic: for divination he uses water, spheres, air, stars, lamps, basins, axes and many other means; moreover, he communicates with ghosts and those in the lower world. In our generation the Emperor Nero exposed all these practices as fraudulent.

The Magi[1]

16. The Magi have certain subterfuges: for example, the gods neither obey nor appear to those with freckles. Was this perhaps why they stood in Nero's way? Tiridates the Magus had come to Nero bringing captives for the emperor's Armenian triumph over himself and, to this end, put a heavy burden on the provinces.

17. He refused to travel by sea, for the Magi consider it sinful to spit into the sea or defile its nature by any other human function. He brought the Magi with him and initiated Nero into their magic banquets. Yet although Tiridates had given Nero a kingdom, he was unable to teach him the art of magic. This should be sufficient proof that magic is execrable, achieves nothing and is pointless.

18. One might well ask what lies the Magi of old perpetrated. In my youth, I met Apion the grammarian, who informed me that the herb *cynocephalia*, known in Egypt as *osiritis*, was a source of divination and a protection against all black magic, but that if anyone completely uprooted it, he would immediately die. He added that he had summoned ghosts to inquire from Homer his native land and the name of his parents, but did not dare to reveal the answers he had allegedly been given.

1. The Magi were members of an ancient Persian clan specializing in cult activities. It appears that they constituted a priesthood serving religions during the Seleucid, Parthian and Sassanian periods. Whether they employed magic is open to question. Herodotus (VII, 191), among others, mentions their use of incantations, but Aristotle (Fragment 36) denies that they employed any form of magic.

BOOK XXXI
WATER

Water, sea and creatures whose habitat is water

1. Nature employs her tireless force nowhere with greater power than in the waves, the ocean swell, the tides that ebb and flow, and – if we admit the truth – in the rapid river currents, since this element commands all others.

2. Water engulfs lands, quenches flames, climbs aloft and lays claim even to the sky, and by a covering of clouds chokes the life-giving spirit that forces out thunderbolts, as the world wages war with itself. What could be more amazing than water standing in the sky? But as though it were a mere trifle to reach such a great height, the water sucks up with itself shoals of fish and often stones as well, carrying more than its own weight aloft.

3. This same water falls back to earth and, if one considers the matter, becomes a source of all things that grow out of the earth – thanks to the miracle of Nature. In order that crops may grow and trees and shrubs live, water travels up to the sky and from there brings the breath of life to plants. Consequently we are bound to admit that all earth's powers are due to the gift of water. I shall therefore start by giving examples of the power of water; for what man could enumerate them all?

Different kinds of water

4–5. Nowhere are healing waters in more plentiful supply than in the Bay of Baiae, nor does any other water afford a greater variety of relief. Some water can have the properties of sulphur or alum or soda or bitumen; some waters are a mixture of acid and alkali. Some water is beneficial because of its steam, whose power is so great that it heats baths and even makes cold water boil in the ground.

Unusual properties of water

21. In this sphere Nature is at her most marvellous. Ctesias records

that in India there is a stretch of stagnant water called Silas in which nothing floats but everything sinks to the bottom. Coelius says that in Lake Avernus even leaves sink, and Varro observes that birds which fly towards it die. The reverse is true with regard to the African Lake Apuscidamus.

23. The River Alcas in Bithynia flows by Bryazus – this is the name of a god and of his temple. Perjurers are said to be unable to endure the waters of this river, as they burn like a flame. In Cantabria the springs of the River Tamaris are thought to be prophetic. They are three in number, 8 feet apart, and unite in one channel to form a huge river.

24. Each spring dries up separately for twelve or occasionally twenty days, and there is not the slightest trace of water although there is a spring near by with an abundant and steady flow. If those who wish to consult these rivers see them stagnant, it is a bad omen, as the governor Larcius Licinius recently discovered after seven days. In Judaea there is a river that dries up every Sabbath.

Poisonous waters

25. But some wondrous things are deadly. Ctesias writes that there is a spring in Armenia inhabited by dark fish and that if these are eaten they cause instant death.

26–7. In Arcadia, near the River Pheneus, there flows from the rocks a stream called Styx which proves instantly fatal to life, but Theophrastus tells us that it contains small fish which are equally deadly; there is no other kind of poisonous spring like this. Theopompus states that near Cychri, in Thrace, there are waters that cause death, and Lycus says that at Leontini there is water that has fatal consequences two days after one has drunk it.

Waters with petrifying properties

29. At Perpenna a spring turns to stone any land it irrigates, and the thermal waters of Aedepsus, in Euboea, have the same effect. Whatever rocks the stream reaches increase in size. At Eurymenae garlands thrown into a spring turn to stone. Bricks thrown into the river at Colossae are found to be of stone when retrieved. At the mine on Scyros all the trees washed by the river are petrified, branches and all.

30. At Mieza, in Macedonia, drops of water form stalactites hanging from arched roofs, while in a cave at Corinth they become

stalagmites after falling. In certain caves the water forms stalactites and stalagmites that join to make pillars, as at Phausia on the peninsula facing Rhodes; these pillars are of different colours.

Waters with beneficial properties

31. Doctors investigate which kinds of waters are beneficial. They rightly condemn stagnant, sluggish waters, and regard running water as more beneficial because it is rendered fine and healthy by the agitation of the current. I am, therefore, surprised that some doctors very highly recommend water from storage-tanks.

37. It is agreed that rain-water ought to be most like air. In the whole world only one spring is said to have a pleasant smell and that is the one at Chabura, in Mesopotamia; the legendary explanation is that Juno bathed in it. Wholesome waters should also be without taste or smell.

38. Some judge the wholesomeness of water by weighing it in a balance. This is pointless, for it is exceedingly rare for one water to be lighter than another. A more reliable and subtle indicator – all things being equal – is that better water warms up and cools down more quickly.

41. The top prize for cool, wholesome water goes, according to the town-crier of Rome, to the Aqua Marcia[1] – a gift (among others) of the gods to our city. It rises at the far end of the Paelignian Mountains, crosses the territory of the Marsi and the Fucine Lake, making directly for Rome. Then it sinks into caves at Tibur, reappears, and completes the last 9 miles along an aqueduct. The first person to begin to bring this water to Rome was Ancus Marcius, one of the kings; Quintus Marcius Rex, when praetor, later carried out repairs, as did Marcus Agrippa.

Prospecting for water

43. It would not be inappropriate to include the method of locating water, which is found mainly in enclosed valleys.

44. Signs of water are the presence of bulrushes – a plant about which I have spoken – and frogs squatting in unusual numbers for any one locality. Willow, alder, vitex, reed, or ivy, which need no

1. In 144 BC the praetor Quintus Marcius Rex provided for the construction of Rome's first high-level aqueduct, the Aqua Marcia, which conveyed the city's purest supply of water from the head of the Anio valley over a distance of thirty-five miles.

special encouragement to grow, and accumulations of rain-water flowing from higher regions to lower, are not reliable indications of water. A much more dependable sign is a misty haze seen from a distance before sunrise; some water-diviners look out for this from a height, lying flat out with their chins touching the ground.

45. There is also a special method of dowsing known only to experts and employed by them in the hottest season and in the fiercest heat of the day, namely, to examine each locality for surface reflections. For if the earth is parched and one spot remains moist, this is a sure sign.

46. But the eyes need to concentrate so much that they become painful, and to avoid this men have recourse to other tests. They dig a hole 5 feet deep and cover it with unfired clay, or else with a well-oiled bronze basin and a burning lamp, covered by a dome of foliage and earth. If the clay is found to be wet or broken, or if moisture covers the bronze or the lamp has gone out before all its oil is used up – or if a woollen fleece becomes wet – then these are incontestible signs of water. Some diviners first dry out the hole by lighting a fire, and this makes the evidence of the vessels even more conclusive.

Wells

48–9. As workmen dig wells, the clods of earth should become damper as the work progresses and the spades should go down more easily. When wells have been sunk to a certain depth, the diggers may encounter sulphurous or aluminous fumes: these are fatal. A lamp that goes out when lowered into the well indicates this danger. Holes are then dug on the right and left of the well to take away the heavier vapour. Apart from these noxious gases, the very depth renders the air oppressive, and the workmen circulate it by continuous fanning with linen cloths. When they reach water, the well is built up without cement so that the inlets are not blocked.

Pipes and methods of carrying water

57. The best way to bring water from a spring is by means of earthenware pipes about 2 inches thick; the connections are made by fitting pipes into one another – so that the upper pipe fits into the lower – and then caulked with quicklime and oil. The water should fall at least $\frac{1}{4}$ inch every 100 feet. If it flows through a tunnel there should be vents every 240 feet. When the water is required to

rise, lead pipes should be used. Water rises as high as its source. If it comes from a long distance away, the pipe should frequently rise and fall so that no pressure is lost.

58. The usual length for a section of pipe is 10 feet. A section with a diameter of approximately 5 inches[1] should weigh 60 pounds; with one of 8 inches, 100 pounds; and with one of 10 inches, 120 pounds; and so on proportionately. Reservoirs must be constructed as circumstances require.

Hot springs and medicinal uses of water

59. I am surprised that Homer did not mention hot springs, although in other contexts he speaks of hot baths. The reason is that in his time medicine did not use hydrotherapy as we do today. Sulphur-impregnated waters, however, are good for muscles, and waters with an alum content for paralysis and similar cases of collapse; waters with bitumen and soda, such as those of Cutilia, are good for drinking and for internal irrigation.

60. A good many people boast about how they can endure the heat of these sulphurous waters for many hours; this is a very dangerous practice, for one should remain in them only a little longer than in the bath and should afterwards rinse oneself in cool, fresh water and not leave without being rubbed with oil.

62. Sea-water can be used for treatment in the same way. Hot sea-water is used for muscle pains, for knitting fractured bones together and for bruised bones. There are many additional uses, the chief being a sea voyage for those with tuberculosis.

63. Egypt is a popular destination not for its own sake but because of the length of the voyage. Furthermore, even sea-sickness caused by a ship's rolling and pitching is beneficial for many conditions of the head, eyes and chest, as well as for all complaints for which hellebore is administered.

Desalination of sea-water

70. Those at sea often suffer from a deficiency of fresh water, and so I shall describe ways of combating this problem. Fleeces, spread

---

1. Pliny describes this as a 'five-finger' pipe – that is, one made from a sheet of lead 15 inches wide. When bent over to form the pipe this gives a diameter of approximately 5 inches. The other figures have similarly been given as diameter measurements.

round a ship, become damp by absorbing sea-spray: fresh water can then be squeezed out of the fleeces. Similarly, hollow balls let down into the sea in nets, or empty containers with their openings sealed, collect fresh water inside.[1] On land salt water is turned into fresh by filtration through clay.

Salt: natural and artificial

73. Salt both occurs naturally and is manufactured. Each type is formed in several ways, but there are two main agents involved: condensation and evaporation. In summer salt is evaporated in the Tarentine Lake by the sun: the whole expanse of water, always shallow and never more than knee-deep, becomes salt. The same happens in Sicily, at Lake Cocanicus and at another lake near Gela, but in these two cases only the water at the edges evaporates. In Phrygia, Cappadocia and at Aspendus the process goes further; in fact, evaporation spreads as far as the centre. The amazing thing is that however much is removed during the day, that amount is made good overnight. All the salt from these pans is in the form of fine powder and not blocks.

74. Another kind of salt spontaneously produced from sea-water is foam left on the edge of the shore and on the rocks by the sea. Three different kinds of natural salt occur: in rivers, lakes and hot springs.

77. There are also mountains of natural salt such as at Oromenus in India, where it is cut out like blocks of stone from a quarry and even replenishes itself; this provides more revenue for the rulers than gold and pearls. It is also dug out of the earth in Cappadocia, where it has evidently been formed by the evaporation of moisture. There it is split into sheets like mica.

78. At Gerra, a town in Arabia, walls and houses are constructed of blocks of salt cemented together by water.

79. The region of Cyrenaica is famous for Hammoniac salt, so called because it is found under the sand.[2] In colour it is like the alum called *schistos* and consists of long opaque blocks.

82. In the provinces of Gaul and Germany they pour salt water on burning logs.

---

1. The water is collected by a kind of osmosis, the walls of the containers acting as a permeable 'membrane'.
2. *hammos.*

Soda and salt

106. No one has given a more detailed description of soda than Theophrastus. A small amount occurs in Media in valleys that are bleached by drought; the locals call this *halmyrax*. It is also found in Thrace, near Philippi, but in smaller quantities and contaminated with earth; this is called *agrium*.

109. Such soda is natural, but in Egypt it is artificially produced in large quantities. This type is inferior in quality, being dark and stony. Soda is made in almost the same way as salt, except that whereas sea-water is poured into the salt-pans, the soda-beds are filled with water from the Nile.

Sponges

123. Sponges are whitened artificially. Fresh sponges of the softest kind are soaked in salt foam throughout the summer and are then laid upside-down, open to the moon and the hoar-frosts. Sponges are animal and even have blood.[1]

124. Some authorities assert that they are guided by a sense of hearing, and contract in response to sound, expelling a lot of moisture. They add that they cannot be torn from rocks and so are cut off with a resulting discharge of a gory liquid.

---

1. Pliny is incorrect: the sponge, although an invertebrate animal, does not have blood vessels.

BOOK XXXII
FISH AND AQUATIC CREATURES

The forces of Nature

1. I come now to the greatest achievement of Nature and am met by a manifest, overwhelming proof of hidden power. For what is more violent than sea, winds, whirlwinds and storms? Has man shown greater skill in assisting Nature in any other sphere than by his use of sails and oars? And one must add the indescribable power of the tides' ebb and flow, by which the whole sea is turned into a river.

The goby[1]

2. Yet even though all these forces may drive in the same direction, they are checked by a single goby, a very small creature. Gales may blow and storms rage but this fish controls their fury, restrains their tremendous force, and compels ships to stop – a thing unachievable either by hawsers or even by dropped anchors, which cannot be drawn back because of their weight. The goby holds off their attacks and tames the fury of the universe with no effort or resistance of its own other than suction. This diminutive fish is strong enough – against all those forces – to prevent vessels from moving. War fleets carry towers on their decks so that men may fight as if from city walls even when at sea. How pathetic men are, when one considers that those rams, equipped with bronze and iron for striking, can be held fast by a little fish some 2 inches long. It is said that at the Battle of Actium this fish brought Antony's flagship to a halt when he was touring his fleet to encourage his men – until, that is, he changed his ship for another. For this reason Octavian's fleet straight away made a more concerted attack. Within living memory a goby stopped the Emperor Gaius' ship on his voyage back from Astura to Antium.

4. In the event this little fish proved to be an omen, for very

---

1. A small carnivorous fish about two inches long. The powers with which it is credited by Pliny are, of course, completely fictitious.

soon after Gaius' return to Rome he was struck down by his own men. His delay occasioned surprise only for a little while, since the cause was immediately discovered. As his quinquereme alone of all the ships in the fleet was not making any headway, men immediately dived overboard and swam round the ship to ascertain the cause. They found a goby attached to the rudder and showed it to Gaius who was furious that such a thing had held him back and prevented 400 rowers from obeying his orders.

5. Accounts agree that what particularly annoyed him was that the fish had stopped him by attaching itself to the outside of the ship, yet did not have the same power when brought inside. Those who saw the goby then, or subsequently, say that it looked like a large slug.

The sting-ray

7. Yet even without the example of the goby, surely the sting-ray, also a sea-creature, would offer proof enough of the power of Nature? Even from a distance – indeed a long distance – if it is touched with a spear or rod, it numbs the strongest and paralyses the swiftest feet.

Ovid, On Fishing

11. I find the characteristics of fish, as explained by Ovid in his On Fishing,[1] quite amazing. The wrasse, when caught in a wicker basket with a narrow neck, does not burst out through the front or thrust his head through the osiers that imprison him, but he turns round, widens the gaps with repeated blows from his tail and creeps backwards. If another wrasse outside happens to see him struggling, he seizes the other's tail with his teeth and helps his efforts to break out.

15. Trebius Niger tells us that the swordfish has a pointed beak with which he pierces and sinks ships.

17. At Myra in Lycia, in the spring of Apollo called Curium, when summoned three times by a pipe the fish come to give oracular responses. If the fish seize the meat thrown to them, this is a favourable response for the inquirers; if they reject it, this is an omen of disaster.

1. *Halieuticon.*

Coral

21. Coral is as highly valued among the Indians as Indian pearls. It is also found in the Red Sea, but there it is darker in colour. The most prized is found in the Gallic Gulf around the Stoechades Islands, in the Sicilian Gulf around the Aeolian Islands, and around Drepanum.

22. Coral is like a shrub in shape and its colour is green. The berries are soft and white under water. When taken out they harden immediately and become pink, being similar in appearance and size to berries of the cultivated cherry. Men say that live coral petrifies at a touch. It is therefore quickly seized and pulled away in nets or cut off by a sharp iron knife. This is the explanation for its name, which comes from the Greek verb.[1] Coral that has a very red hue and several branches, and is neither rough nor stony nor empty nor hollow, is the most valuable.

23. Coral-berries are no less valued by Indian men than specimen Indian pearls by Roman ladies. Indian soothsayers and seers believe that coral is potent as a charm for warding off dangers. Accordingly they delight in its beauty and religious power. Before this became known, the Gauls used to decorate their swords, shields and helmets with coral. Now it is very scarce because of the price it commands, and is rarely seen in its natural habitat.

The tortoise

32. Like the beaver, the tortoise[2] is amphibious. It has the same medical properties, and is notable for the high price it can fetch and for its peculiar shape. Its blood improves the eyesight and checks cataracts; it is an antidote for the poisons of all snakes, spiders and similar creatures. The flesh of the tortoise is reported to be useful for fumigation and for countering magic tricks and poisons. Tortoises are common in Africa.

Oysters

59. Oysters have long since taken pride of place as a delicacy for our tables. They love fresh water and places where there is an inflow from many streams. Deep-sea oysters, therefore, are small

1. *keirein*: 'to cut'.
2. Pliny surely means turtle here.

and rare. Oysters also breed in rocky places that lack fresh water, for example, round about Grynium and Myrina. Their growth corresponds very closely to the waxing of the moon.

60. Oysters vary in colour; they are red in Spain, tawny in Illyricum and black in flesh and shell in Circeii.

61. Experts mention this mark of distinction: if a purple line encircles the beard they consider such oysters to be of a nobler variety. Oysters like to travel and to be transferred to unfamiliar waters. Thus oysters from Brundisium that have been fattened in Lake Avernus are believed to retain their own taste as well as acquiring that of the local variety in the lake.

62. I shall now speak about countries where oysters are bred so that those shores should not be cheated of their due fame – but I shall quote the words of another authority, one who is the greatest expert of our time. This then is what Mucianus has to say:

Oysters from Cyzicus are larger than those from the Lucrine Lake, fresher than the British variety, sweeter than those from Medullae, sharper than oysters from Ephesus, fuller than those of Spain, less slimy than ones from Coryphas, softer than those from Histria and whiter than oysters from Circeii.

None are fresher or softer than the last named.

63. The historians who chronicled Alexander's expedition inform us that oysters a foot long are found in the Indian Ocean, and among ourselves a gourmet coined the name *tridachna*[1] for these, intending it to be applied to those which are so big that they require three bites.

Medical uses for oysters

64. Oysters are singularly good for settling the stomach and restoring appetites. Luxury has provided the extra feature of coldness by having oysters buried in snow, thus linking the tops of mountains to the depths of the sea. Oysters are a mild laxative and, if boiled with honey-wine, cure bowel irritation and straining with motions, provided there is no ulceration. They also cleanse an ulcerated bladder, and boiled intact in their shells, as gathered, they are extremely good for bad colds.

1. From *treis*, 'three', and *daknein*, 'to bite'.

Leeches

123–4. Leeches, also known as 'bloodsuckers', are used to let blood. They are supposed to have the same function as cupping-glasses, namely to relieve the body of excess blood and open the pores of the skin. There is a snag, however: once applied, their use is addictive and patients want the same treatment every year about the same time. Many think that leeches should be applied for gout. When they have had their fill the leeches either drop off from sheer weight of blood or are detached by a pinch of salt. Sometimes, however, they leave their heads embedded in the skin and so cause incurable wounds which have often caused death. An example is the case of Messalinus, a patrician of consular rank, who applied leeches to his knee; this remedy turned into a deadly poison. Red leeches are especially feared, so men cut them off with scissors while they are sucking and the blood runs down as though through tubes. As they die their heads gradually shrink and are not left in the skin.

An inventory of sea-creatures and fish

142. I have now finished my account of the natural qualities and properties of aquatic animals and plants. It does not seem outside the scope of this work to point out that in all the seas there are altogether 144 named species; this cannot be matched in the case of land-based creatures and birds.

144. On the other hand, believe me, in the sea, immense though it is, and in the ocean, the number of living creatures is known. We may well wonder that we are better acquainted with the things that Nature has submerged in the depths!

MINING AND MINERALS

BOOK XXXIII

GOLD AND SILVER

Man's greed and exploitation of the earth's resources

1. I shall now discuss metals and the natural resources we use to pay for goods. We search for these deep within the earth, diligently and in a number of ways. In some places we dig for riches, when our life-style requires gold, silver, electrum and copper; and in others out of sheer self-indulgence, when gems and pigments for wall-painting are required; and in yet other places we dig with sheer recklessness when iron is needed – a metal even more welcome than gold amid the bloodshed of war. We pursue all the lodes in the earth, live above mined areas, and then are amazed that sometimes the earth gapes open or begins to tremble, unwilling to believe that this might be our holy parent's way of expressing her indignation.

2. We search for riches deep within the bowels of the earth where the spirits of the dead have their abode, as though the part we walk upon is not sufficiently bountiful and productive. And in doing this our least important goal is the discovery of remedies for diseases. Indeed, how many men dig for medicinal purposes? Yet Earth provides us with medicines on her surface, just as she provides corn – generous and unstinting indeed with all things that help us.

3. But what she has hidden and kept underground – those things that cannot be found immediately – destroy us and drive us to the depths. As a result the mind boggles at the thought of the long-term effect of draining the earth's resources and the full impact of greed. How innocent, how happy, indeed how comfortable, life might be if it coveted nothing from anywhere other than the surface of the earth – in brief, nothing except what is immediately available!

4. Gold is mined and, with it, malachite[1] which jealously guards its name, derived from gold,[2] to make it appear more valuable. Greed hankered after silver, but was grateful in the mean time to have discovered cinnabar and to have invented a use for red earth. Alas for our fertile ingenuity! Think of the many ways in which we have inflated the value of objects! The art of painting has

1. Copper carbonate.
2. *Chrysocolla.*

arrived on the scene, and so we have made gold and silver more expensive by chasing their surface. Man has learnt to challenge Nature. The incentive of vice has increased the range even of art. Once we were content to engrave permissive scenes on our cups and to take our drink through obscene figures.

5. Then, with an excess of gold and silver, these artistic efforts were tossed aside and considered paltry. Out of this same earth we dug fluorspar[1] and rock-crystal which owe their value to their very brittleness. It became a symbol of wealth, the true cachet of luxury, to possess an object that could be completely destroyed in a second. But even then we were not satisfied. We drink from an array of gemstones, covering our cups with emeralds and desire to possess India for the sake of our drinking. Nowadays gold is a mere accessory.

6. If only gold could be completely banished from life, reviled and abused as it is by all the worthiest people, a discovery whose only purpose was the destruction of human life! What a far happier age it was when goods were bartered, as, following Homer, we must believe was the practice in Trojan times.[2] In my opinion this was how trade came into being – in the quest for the necessities of life. Homer records that some people bought things with ox-hides, others with iron and prisoners, and, although he himself was an admirer of gold, he reckoned the value of goods in cattle, observing that Glaucus exchanged gold armour worth 100 oxen with that of Diomedes, worth 9 oxen.[3] As a result of this practice, even at Rome a fine under the old laws is assessed in cattle.

8. The first person to put gold on his fingers committed the worst crime against human life. There is no record, however, of who this was. For although antiquity endowed Prometheus with an iron ring, intending this to be understood as a fetter and not an ornament, I believe the whole story to be a myth.[4] At any rate, Midas' ring, which when turned round rendered the wearer invisible, is undeniably even more of a myth.[5] It was the hand –

1. Calcium fluorite; also known as Blue John, and at one time extensively mined in Derbyshire.

2. *Iliad* VII, 472 ff.

3. *Iliad* VI, 234–6.

4. Prometheus gave men the gift of fire in a fennel stalk and was punished for this by Zeus.

5. Midas, king of Phrygia, was renowned for his immense riches. Dionysus granted his wish that everything he touched should turn to gold. When he implored Dionysus to take back this gift, Dionysus ordered him to bathe in the sources of the River Pactolus which thenceforward had an abundance of alluvial gold. Small quantities of gold are still found in the river.

indeed, the left hand – that brought gold its very great esteem, but not a Roman hand, since Romans were accustomed to wear an iron ring as a symbol of courage in war.

Gold at Rome

14. For a long time gold was found at Rome only in very small quantities. At any rate, when peace had to be purchased after the city's capture by the Gauls, no more than 1,000 pounds of gold could be produced. I know, of course, that in Pompey's third consulship[1] there was lost from the throne of Jupiter Capitolinus some 2,000 pounds of gold that had been stored there by Camillus.[2] This was why most people thought that 2,000 was the actual sum total. But the additional amount was part of the booty taken from the Gauls, who had removed it from the temples in the part of Rome which they had captured.

15. The case of Torquatus is proof that the Gauls used to wear gold in battle.[3] So it seems that the gold belonging to the Gauls and to the temples reached a total weight of 2,000 pounds and no more. This was to be the meaning of the augury, when 'Jupiter Capitolinus had repaid twice'. Since I was led on to the topic of gold from my discussion of rings, it is not inappropriate, incidentally, to record that the official in charge of the Temple of Jupiter Capitolinus, when arrested, broke the stone of his ring in his mouth and died immediately so that any proof of embezzlement was destroyed.

16. It follows that there were only 2,000 pounds of gold, at most, when Rome was captured in AUC 364 although the census showed that there were already 152,573 free citizens. The gold that Gaius Marius the Younger had taken from Rome to Praeneste 307 years later[4] from the burning temple on the Capitoline and from all the other shrines in Rome amounted to 14,000 pounds; this was carried in Sulla's triumph under a placard publicizing the amount, together with 6,000 pounds of silver. On the previous day Sulla had displayed 15,000 pounds of gold and 115,000 pounds of silver, acquired as a result of his other victories.

20. There is further evidence that rings were commonly worn at

1. 52 BC.

2. Camillus, appointed dictator, defeated the Gauls in 390 BC.

3. Titus Marcus Torquatus derived his surname from the gold torque or necklace that he took from a Gaul whom he killed in single combat (360 BC).

4. 390 BC.

the time of the Second Punic War: for had this not been the practice, it would have been impossible for Hannibal to send 3 modii[1] of rings to Carthage. Also the hostility between Caepio and Drusus, which was the main cause for the outbreak of the Social War and the disasters that ensued, arose from a ring put up for auction.

21. Even then not all senators had gold rings. Many men, within our grandfathers' memory, had held the office of praetor and had worn an iron ring right into old age, as Fenestella records: for example, Calpurnius, Manilius[2] – the latter had been a general under Gaius Marius[3] in the Jugurthine War[4] and, according to many authorities, the Lucius Fufidius to whom Scaurus dedicated his *Autobiography*. Another piece of evidence is that in the family of the Quintii it was not customary even for women to have a gold ring, and the majority of the peoples who live under our Empire at the present time possess no gold rings at all. Even now the East and Egypt do not seal documents and are content with only a signature.

The history of the equestrian order

29. As soon as rings came into common use, they were used to distinguish the second order, the knights, from the plebeians – just as a tunic distinguished senators from those who wore the ring, although this distinction came later. We find that even heralds wore a wider purple stripe on their tunics, as did, for example, the father of Lucius Aelius Stilo Praeconinus who derived his surname from his father's office.[5] So rings clearly introduced a third order between plebeians and the Senate; the title of 'knight', formerly conferred if one could provide a horse for war, is now awarded on the basis of wealth. This is a recent innovation.

30. When the late Emperor Augustus appointed judicial panels, the majority of those empanelled belonged to the class who wore an

1. Three-quarters of a bushel.
2. Manilius (born 138 BC) served under Marius in Africa (107 BC) and was later appointed dictator.
3. Marius (157–86 BC) established himself as a leading power in Rome with the capture of Jugurtha in 106 BC and victory over the Germans in 102. He was consul seven times and responsible for reorganizing the Roman army. His bid to secure from Sulla command of the campaign against Mithridates in the East led to civil war.
4. 112–106 BC.
5. As a herald (*praeco*).

iron ring, and were called not knights but judges; the term 'knights' was still reserved for cavalry squadrons financed by the state. At first there were only four panels of judges and scarcely a thousand names were listed on each, since the provinces had not yet been admitted to this office; the regulation prohibiting people who have recently acquired citizenship from acting as judges on any of the panels has remained in force until the present time.

31. The panels themselves were also distinguished by different titles according to whether they consisted of so-called treasury officials,[1] chosen members[2] or judges.[3] In addition there was a group called the Nine Hundred, chosen from the whole body as inspectors of the ballot-boxes at elections. Pride in these titles caused divisions within this order also, one calling himself a member of the Nine Hundred, another one a chosen member, yet another a treasury official.

32. Eventually, in the ninth year of Tiberius' principate,[4] the equestrian order was incorporated as a single body. In the consulship of Gaius Asinius Pollio and Gaius Antistius a regulation was brought in specifying who should be allowed to wear rings. The trivial and curious reason for this was in effect that Gaius Sulpicius Galba had as a young man tried to curry the emperor's favour by establishing a tax for keeping eating-houses and had complained in the Senate that proprietors had protected themselves from this charge by means of their rings. A rule was therefore imposed that no one should have the right to wear a ring unless he himself as well as his father and grandfather, was free-born, had capital assets amounting to 400,000 sesterces, and, under the Julian law relating to the theatre, possessed a seat in the first fourteen rows.

33. After this people began to apply in droves for this mark of distinction, and, because of the disputes that arose as a consequence, the Emperor Gaius Caligula added a fifth panel. This provoked such arrogance that panels which could not be filled in the principate of the late Emperor Augustus, will not hold members of that order, and everywhere there are examples of freed slaves making the leap to these privileges, a thing that never occurred in former times since the iron ring was the distinctive symbol of knights and judges. And this practice began to be so commonplace that during the censorship

1. *tribuni aerarii.*
2. *selecti.*
3. *iudices.*
4. AD 22.

of the Emperor Claudius[1] a knight named Flavius Proculus impeached 400 people for fraudulently obtaining these privileges. The result is that an order intended to distinguish the holder of the title from other men of free birth has been shared with slaves.

34. The Gracchi were the first to glorify the equestrian order by the title 'judges', seditiously currying favour with the plebeians to achieve the humiliation of the Senate. But soon the authority vested in the title of 'knight' ended with the development of different factions and came to be associated with tax-farmers, and so for some time the equestrian order became the third rank. Finally, Marcus Cicero, as a result of the Catilinarian conspiracy,[2] put the title of 'knight' on a secure basis during his consulship, boasting that he himself had sprung from that order; he used his own claim to popularity as a means of bidding for their influential support. From that time forward the knights became a third body in the state and the name 'equestrian order' was added to the formula: 'The Senate and the People of Rome'. Since this is the most recent addition, it is even now written after 'People'.

35. The knights' very name has often been altered, even when those concerned were genuine cavalrymen. Under Romulus and the Kings they were called the 'Celeres', then the 'Flexuntes', and later the 'Trossuli' because they captured a town of that name in Etruria, nine miles this side of Volsinii, without any help from the infantry. That name survived until after the time of Gaius Gracchus.

36. At any rate, Junius, who was called Gracchanus from his friendship with Gaius Gracchus, wrote as follows: 'Concerning the equestrian order, they were previously called the Trossuli but are now simply known as the knights because people do not know the meaning of the word "Trossuli", and many of them are ashamed of being called by this name.'

Gold

37. There are some facts concerning gold that I should include. The state gave gold torques to auxiliaries and foreign soldiers, but Roman citizens received only silver ones. Also it gave bracelets to citizens, but not to foreigners.

38. What is even more surprising for us is that gold crowns were awarded to citizens. I myself have been unable to discover who was

1. AD 47–8
2. 63 BC.

the first to receive a gold crown. Lucius Piso records that the first
person to make such an award was the dictator Aulus Postumius;
after the storming of the Latins' camp at Lake Regillus,[1] he gave a
gold crown to the soldier who had been mainly responsible for its
capture. The crown was made of gold taken from the spoils and
weighed two pounds.

The introduction of coinage at Rome[2]

42-3. The second crime against mankind was committed by the
person who first struck a gold denarius; the crime itself is not
recorded and its author unknown. The Roman people did not use
struck silver coinage before the defeat of King Pyrrhus. The *as*
weighed one pound – hence the 'little pound' and the 'two-
pounder'. This is the reason why a fine is levied in terms of 'heavy
bronze',[3] and why in accounts outgoings are called 'sums weighed
out',[4] likewise interest, 'weighed on account',[5] and payment, 'weighing down'.[6] Moreover, it explains the term 'soldiers' stipend',[7] that
is, 'weighed-out heaps of coins'. King Servius was the first to imprint
a type on money; previously, so Timaeus records, plain metal was
used at Rome. The type was of an ox or a sheep, *pecus*, which
explains the origin of the term for money, *pecunia*. The highest
property-rating in the reign of King Servius was 120,000 *asses* and
this was the qualification for the first class of citizens.

Silver coinage

44. Silver was first struck in AUC 485[8] in the consulship of Quintus
Ogulnius and Gaius Fabius, five years before the First Punic War. It
was decided that the value of the denarius should be 10 pounds of
bronze; that of a half-denarius, 5 pounds; and that of a sesterce, 2½
pounds. The weight of the standard pound of bronze, however, was
reduced during the First Punic War when the state could not meet

1. 497 BC.
2. Pliny's account of the development of early Roman coinage is unsound. See
M. H. Crawford, *Roman Republican Coinage* (Cambridge, 1974).
3. *aes grave.*
4. *expensa.*
5. *impendia.*
6. *dependere.*
7. *stipendiumo.*
8. 269–268 BC.

its commitments, and it was decreed that the *as* should be struck weighing 2 ounces. This produced a saving of five-sixths and the state's debt was wiped out. This bronze coin had the type of a Janus on the obverse, and the ram of a warship on the reverse. The third of an *as* and the quarter of an *as* both had a ship.[1]

45. In soldiers' pay 1 denarius has always been the equivalent of 10 *asses*. The types on silver were of a two- and a four-horsed chariot, which led to the coins being called *bigati* and *quadrigati* respectively. Next, under the law of Papirus,[2] *asses* weighing half an ounce were struck. In the tribunate of Livius Drusus,[3] the silver was alloyed with an eighth part of copper. A coin known as the victoriate was struck under the law of Clodius.[4] A similar coin had formerly been imported from Illyria and regarded as legal tender; its type was a figure of Victory, hence the name.

Gold coinage

47. The first gold coin was struck fifty-one years after the first silver coin; a scruple of gold was equivalent to 20 sesterces at 400 to the pound of silver at the then prevailing rate for the sesterce. It was later decided to strike denarii at 40 to the pound of gold and the emperors gradually reduced the weight of the gold denarius; most recently Nero devalued it to 45 denarii to the pound.

Greed and its effect on Roman character

48. Money was the underlying cause of greed: usury was invented and with it a means of making profit without effort. In rapid stages mere greed gave place to gold fever, which flared up with a kind of frenzy. For example, when Septumuleius, a friend of Gaius Gracchus, heard that a price equivalent to its weight in gold had been set on Gracchus' head, Septumuleius cut it off and brought it to Opimius. He had put lead in the head's mouth and so not only added to the horror of the crime, but cheated the state into the bargain. And it was not a Roman citizen, but King Mithridates, who, to the discredit of the reputation of Romans generally, poured molten gold into the mouth of the general Aquilius whom he had captured. These

1. On the reverse.
2. 89 BC.
3. 123 BC.
4. 104 BC.

outrages are the outcome of an acquisitive, materialistic attitude.

49. One has a sense of shame on seeing the novel names that are sometimes derived from Greek words for silver vessels with filigree or gold inlay. Because of these devices, gilded vessels fetch more than pure gold. But we know that Spartacus forbade anyone in his camp to possess gold or silver, so much stronger at that time was the moral fibre of our runaway slaves. The orator Messala has recorded that the triumvir Antony used gold chamber-pots for all the calls of nature, a charge that would have shamed even Cleopatra. Previously, foreigners had held the record for extravagance. King Philip had the habit of sleeping with a gold goblet under his pillow; Hagnon of Teos, a general of Alexander the Great, had his sandals soled with gold nails.

51. For my part I am surprised that the Roman people always imposed tribute on defeated nations in silver rather than in gold. For example, the tribute imposed on Carthage, when it fell after the defeat of Hannibal, was 800,000 pounds of silver, spread over fifty years – but no gold. Nor can this be ascribed to the world's poverty, for Midas and Croesus had already possessed limitless wealth, and Cyrus, after his conquest of Asia Minor, found booty comprising 24,000 pounds of gold, besides golden vessels and articles, including a throne, a plane-tree and a vine. By this victory he won 500,000 talents of silver and a crater that had belonged to Semiramis,[1] weighing 15 talents.

52. Marcus Varro records the weight of the Egyptian talent as 80 pounds of gold. Saulaces, a descendant of Aeetes, was reigning in Colchis when he is said to have stumbled upon virgin soil in the country of the Suani and elsewhere and to have dug up a great amount of gold and silver; Saulaces' kingdom was, moreover, famous for golden fleeces. Stories are told of his gold-vaulted ceilings and silver beams with columns and pilasters; these originally belonged to Sesostris,[2] king of Egypt, whom he conquered. Sesostris was so proud a monarch that every year he used to yoke various kings, chosen by lot from among his subjects, to his chariot for a triumphal procession.

53. Some of our own deeds will be considered legendary by our descendants. Caesar, the future dictator, was the first aedile to use exclusively silver equipment in the arena at the funeral games given

---

1. Semiramis and Ninus, her husband, were the mythical founders of the Assyrian empire of Nineveh.

2. A semi-legendary king of Egypt at the beginning of the second millennium BC.

in honour of his father;[1] this was also the first time that criminals fought wild beasts with silver weapons, a practice currently matched even by our municipal towns. Gaius Antonius presented plays on a silver stage and likewise Lucius Murena. The Emperor Gaius Caligula brought into the circus mobile scaffolding bearing 124,000 pounds of silver. His successor Claudius, when celebrating his triumph after the conquest of Britain, publicized on placards the fact that there were 7,000 pounds of gold crowns from Hither Spain, and 9,000 from Gallia Comata. Nero filled Pompey's theatre with gold for one day when he showed it to the king of Armenia, Tiridates. And yet how small the theatre was compared to Nero's Golden House which dominates Rome!

57. Ceilings, such as the gold-covered ones we now see in private houses, were first gilded on the Capitol when Mummius was censor after the destruction of Carthage.[2] The practice of gilding passed from ceilings to vaulted roofs and walls, which are now covered with gold as if they were pieces of plate. However, Catulus' contemporaries were divided in their opinions of his gilding of the bronze roof-tiles of the Temple of Jupiter on the Capitol.[3]

The physical properties of gold and the reasons for its popularity

58. The main reason for the popularity of gold is not, in my view, its colour; silver is lighter and more like daylight: that is why it is more commonly employed for military standards, since its sheen is visible from further away. Those who think that gold is popular because of its starlike colour are clearly wrong because, compared to gems and other things, gold is not conspicuous.

59. Gold is preferred to other metals not for its weight and malleability – since in respect of both properties it takes second place to lead – but because, remaining unscathed in conflagrations and on funeral pyres, it is the only metal that loses nothing by contact with fire. Indeed, the opposite is the case: the quality of gold is enhanced the more it is subjected to fire. Fire also provides a test of quality, making the gold take on its own colour and glow red; this process is known as 'assaying'.

60. That gold is not easily affected by fire is the first proof of its

1. 65 BC.
2. 146 BC.
3. 79–60 BC.

quality. But it is also remarkable that, although gold is resistant to charcoal prepared from the hardest wood, it very quickly becomes red-hot when subjected to a fire of chaff and is cupelled with lead to purify it. Another important determinant of its value is that it suffers very little wear in use; with silver, copper and lead, however, smear lines may be drawn, and the substance that rubs off these metals soils the hands.

61. No other metal is as malleable as gold or able to be divided into so many portions; thus an ounce of gold can be beaten into upwards of 750 leaves four inches square. The thickest kind of gold leaf is called Praenestinian, retaining a name derived from the superbly gilded statue of Fortune at Praeneste.

62. The leaf next in thickness is called Quaestorian. In Spain tiny pieces of gold are called 'scrapers'. More frequently than any other metal, gold is found in nuggets or as placers. Whereas other ore-minerals from mines are refined by some heat process, gold is pure when found and already in a finished state. In addition to native gold there is another type obtained by mechanical means which I shall describe. Gold surpasses any other substance in its resistance to rust, verdigris, or any other condition that corrupts it or diminishes its weight. It consistently withstands brine or vinegar which destroy other materials, and, what is more, can be spun into thread and woven into a wool-like fabric without any addition of wool itself.

63. Verrius tells us that Tarquinius Priscus celebrated a triumph wearing a golden tunic. I have seen Agrippina, the wife of the Emperor Claudius, at a show where he was presenting a naval battle, seated by him, wearing a military cloak made entirely of gold cloth. Gold has long been woven into the fabric called Attalus cloth; this was invented by the kings of Asia.

Sources of gold and Roman mining techniques

66. Leaving aside tales of Indian gold obtained by ants, or the gold dug up by griffins in Scythia, gold is obtained in our part of the world in three ways. First, it can be obtained from placers in rivers, for example, the Tagus in Spain, the Padus in Italy, the Hebrus in Thrace, the Hermus in Asia Minor and the Ganges in India; this is gold in its most perfect state, thoroughly polished by the friction induced by the current. A second method is to sink shafts. And thirdly it is sought for in the debris of collapsed mountains. Let me describe each method.

67. Gold prospectors begin by gathering earth that indicates the presence of gold. The auriferous material of the deposit is washed and the sediment allows an analysis to be made of the parent ore body. Sometimes by a rare stroke of luck a pocket is found on the surface, such as the one recently found in Dalmatia during Nero's principate which yielded fifty pounds of gold in a day. Gold occurring in this manner is called an outcrop if there is also auriferous earth beneath. The otherwise parched and barren mountains of Spain, in which nothing else is produced, are compelled to yield a harvest of this commodity.

68. Gold mined by means of shafts is called 'channelled' or 'trenched'; it is found adhering to quartz, not in the way it shines in lapis lazuli in the East, or in the granite of Thebes or in other precious stones, but glistening as nodules in a quartz matrix. Traces of veins appear here and there along the walls of the underground galleries and the overburden is supported by wooden props.

69. The mineral is mined, crushed, washed, fired and ground to a fine powder. Ground ore is known as *scudes*, and the silver when tapped is called 'sweat'. The waste product from the furnace, regardless of the ore, is called 'slag'. In the case of gold, the slag is pounded and fired a second time; the crucibles used for this operation are made of a white fire-clay. This alone can withstand the forced draught and intense heat of both furnace and molten metal.

70. The third method used for extracting gold rivals the achievements of the Giants. By the light of lamps, long galleries are cut into the mountain. Men work in long shifts measured by lamps, and may not see daylight for months on end. Local people call these mines 'deep-vein'. The roofs of these are liable to give way and crush the miners, which makes diving for pearls or getting purple-fish from the depths of the sea seem comparatively safe. So much more dangerous have we made the earth. Arches are left at frequent intervals to support the mountains above.

71. In open-cast and deep-vein mining masses of flint are encountered. These can be split by fire-setting which involves the use of vinegar. Fire-setting in galleries usually makes them suffocatingly hot and smoke-filled. Instead, therefore, the rocks are split by means of crushers which carry 150 pounds of iron. The miners then carry the ore out of the workings on their shoulders, each man forming part of a human chain working in the dark; only those at the end of the line see daylight. If the bed of flint seems too long, the miners bypass it. Yet flint is considered relatively easy to work.

72. There is a kind of earth consisting of clay and gravel called

'conglomerate' which is almost unworkable. The method used in this case is to attack it with iron wedges and crushers. Conglomerate is considered the hardest of all things – except the greed for gold which is even more stubborn. When the work is completed the miners cut through the tops of the arches, beginning with the last. The opening of a fissure gives warning of the impending collapse, but this is only seen by a watchman perched on the top of the mountain.

73. With a shout or a wave the look-out gives the order for the miners to be called off and at the same time rushes down from his vantage-point. The ruptured mountain falls asunder with an unimaginable crash, and is accompanied by an equally incredible blast of air. Like conquering heroes the miners contemplate their triumph over Nature. Yet even at this juncture there is no gold visible, nor did they have any positive indication that there was any when they began to dig. The mere hope of obtaining their coveted prize was sufficient reason for embarking on such danger and expense.

74. Another equally laborious and even more expensive task involves the feat of bringing streams along mountain-ridges – often a distance of 100 miles – to wash away the debris from mining operations. The miners call these water-channels *corrugi*, a term derived from the word 'confluence',[1] and they involve countless problems. The incline must be steep to produce a surge rather than a steady flow of water and consequently high-level sources are required. Gorges and crevasses are bridged by aqueducts. Elsewhere, impassable rocks are cut away to allow space for hollow wooden troughs.

75. The workmen cutting out the rock hang suspended by ropes, so that viewed from a distance the operation seems to involve not so much a species of strange animals as of birds. Most hang suspended as they take the levels and mark out the route – man leads rivers to run where there is no place for him to plant his own footsteps. The washing of the ore is spoilt if the current of the stream brings silt, or *urium* as this residue is called. To avoid this the water is guided over flint or pebbles. On the ridge above the head of the mine reservoirs are excavated, measuring more than 200 feet each way and 10 feet deep. Five sluices each about 3 feet square, are constructed in the walls. When the reservoir is full, the sluices are knocked open so that the violent downward surge of water is sufficient to sweep away the rock debris.

---

1. *contrivatio.*

76. The next process takes place on level ground. Water conduits, the Greek name for which is *agogae*, are cut in steps and their floors covered with gorse – a plant resembling rosemary – which is rough and so traps the gold particles. The conduits are boarded with planks and carried on arches over steep pitches. Thus the tailings flow down to the sea and the shattered mountain is washed away. It is because of this that Spain has now encroached substantially on the sea.

77. The material extracted by such enormous labour in deep-vein mining is, by this latter process,[1] washed out so as not to fill up the shafts. Gold mined from deep veins does not require heat treatment but is pure gold. Nuggets are found in such mines, and similarly in pits; some weigh more than 10 pounds. These are called *palagae* or *palacurnae*, while gold-dust is also known as *balux*. The gorse is dried and burnt and its ash is washed on a bed of grassy turves so that the gold is deposited on these.

78. According to some sources, Asturia, Gallaecia and Lusitania produce 20,000 pounds of gold in a year; Asturia supplies the largest amount. Spain has long been the main gold-producing area in the world. By an old senatorial decree, Italy is protected from over-exploitation; otherwise, it would have been the most productive region for ore-minerals, as it is for agricultural produce. There is in existence a decree of the censors concerning the gold mines of Victumulae in the region of Vercellae, which prohibits contractors from employing more than 5,000 men in mining operations.

Gold produced from orpiment

79. There is a method of making gold from orpiment which is mined in Syria for painters; it is found on the surface and has the colour of gold, but is brittle and like selenite. Its potential attracted the Emperor Gaius Caligula who was obsessed with gold. He ordered a great weight of orpiment to be melted; and certainly it produced excellent gold, but the yield was very low and so, although orpiment sold for 4 denarii a pound, he lost out by the experiment which his greed had led him to initiate. The experiment was not subsequently repeated by anyone else.

80. All gold contains a varying proportion of silver – some a tenth, some an eighth. In one mine only – Albucrara in Gallaecia – the proportion of silver found is a thirty-sixth, which makes this

1. Hushing.

gold more valuable than the rest. Where the proportion of silver is at least one-fifth, the ore is called electrum; grains of this are found in 'channelled' gold. An artificial electrum alloy is also made by adding silver to gold. If the proportion of silver exceeds one-fifth, the electrum offers no resistance to the anvil.

81. At Lindos on the island of Rhodes there is a temple of Athena which possesses an electrum cup that was dedicated by Helen; history further adds that it is the same size as her breasts.

Gold statues

82. The first statue made from solid gold is said to have been set up in the Temple of Anaitis in Anaitica, which we have so named because of the local population's particular devotion to that goddess.

83. This gold statue was taken as booty during Mark Antony's campaigns in Parthia. There is a story told of a witty exchange involving one of our army veterans who was being entertained as a guest to dinner at Bononia by the late Emperor Augustus. Asked whether it was true that the man who first committed this sacrilege against Anaitis had been struck blind, was paralysed and had died, the veteran replied that the emperor was eating his dinner off one of the goddess's legs and that he himself had committed the sacrilege and owed his entire fortune to that plunder. The first solid gold statue of a man was one set up by Gorgias of Leontini in the Temple of Apollo at Delphi.[1] Such was the financial reward from teaching oratory!

Gold-refining

84. Gold is first heated with twice its weight of salt, three times its weight of copper pyrites, and then heated again with two parts of salt and one of alum. This process removes the impurities when the other substances have been burnt away in an earthenware crucible; the gold itself is left behind pure and uncorrupted.

Silver

95. Let me speak now about silver, the next source of madness in men. Silver is found only in deep shafts. It does not advertise its existence, having no shiny particles such as are seen in the case of

1. In about 500–497 BC.

gold. Its ore is sometimes red, sometimes the colour of ash. It can be smelted only with lead[1] or the lead mineral called galena, which is found mostly mixed with veins of silver. When cupelled, part of the ore precipitates as lead, while the silver floats on the surface like oil on water.

96. Silver is found in very nearly all the provinces, but the best comes from Spain, where together with gold it occurs in barren ground and even in the mountains. Wherever one vein is found, another is subsequently discovered not far away. Indeed, this happens with almost every metal and is apparently the reason for the Greek use of the term *metalla*.[2] It is a remarkable fact that shafts begun on Hannibal's initiative all over the Spanish provinces are still in existence; they are named after their discoverers. One such mine, known today as Baebelo, provided Hannibal with 300 pounds of silver a day. The galleries ran for between 1 and 2 miles into the mountain, and along the whole of this distance water-men stood day and night in shifts measured by lamps, bailing out water and making a stream.

98. The vein of silver nearest the surface is called 'raw'. In earliest times digging stopped when alum was discovered and no further search was made, but recently a vein of copper was found below the alum and this revived men's hopes. The fumes from silver mines are dangerous to all animals but especially to dogs. The beauty of gold and silver is in proportion to their softness. Most people are surprised that black lines can be drawn from silver.

Mercury

99. There is also a mineral found in these veins of silver that contains a distillation – always in liquid form – which is called mercury.[3] It is universally poisonous and destroys any container, corroding it with an invasive process of disintegration. All substances float on its surface except gold, which is the only metal that it attracts to itself. Mercury is thus very good for refining gold, since, if the two are repeatedly shaken together in earthenware vessels, the mercury draws out all the impurities in the gold. After the impurities have been driven out, separation of the mercury from gold is achieved by pouring both on to well-dressed hides; the mercury is exuded through the hides like a kind of sweat and the gold is left pure.

1. Cupellation.
2. *ta met' alla*: 'one after another'.
3. *argentum vivum*.

The gilding of copper substrates

100. When articles made of copper are gilded, mercury is applied under the gold leaf and this keeps the leaf very firmly in place. But if the gold leaf is applied in one layer, or is very thin, its pale colour betrays the mercury backing.

101. In the same silver mines is found what may properly be called a rock that is composed of white and shiny, but not transparent, froth. Some people call antimony *stimi* or *stibi*, others *alabastrum* or *labarsis*. There are two kinds: male[1] and female.[2] The latter is preferred, since the male kind is more uneven and rougher to the touch, as well as lighter in weight, less brilliant and more like sand. Pure antimony, however, is sparkling, friable and splits into thin layers rather than into globules.

102. Antimony is an astringent and a coolant, but is mainly used round the eyes. This is why it is generally known by a Greek name meaning 'wide-eyed',[3] because it is used as an eyebrow cosmetic to dilate women's eyes.

Some medical uses of metals

105. The Greeks call the silver slag 'dross'. It has an astringent and cooling effect on the body and, like lead sulphide – which I shall discuss under 'lead' – it heals when used in plasters and is especially effective in causing wounds to close over. Likewise it is a very effective means, when administered together with oil of myrtle in enemas, of counteracting both constipation and dysentery. People add this dross to preparations called 'emollient plasters' which are used on the raised edges of ulcers, on sores caused by rubbing and on suppurating head sores.

106. In the same mines occurs the mineral called litharge.[4] There are three kinds, with Greek names: the best-known is *chrysitis*, 'golden'; then, *argyritis*, 'silver'; and finally *molybditis*, 'leaden'. Generally speaking, all these colours are found in the same sample. The Attic kind is the most sought after, then the Spanish. The 'golden scum' is obtained directly from the mineral ore, the 'silver' from the silver, the 'leaden' from smelting the lead itself, which is

1. Stibnite, or antimony sulphide.
2. Metallic antimony.
3. *platyopthalmon.*
4. *spuma argenti*: lead monoxide.

processed at Puteoli – hence its name *argyritis Puteolana*. Each kind of litharge is made by heating the ore until it liquefies; it then flows from an upper into a lower vessel and is taken out of that by means of small iron spits. It is then directly exposed to the flame on a spit to make it of moderate weight.

Cinnabar

111. Cinnabar[1] is also found in silver mines. Nowadays it is of great importance among pigments; formerly it was not only very important but also sacred. Verrius lists several writers who say – and one must believe them – that on holidays the face of the statue of Jupiter used to be covered with cinnabar, as well as the bodies of those in triumphal processions. They add that Camillus was so adorned at his triumph and that in accordance with this ritual, even in their time, cinnabar was added to the perfumes used at the banquet that follows a triumph. This explains why one of the most important duties of the censors was to arrange a contract for covering Jupiter with cinnabar. I myself am at a loss as to the origin of this custom, although nowadays cinnabar is in demand among the peoples of Ethiopia, whose leaders cover themselves all over with this; all their cult statues are of the same colour. I shall accordingly expound all the facts concerning cinnabar in some detail.

113–4. Theophrastus records[2] that cinnabar was discovered by an Athenian named Callias some ninety years before the archonship of Praxibulus at Athens, that is, AUC 349.[3] He goes on to state that Callias had hoped that gold could be extracted, by smelting, from the red sand found in silver mines and that this was the origin of cinnabar. This mineral was, even then, found in Spain, but there it was a hard, sandy type. It also occurred in Colchian territory on a certain inaccessible rock from which the natives brought it down by throwing spears. This cinnabar, however, was of an impure quality; the best is found in the territory of the Cilbi to the east of Ephesus where the sand is the red colour of cochineal beetles. This sand is ground up, explains Theophrastus; the powder is washed and the sediment rewashed. Skill makes a difference: some workers produce cinnabar after one washing, while with others the cinnabar is rather weak and is improved by a second washing. The importance of the

1. Mercuric sulphide.
2. *On Stones*, 58–9.
3. 405 BC.

colour comes as no surprise to me, for in Trojan times red ochre was valued highly, as Homer testifies (*Il.* II, 637) he commends red-coloured ships, but otherwise he seldom refers to colours and paintwork.

118. Juba records that cinnabar is also produced in Carmania, and Timagenes that it is found in Ethiopia as well; but it is not exported to us from either place and indeed from hardly anywhere other than Spain. The most famous cinnabar mine providing revenue for the Roman people is that of Sisapo in Baetica. The security precautions are second to none. Smelting and refining of the ore are not allowed locally, but as much as 2,000 pounds of crude ore a year are sent to Rome under seal and there purified. To prevent the price going sky-high, it is fixed by law at 70 sesterces for about a pound. The mineral is adulterated in many ways – a source of illegal profit for the mining company.

119. There is another kind of 'cinnabar'[1] found in almost all silver mines, and, similarly, lead mines; this is produced by smelting rock that is veined with metal, and is not obtained from the rock that, when smelted, produces the round drops called mercury, but from other minerals found alongside. In the cinnabar mines of Sisapo the vein of sand contains no silver. It is smelted like gold and assayed by means of gold brought to red heat. If it has been adulterated it turns black, but if genuine it retains its colour.

122. Those who polish cinnabar in workshops tie loose masks made of bladder-skin over their faces to prevent inhalation of dust as they breathe; the dust is a very serious health hazard. The masks, however, allow them to see over the top. Cinnabar is used for lettering in books and it makes more colourful lettering for inscriptions on walls, marble or even tombs.

The touchstone

126. Together with the mention of gold and silver belongs an account of the stone known as the touchstone, usually found only in the River Tmolus, according to Theophrastus, but now known to occur in a number of places. Some call this 'Heraclian' stone, others 'Lydian'. The pieces are of medium size not exceeding 4 inches in length and 2 inches in width. The side of the stone exposed to the sun is better than the underside. Experts use the touchstone like a file to take a scraping from an ore; they can tell to a scruple how

1. Red lead or lead carbonate.

much silver or copper is present, and their amazing calculation is absolutely without error.

127. Silver may differ in two respects. A shaving of silver that remains white when placed on a white-hot iron shovel passes the test of purity. If it turns red, it is of the second quality; if black, it is worthless. But even this test is open to manipulation. If the shovel is dipped in human urine, the silver shaving is stained as it is heated and gives a fraudulently white appearance. A further test for polished silver is if it condenses the vapour from a person's breath and precipitates moisture.

Mirrors

128. It used to be believed that only the best silver could be beaten into plates to produce a reflected image. This was once true, but now even this test is open to fraud. However, the power to reflect images is marvellous. It is generally agreed that the phenomenon is due to the air bouncing back and making contact with the eyes. Similarly, by using a mirror in which the thickness of the metal has been polished and beaten into a slightly concave shape, objects are greatly magnified. Such a big difference does it make whether the surface absorbs the air or reflects it.

129. It is also possible to make goblets with a number of mirrors beaten, as it were, outwards from the inside in such a shape that a single face sees itself reflected as a multiplicity of images equal in number to the reflecting surfaces. Vessels, like those dedicated in the Temple of Smyrna, are devised that achieve strange effects.

130. To conclude the topic of mirrors, the best examples in our ancestors' time were those manufactured at Brundisium from an alloy of tin and copper.[1] Silver mirrors are now extremely fashionable; they were first made by Pasiteles in the time of Pompey the Great. Recently the belief has grown up that a better reflection is achieved by applying gold to the back of glass.

The debasement of silver coinage and a digression on wealth

132. The triumvir Antony alloyed the silver denarius with iron, and forgers alloyed silver coins with copper. Others used a reduced standard of coinage – 84 denarii from a pound of silver. So a method was invented for assaying the denarius under a law that was

1. Bronze.

so popular among the common people that district by district and with one accord they voted statues to Marius Gratidianus. Indeed, it is a remarkable fact that in this field alone bogus specimens are objects of study, so that an example of a forged denarius is subjected to close scrutiny and an adulterated coin costs more than genuine ones.

133. In past ages there was no numeral representing a sum greater than 100,000, and so even today we employ multiples of that number, using the expression 'times', as in 10 × 100,000, or larger multiples. This arose from the practice of usury and the issue of struck coinage. So even now we use the expression 'someone else's copper'[1] for a debt. Subsequently 'Rich'[2] became a surname, although it must be observed that the first man to be so dubbed beggared his creditors.

134. Later Marcus Crassus, who belonged to the Rich family, used to say that only a man who could maintain a legion of soldiers on his yearly income was wealthy. He owned estates worth 200 million sesterces, being the richest citizen after Sulla. Nor would he have been satisfied until he had obtained the whole of Parthia's gold as well. It is true that he was the first to win lasting fame for his wealth, but it is with pleasure that I censure insatiable greed of that kind. I have known of many freed slaves since his time who have been richer; not long ago there were three at the same time when Claudius was emperor,[3] namely, Callistus, Pallas and Narcissus.

135. There is Gaius Caecilius Isidorus, the ex-slave of Gaius Caecilius, who in the consulship of Gaius Asinius Gallus and Gaius Marcius Censorinus[4] drew up his will, dated 27 January. In it he stated that in spite of his many losses in the Civil War, he left 4,116 slaves, 3,600 pairs of oxen, 257,000 head of other cattle and 60 million sesterces in money. He ordered that 1,100,000 sesterces be spent on his funeral.

136. However enormous the riches that may be amassed by other men, these can be only a fraction of the riches amassed by Ptolemy, who, according to Varro, at the time of Pompey's campaigns in the regions adjoining Judaea, maintained 8,000 cavalry at his own expense and gave a feast to 1,000 guests with 1,000 gold goblets which were changed for every course!

1. *aes alienum.*
2. *Dives.*
3. AD 41–54.
4. 8 BC.

137. And Ptolemy's estate can have been only a fraction – and now I am not speaking about kings – of that of Pythes[1] from Bithynia who presented the famous gold plane-tree and vine to King Darius[2] and gave a banquet to Xerxes' forces (some 788,000 men) with the promise of five months' pay and corn on condition that one at least of his five children should be left to cheer his old age. Also compare even Pythes himself with King Croesus! What madness to waste our lives coveting something that has fallen to the lot of slaves and with which even kings cannot be sated.

Famous objets d'art *made of silver*

154. It is remarkable that no one has achieved fame in the art of engraving gold, while many have become famous for chasing silver. The most acclaimed craftsman was Mentor; he only ever made four goblets and, it is said, not one survives because the Temple of Diana at Ephesus and the Capitol at Rome were destroyed by fire.[3] Varro says that he possessed a bronze statue by Mentor. Next in esteem were Acragas, Boethus and Mys. Works by all these artists survive today on the island of Rhodes. In the Temple of Athena, at Lindos, there is a work by Boethus; and in the Temple of Bacchus in Rhodes town, there are some goblets engraved with Centaurs and Bacchantes by Acragas, as well as goblets with Sileni and Cupids by Mys. Hunting scenes on goblets engraved by Acragas enjoyed a great reputation.

156. Zopyrus engraved scenes of the Athenian council of the Areopagus and the trial of Orestes on two goblets valued at 10,000 sesterces. There was also Pytheas, one of whose works fetched 10,000 denarii for a weight of two ounces. This was an embossed base of a bowl showing Odysseus and Diomedes stealing the Palladium. The same artist also carved some tiny drinking cups in the shape of cooks known as 'Little Chefs' of which no one was allowed to make casts because the fine work was extremely fragile. Likewise Teucer, an artist who specialized in embossed work, attained fame; but suddenly this kind of art went out of fashion, so that it is now only sought after in antique specimens – value attaches to engraved items worn by use even if the design has become indistinguishable.

1. The riches of Pythes are described by Herodotus (VII, 27–9 and 38–9).
2. 528–485 BC.
3. The Temple of Diana in 356 BC, and the Capitol in 83 BC.

Changing prices

164. The prices which I have often given, vary from place to place and almost from year to year; they fluctuate because of the cost of shipping or because of the amount paid by a particular merchant, or because some powerful bidder whips up the selling price at auction. I well recall that during Nero's principate Demetrius was prosecuted by the whole district of Seplasia in Capua in a case heard by the consuls, but I have had to state prices generally current at Rome in order to give a standard value for commodities.

BOOK XXXIV

COPPER AND BRONZE SCULPTURE; TIN, LEAD AND IRON

1. I shall now discuss copper-ores, which are the next in value.

2. Copper is produced from an ore that the Greeks call *cadmea*,[1] a highly reputed variety. Cadmea comes from overseas; but formerly it was found in Campania and now it occurs in the region of Bergamum on the very border of Italy; it has also recently been reported in the province of Germany. In Cyprus, where copper was first discovered, it is obtained from copper-stone; this ore was of a particularly low quality and better sources, especially *aurichalcum*,[2] were found in other countries.

4. The highest reputation is now enjoyed by Marian copper, also called Corduba copper.

5. At one time bronze[3] used to be alloyed with gold and silver, but the craftsmanship was valued more highly than the metal. Nowadays it is difficult to know whether the craftsmanship or the metal is worse. However, like everything else, art is undertaken for financial gain, not fame as in former days.

6. Corinthian is the most highly praised bronze and was renowned in earlier times. This alloy was created accidentally when Corinth was burned in the course of being captured. Many people have been affected by a mania for this bronze. In fact it is recorded that Verres, who had been successfully prosecuted by Marcus Cicero, met the same fate as Cicero, being proscribed by Antony because he refused to give up some Corinthian pieces. It seems that most collectors only pretend to be knowledgeable, so as to set themselves apart from the man in the street, and do not have any genuinely more refined insight into the subject.

---

1. A mixture of zinc carbonate and zinc silicate.

2. Copper pyrites.

3. The term *aes* usually refers to bronze (an alloy of copper and tin) but is occasionally used for brass (copper and zinc). Ancient copper also contained traces of lead and arsenic. Pure copper is known as *aes Cyprium*.

Bronze candelabra and the story of Gegania

11–12. Aegina specialized in producing the upper parts of candelabra, and Tarentum the stems. Credit, therefore, for the manufacture of these articles is shared between the two places. People find no shame in buying them for the equivalent of a military tribune's pay. At the sale of such a candelabrum of this kind the fuller Clesippus, an ugly hunchback, was thrown in as part of the lot on the instruction of Theon the auctioneer. A woman named Gegania bought the lot for 50,000 sesterces. She hosted a party to show off her purchases, and for the amusement of those present the man appeared in the nude. Gegania became shamelessly passionate and took him to bed. She later mentioned him in her will. Thus, having become very rich, he worshipped the candelabrum as a deity. However, morality was vindicated in that he put up a fine tombstone to perpetuate the memory of Gegania's shame throughout the world.

Bronze statues

15. Bronze-working came generally to be associated with statues of gods. I find that the first image cast in bronze at Rome was that of Ceres; it was paid for out of the property of Spurius Cassius, who was executed by his own father when he tried to make himself king. The art then passed from representations of gods to statues and likenesses of men in a variety of forms. In early times people used to stain statues with bitumen, and it is thus remarkable that subsequently they should have preferred to cover them with gold. This may have been a Roman invention, but even so it does not have any long history at Rome.

16. Likenesses of men were not usually made unless they deserved lasting commemoration for some outstanding reason, such as victory in the sacred games – particularly those held at Olympia where it was customary to dedicate statues of all winners. When an individual had won three times, exact likenesses were made of him and these were known as 'portrait statues'.[1]

17. The first portrait statues erected at public expense in Athens were probably those of the tyrannicides Harmodius and Aristogiton. This happened in the same year as the expulsion of the kings at Rome. Out of a most civilized sense of rivalry, the setting up of statues was afterwards adopted by the whole world. The custom

1. *iconicae.*

arose of having statues adorn the forums in all municipal towns, and
– so that such records should not be seen only on tombs – the
memory of men was perpetuated by inscribing rolls of honour on
statue bases to be read for all time.

Greek and Roman styles of sculpture

18. In the old days statues were dedicated portraying the subject
clad in a toga, and naked figures holding spears were also popular:
these were modelled on young Greek men in the gymnasia and
were known as 'figures of Achilles'. Greek statues are usually of
nudes, whereas Roman statues have breastplates. The dictator Caesar
gave permission for a statue wearing a cuirass to be put up in the
Forum in his honour.

19. Equestrian statues are certainly popular at Rome; the fashion
for these was no doubt derived from Greece. The Greeks, however,
used only to put up statues of winners of horse races at their sacred
games, although subsequently they also set up statues of winners in
two- and four-horse chariot races. Such is the origin of our dedication
of chariots in honour of those who have celebrated a triumph. This,
however, is a late innovation and it was not until the time of the
Emperor Augustus that chariots with six horses were found, or
indeed with elephants.

Famous Roman statues

20. After defeating Antium,[1] the Roman people fixed on the Rostra
the beaked prows of ships captured in their victory. Similarly, the
statue of Gaius Duilius, the first to achieve a naval victory over the
Carthaginians, still stands in the Forum.

22. There was a different but very good reason for the statue of
Marcus Horatius Cocles which still survives: it was erected because
single-handed he kept the enemy from crossing the Sublician Bridge.

26. I also find that statues of Pythagoras and Alcibiades[2] were set
up in a corner of the Comitium when, during the Samnite Wars,[3]

1. 338 BC.

2. Alcibiades (*c.* 450–404 BC) led a notoriously chequered career. He was commander of the ill-fated Athenian expedition to Syracuse (415 BC). Back in Athens he
was charged with being the ringleader in the mutilation of the statues of the Hermae
and took refuge with Tissaphernes. He returned to Athens to be appointed
commander-in-chief. Defeated at Notium, he went into exile and was killed shortly
after.

3. 343 BC.

Pythian Apollo had ordered a statue of the bravest man of the Greek race to be erected in some conspicuous place, as well as one of the wisest man. These remained until the dictator Sulla built the Senate House on the site. It is cause for surprise that our illustrious senators rated Pythagoras above Socrates[1] whom Apollo had ranked above all the rest of mankind for wisdom, or Alcibiades above any number of men for courage, or anyone above Themistocles[2] for wisdom and courage combined.

27. The reason for placing statues on top of columns was to raise them above all other mortals. This is also the significance of the new invention of the arch. This honour originated with the Greeks. I do not think that anyone ever had more statues than did Demetrius of Phaleron at Athens;[3] the Athenians put up 360 statues in his honour at a time when the year did not exceed that number of days. However, the Athenians soon destroyed them.

36. When Marcus Scaurus was aedile, 3,000 statues appeared on the stage of what was only a temporary theatre. Mummius filled Rome with statues after his conquest of Achaia, although on his death his estate was too small to provide his daughter with a dowry.

Colossi

39–40. We see enormous statues raised – as tall as towers – which are known as colossi. Such is the statue of Apollo on the Capitol, brought over by Marcus Lucullus from Apollonia in Pontus. This was about 45 feet high and cost 500 talents to make. Another example is the statue of Jupiter, dedicated to the Emperor Claudius in the Campus Martius, which is overshadowed by the nearby Theatre of Pompey. Then there is the statue at Tarentum, made by Lysippus,[4] which is 60 feet high. The remarkable thing about this statue is that, although it can be rocked by the hand, it is said to be

1. Socrates (469–399 BC) left no writings, and is known mainly through the dialogues of Plato and the *Memorabilia* of Xenophon. The celebrated Athenian philosopher's teaching led to his unpopularity, indictment on trumped-up charges and execution.

2. Themistocles (c. 528–462 BC) used the silver revenue from the rich Third Strike (c. 483 BC) at Laurion to finance an Athenian navy, on the advice of the Delphic oracle. He defeated the Persians in the decisive naval battle at Salamis (480 BC).

3. An Athenian orator and statesman (c. 345–282 BC), Demetrius was exiled in 307 BC, after ten years as tyrant.

4. Lysippus (*fl.* 328 BC), from Sicyon, was a sculptor famous for the slender proportions of his figures and the precision of their detail.

so finely balanced that it cannot be blown down by any storms. The sculptor provided against this eventuality by erecting a column a short distance from it on the side where it would break the force of the wind. Because of the size of the statue and the difficulty of moving it, Fabius Verrucosus left it behind when he transported the statue of Hercules from the Campus Martius to the Capitol, where it still stands.

The Colossus of Rhodes

41. No statue has commanded greater admiration than the Colossus of Rhodes[1] made by Chares of Lindos, the pupil of Lysippus. It was about 105 feet high. Sixty-six years after its erection the Colossus was toppled by an earthquake, but even lying on the ground it is amazing. Few people can make their arms meet round its thumb, and the fingers alone are larger than most statues. Where the limbs have been broken off, there are huge, gaping holes, and inside can be seen large, heavy pieces of rock that the sculptor used to steady the Colossus when set up. Records show that it took twelve years to complete, at a cost of 300 talents; this money came from the sale of the siege-engines abandoned by King Demetrius when he became tired of the long-drawn-out siege of Rhodes.[2] There are a hundred other large-scale statues in Rhodes, which, although smaller than the Colossus, would each have brought fame to wherever they stood; and in addition there are the five colossi of gods made by Bryaxes.

43. Large-scale statues also used to be made in Italy. We can see the Tuscan Apollo in the library of the Temple of Augustus. It is debatable whether the statue – which measures 50 feet from head to toe – is more remarkable for the quality of its bronze or for its beauty. After he had defeated the Samnites, in a war fought under a solemn oath,[3] Spurius Carvilius made the statue of Jupiter that stands in the Capitol from captured breastplates, greaves and helmets. The figure is so large that it can be seen from the Temple of Jupiter Latiaris on the Alban Mount. From the bronze filings left over, Spurius Carvilius made a statue of himself which stands at the feet of the Jupiter. But of all such colossi the Mercury made by Zenodorus for the Arverni, in Gaul, has pride of place. It took ten

1. A statue of the sun-god – one of the Seven Wonders of the World.
2. 305–304 BC.
3. 293 BC.

years to complete, at a cost of 40 million sesterces. When he had given sufficient proof of his skill in Gaul, Zenodorus was summoned to Rome by Nero and there he cast the gigantic statue – about 106 feet in height – intended as a likeness of Nero, but now dedicated to the Sun after the condemnation of Nero's crimes; it is a breathtaking sight.

46. In his studio we used to admire not only the extraordinary verisimilitude of the clay model but also the framework of small timbers that formed the armature for the work in hand. This statue shows that erstwhile expertise in bronze-casting has not gone by the board.

Famous Greek sculptors

49. An almost countless collection of artists has achieved fame from smaller representations and statues. Outstanding is the Athenian Phidias,[1] famous for his statue of Olympian Zeus, which was made of ivory and gold, although he also cast statues in bronze.

54. In addition to that statue, which no one has rivalled, Phidias made the chryselephantine statue of Athena which stands in the Parthenon. Besides an *Amazon* he made a bronze *Athena* of such remarkable beauty that it is called 'the Fair'. In his works, Phidias is justly considered to have been the first to reveal the potential of sculpture.

55. Polyclitus[2] of Sicyon, a pupil of Hagelades, made a statue of a youth binding his hair, the *Diadumenos* – a young man with soft features, and renowned for costing 100 talents; a statue of a man carrying a spear, the *Doryphoros* – a youth, but of rugged appearance; and a statue that artists call the *Canon*, since they draw their outlines from it as if from a sort of standard. In fact, thanks to this one work, Polyclitus, alone of all men, is considered to have created the very art of sculpture. He also made a statue of a man using a strigil, the *Apoxyomenos*; a nude statue of a man attacking with a spear; and two boys playing dice, the *Astragalizontes* – also in the nude; the latter stands in the forecourt of the Emperor Titus' house and is generally considered to be the most perfect surviving work of his art.

1. The Athenian sculptor (born *c.* 490 BC) responsible for the overall execution of the Parthenon sculptures.
2. Polyclitus (*c.* mid-fifth century BC) was famed for his intellectual absorption with the mathematical proportions of the human form. He established criteria for the proportions of free-standing statues.

56. Polyclitus is considered to have perfected the art of sculpture, which Phidias had revealed. Polyclitus' very own discovery was the art of making statues with their weight concentrated on one leg. Varro says that these statues are of square build and generally follow a single model.

57. Myron, also a pupil of Hagelades, was born at Eleutherae.[1] His statue of a heifer brought him particular fame; this was celebrated by some verses, since the majority of men owe their reputation to someone else's talent rather than their own. His other works include *Ladas, the Discus-thrower* (the *Discobolos*), *Perseus and the Sawyers*, and the group of *Marsyas Wondering at the Flute in the Company of Athena*.

58. Myron seems to be the first sculptor to have extended the scope of realism; there was more harmony in his art than in Polyclitus', and he exercised more care with regard to proportions. Yet although Myron took pains in his representation of the body, he did not express the feelings of the mind. Nor did he render the hair, or the pubic hair, with any more accuracy than the unsophisticated sculptures of olden times.

59. Pythagoras of Rhegium in Italy outclassed Myron with his figure of a pancratiast which stands in Delphi; with this work he also defeated Lentiscus. He was the first sculptor to show the sinews and veins and to give a lifelike rendering of the hair.

61. Lysippus of Sicyon was self-taught, according to Duris, but was originally a bronze-smith who hit on the idea of becoming a sculptor as a result of a remark made to him by the painter Eupompus. He asked Eupompus on which of his predecessors he modelled himself. Eupompus pointed to a crowd of people and said: 'You should imitate Nature, not any other artist.'

62. Lysippus was a most prolific sculptor, and surpassed all other artists in the sheer quantity of his output, which included a statue of a man using a strigil, the *Apoxyomenos*, dedicated by Marcus Agrippa in front of his Baths; Tiberius also much admired this statue. Although he expressed some control over his feelings at the beginning of his principate, Tiberius could not restrain himself in this case and removed the *Apoxyomenos* to his bedroom, substituting a copy. But the people of Rome were so indignant about this that they staged a protest in the theatre, shouting, 'Bring back the *Apoxyomenos!*' And so despite his passion for it, Tiberius was obliged to replace the original statue.

1. *fl. c.* 480–440 BC.

65. Lysippus is said to have made a substantial contribution to the art of bronze sculpture by achieving a detailed rendering of the hair, and by making the head smaller than had previous sculptors, and the body more slender and solid, so that his statues appeared to be taller. He carefully preserved 'symmetry' – there is no equivalent term in Latin – by the new and untried method of modifying the squareness of the figure that the old generation of sculptors had portrayed. He used to say that he made men as he visualized them, whereas his predecessors made them as they were. A special characteristic of Lysippus was his finish that preserved even the smallest details.

69–70. Praxiteles, although more successful and renowned for his sculpture in marble, nevertheless made some very beautiful bronzes, including the *Rape of Persephone*, the *Girl Spinning*, the *Bacchus* with a figure of Drunkenness – and the famous *Satyr*. He also sculpted a youthful Apollo slaying a lizard creeping towards him – the *Sauroctonos*.

The sculptor Perillus

89. Nothing good can be said about Perillus. He was more cruel than the tyrant Phalaris, for whom he made a bull with the promise that if a man were shut inside it and a fire lit beneath, the bull would bellow. Perillus himself was the first to suffer this torture, a cruelty that in his case was fitting, for such were the depths the sculptor plumbed when he turned this most humane art away from images of gods and men. And after his example many other masters of that art laboured at applying it to implements of torture. Perillus' works are preserved for one reason only, that anyone who sees them will hate the hands that made them.

Copper and bronze used in casting

97. The composition of bronze for statues, as well as for sheets of metal, is as follows: the ore is melted and to the melt is added a third part of copper scrap – that is, used, second-hand copper. This scrap contains an intrinsic, seasoned brightness, since it has been subdued by friction and tamed by use. Tin is also alloyed with it, in the proportion of one part of tin to eight parts of copper.

98. Then there is the bronze referred to as 'suitable for moulds'; this is very delicate because a tenth part of lead and a twentieth part of silver-lead is added; it is the best way to impart the colour called

Grecian. The last kind is known as pot-bronze – a name that comes from vessels made of it. The alloy is composed of three or four pounds of silver-lead for every hundred pounds of copper. Lead added to Cypriot copper makes the purple in the borders of robes on statues.

Copper, its compounds and their uses

99. Bronze articles are prone to more rapidly spreading corrosion when polished than when neglected, unless they are smeared with oil. People say that they are best preserved by being dipped in liquid pitch. Bronze was used formerly to ensure that records would last indefinitely; official business was inscribed on bronze tablets.

100. Copper-ores and copper mines provide a variety of medicinal products which quickly cure all types of ulcers. The most useful substance, however, is *cadmea*.[1] This is indeed also a by-product of silver-smelting; in this case it is whiter and not so heavy as *cadmea* occurring in ore-deposits but in no way comparable with that obtained in copper-processing.

Copper slag and copper compounds used in medicine

107. Copper slag is washed in the same way as copper but is a less effective remedy. Flower of copper[2] is also useful as a medicine. To produce it copper is fused and then transferred to furnaces where a more intense, forced draught makes the metal give off layers like scales of millet which people call the 'flower'. When the sheets of metal are cooled in water, they shed further copper scales, of a reddish colour, known as 'husk'. In this way flower of copper can be adulterated, with the husk being sold as a substitute for it. True flower of copper is a scale knocked off bars into which have been welded cakes of the metal, especially in foundries in Cyprus.

108. But doctors, if they will allow me to make this observation, are ignorant of all this. They are misled by names and far removed from the manufacture of drugs that used to be their province. Now, whenever they come upon books of prescriptions, they want to make up medicines out of them, and conduct experiments at the expense of their wretched patients. They rely on chemists' shops in Seplasia which cause untold harm with quack remedies, and for a

1. Zinc carbonate and zinc silicate.
2. Red cuprous oxide.

long time they have been buying plasters and eye-salves over the counter. In such a way are the inferior products and fraudulence of Seplasia advertised.

110. Great use is also made of verdigris, which can be made in several ways. One can scrape it from the ore from which copper is smelted, or drill holes in brass and suspend the scrapings over strong vinegar in sealed containers. The verdigris is of much better quality if scales of copper are suspended instead of scrapings.

112. Verdigris can be detected by papyrus that has been steeped in an infusion of gall. When smeared with genuine verdigris, the papyrus immediately turns black.

113. The potency of verdigris makes it well-suited for eye-salves. Its caustic property induces watering of the eyes, and so it is vital to wash it off with warm swabs until the corrosive effect has ceased.

Copper pyrites

117. The ore from which copper and *cadmea* are obtained by smelting is called copper-stone.[1] *Cadmea* is quarried from rocks above ground, while copper-stone is mined underground and immediately crumbles into small pieces, being of a soft consistency resembling matted down. Copper-stone further differs from *cadmea* in that it contains three kinds of mineral – copper, *misy*[2] and *sory*[3] – each of which I shall describe; the veins of copper in it are oblong in shape. Top quality copper-stone is honey-coloured, streaked with fine veins and liable to crumble, but is not stony.

121. Some authorities have recorded that *misy* is made by burning copper-ore in trenches, its fine yellow powder mixing itself with the ash of the burnt pine-wood. But, although obtained from copper-stone, it is an integral part of that mineral and separated from it only by force. The best kind of *misy* is extracted in copper-works in Cyprus. When broken, its fragments sparkle like gold; when ground, it has a sandy appearance. A mixture of *misy* is used by gold-refiners.

138. I must now give an account of iron mines and ferriferous ores. Iron is the best and worst of life's materials. With iron we plough the ground, plant trees, trim other trees that support our vines, and compel the vines to put out new shoots by cutting out

1. *chalcitis.*
2. Copper pyrites.
3. Decomposing marcasite.

dead wood every year. With it we build houses, quarry rocks and achieve many other useful tasks. But we also use iron for wars and bloodshed and plunder. We use it not only at close quarters, but as winged missiles. Once iron projectiles were hurled by the arm, but now they are actually equipped with wings! This I consider to be the most criminal misuse of man's genius: to make it possible for death to reach human beings more quickly, we have taught iron how to fly and have given it wings!

139. Let man, not Nature, take the blame for this. Men have often tried to make it possible for iron to be without guilt. In a treaty granted by Porsenna to the Roman people after the expulsion of the Kings, we find it specifically stated that iron shall be used only for agriculture. The earliest authors have recorded that in their day it was the practice to write with a pen made of bone. There survives an edict of Pompey the Great, dating from his third consulship,[1] at the time of the disorders in which Clodius was killed, that prohibited the presence of any weapon in Rome.

140. The art of those far-off days provided a gentler purpose for iron. When the sculptor Aristonidas wished to convey the madness of Athamas sitting listlessly and overcome with remorse after he had thrown his son Learchus from a rock, he mixed copper and iron together in order that the redness of shame might be represented by the rust of the iron shining through the bright surface of the copper.

141. This statue still survives in Rhodes. Also there is an iron figure of Hercules, made by Alcon and inspired by the perseverance of that god in his Labours. And at Rome we see iron goblets as votive offerings in the Temple of Mars the Avenger. And Nature's kindness has curtailed the power even of iron by exacting the penalty of rust, by this foresight making nothing in the world more perishable than that which is most hostile to mortals.

Iron-ores and smelting

142. Iron-ores are found almost everywhere and are produced today on the Italian island of Ilva. The ores are easily recognized, being clearly visible because of the earth's colour. The smelting of the ore follows the same technique as that used for copper production. In Cappadocia alone there is a question of whether the iron deposits result from the action of water or the nature of the earth, since that

1. 52 BC.

region supplies iron from its foundries when the earth has been flooded by the River Cerasus, but not otherwise. There are many different kinds of iron, and this variety is first encountered in the nature of the soil or climate: some lands produce a soft iron akin to lead; others, a brittle, coppery kind especially to be avoided for wheels and rails – for this purpose the soft iron is suitable. All the different kinds of iron are called 'edging ores', a term not used in the case of other metals. This term is derived from the expression 'to draw out a sharp edge'.

144. There is also a great difference in the way furnaces are used: by one special process[1] the iron is smelted to give hardness to a blade; by another, to give solidity to anvils and hammer-heads. But the chief difference is the water into which the red-hot metal is at intervals plunged. The water in some districts is more beneficial than in others and has made places famous for their iron, as, for example, Bilbilis and Tarragona in Spain, and Comum in Italy, although there are no iron mines in these places.

145. Of all the various kinds of iron Chinese takes first prize: it is exported to us along with fabrics and skins. The second prize goes to iron from Parthia. These are the only kinds of iron forged from pure metal, all others being alloyed with a softer metal. In our part of the world, in certain places, the lode provides ores of such high tenor, as in Noricum; in other places it is the working, as at Sulmona; and in yet others, as I have said, it is the water.

146. It is usual to quench smaller iron-forgings with oil lest they should be hardened by water and become brittle. It is remarkable that when ore is smelted the iron liquefies and subsequently becomes spongy and brittle.

Lodestone and 'magnetic suspension'

147. I shall defer to its appropriate place discussion of lodestone's attraction to iron.

148. The architect Timochares had begun to use lodestone in the construction of the vaulting of the Temple of Arsinoë[2] at Alexandria, so that the iron statue it housed might appear to be suspended in mid-air, but the project was halted by his death and the death of King Ptolemy who had commissioned this work.

---

1. In this case iron with a high carbon content (*nucleus ferri*).

2. Arsinoë (born *c.* 300 BC) was the daughter of Lysimachus and Nicaea, and the wife of Ptolemy II Philadelphus, king of Egypt (reigned 286–247 BC).

149. Iron-ore occurs more abundantly than all other ores. In the coastal region of Cantabria, washed by the Atlantic, there is a very high mountain which consists – almost incredibly – entirely of iron-ore.

150. Iron can be protected from rust by white lead, gypsum and pitch; the Greeks call rust *antipathia*.[1] It is claimed that this protection may also be achieved through a religious ceremony. In the city of Zeugma on the River Euphrates there is said to be an iron chain used by Alexander the Great in building the bridge there; the links used to repair it have been attacked by rust although the original ones are rust-free.

Lead and tin

156. My next subject is the nature of lead, of which there are two kinds: black and white. White lead[2] is the more valuable and was called *cassiteros* by the Greeks. According to legend, they used to fetch it from islands in the Atlantic[3] and brought it back in wicker-work containers covered with stretched hides. It is now well-known that in Lusitania and Gallaecia it is extracted from surface strata of a sandy black colour.

157. Tin is detected only by its weight. Tiny pebbles of tin appear sporadically, especially in the dried-up beds of what were once fast-moving rivers. The workmen wash the sand and heat the residue in furnaces. Tin is also found in gold mines called *alutiae*, through which a current of water is sent to wash out the black pebbles with their white spots of tin. The pebbles weigh the same as the gold, and so remain with it in the bowls in which it is collected. Afterwards the pebbles are separated in the furnaces and fused into tin.

158–9. There is no lead in Gallaecia, although neighbouring Canta-bria has rich deposits of this and no tin. Tin yields no silver although silver can be obtained from lead. Homer bears witness to the fact that tin was important even in Trojan times and he called it *cassiteros*. There are two sources of lead: either it comes from an ore of its own that produces no other substance, or it occurs with silver,[4] in

1. 'the opposite of iron'.
2. By 'white lead' Pliny means tin.
3. The Cassiterides ('Tin Islands'), generally identified as St Michael's Mount in Cornwall.
4. As in argentiferous galena, an ore containing lead and silver.

which case both metals are smelted together. The metal which liquefies first in the furnace is known as *stagnum*, the next to liquefy is argentiferous lead, and the residue is impure lead which constitutes a third of the original charge. When the impure lead is smelted again, it yields black lead,[1] losing two-ninths of its weight.

164. The lead that we use in the manufacture of pipes and sheets is mined with considerable effort in Spain and the Gallic provinces; in Britain, however, it is found on the surface in such large quantities that there is a law limiting production. The following are the names of the different sources of lead – namely, Oviedo, Capraria and Oleastrum. There is, however, no intrinsic difference between them provided that the slag is carefully removed by smelting. What is amazing is that these mines – but only these – when abandoned, replenish themselves and become productive again.[2]

Some medical uses of lead

166. In the medical field lead is used by itself to remove scars. Because of their coldness, lead plates are applied to the lower regions and the kidneys to check excessive sexual appetite and nocturnal emissions if the latter amount to a clinical disorder in their frequency. By such plates the orator Calvus, it is said, controlled his passions and preserved his physical strength for the hard work involved in study. Nero, whom the gods were pleased to make emperor, used to put a lead plate on his chest when singing loudly and so demonstrated a way of conserving the voice.

167. For medicinal use lead is melted in earthenware vessels with a layer of finely powdered sulphur beneath, and then stirred with an iron rod. While it is being melted the nasal passages must be protected to prevent any inhalation of the injurious, deadly vapour from the lead furnace. These fumes very quickly harm dogs, and the fumes from all metals are harmful to flies and gnats – which is why these annoying insects are not found in mines.

169. Calcined lead is washed like antimony and *cadmea*. It can act as an astringent, arrest bleeding and heal scars. It is just as useful for eye-medicines (especially to prevent the eyes bulging), for healing cavities or lumps left by ulcers, and for haemorrhoids and anal fissures and swellings.

1. Pliny means pure lead.
2. Pliny here subscribes to a commonly held view in the ancient world that mineral deposits were self-regenerating.

BOOK XXXV

PAINTING, SCULPTURE AND ARCHITECTURE

1. Now I come to the various kinds of minerals and stones. These add up to an even larger series of topics, and have been treated separately in many volumes, especially by the Greeks. I shall be as brief as befits my plan, but without omitting anything that is necessary and part of the natural world.

Painting and changing tastes

2. First I shall conclude my remarks on painting. It was once a celebrated art – much in vogue with kings and nations – which brought fame to those it thought worthy of being handed down to posterity. Currently, however, it has been supplanted by marble and gold: slabs of marble not only cover entire walls but are also engraved with patterns and decorated with twisting lines that represent objects and animals.

3. We are no longer content with panels depicting views of mountain-ranges in our bedroom, but have begun even to paint the stonework. This practice dates from Nero's principate.

Portraiture

4. Portraiture, the medium by which exact likenesses of people were handed down through the ages, is completely out of fashion. Bronze shields are set up as monuments; they bear a design in silver with a faint outline of human figures. Heads of statues are interchangeable – indeed, before now, sarcastic verses have gone the rounds on this topic. Meanwhile, people cram the walls of their galleries with old pictures and revere portraits of strangers. As for likenesses of themselves, their concern for honour extends only as far as the price, so that they are quite content for their heir to break up the statue and drag it out of doors by a noose.

5. So no one's likeness lives on, and men leave portraits that represent not themselves, but their money. They decorate their dressing-rooms with portraits of athletes from the palaestra, and display pictures of Epicurus all round their bedrooms and on their persons. They make sacrifices on his birthday and observe his festival (which they call *eikas*) on the twentieth of every month. And these are people who are not concerned for their fame even during their lifetime! Such then is the situation. A fatal malaise has taken hold of the arts, and since our minds cannot be portrayed, our physical features also are neglected.

6. It was quite different with our ancestors. In their halls portraits were objects to be admired. They were not statues by foreign artists, not bronzes, not marbles, but wax masks of members of their family, and these were displayed on individual urns so that their likenesses might be carried in procession at family funerals. For, invariably, when someone died, all the members of his family who had ever existed were present. The family tree was traced by lines connecting the painted portraits.

7. Our ancestors' archive rooms were filled with books, records and written accounts of their achievements while in office. Outside the houses and round the door-lintels were other likenesses of those remarkable men. Spoils taken from the enemy were fastened to their doors and not even a subsequent purchaser of the house was allowed to take these down. Consequently, as they changed owners, houses celebrated an ongoing triumph.

9. I must not fail to include an innovation whereby likenesses — in bronze at least, if not in gold or silver — are put in libraries in honour of those whose immortal spirits speak to us in these places. Indeed even imaginary likenesses are modelled, and our longing for such gives birth to faces that have not been handed down to us, as has happened in the case of Homer.

10. In my view there is nothing more pleasing than that people should always wish to know what kind of man someone had been. At Rome this practice began with Asinius Pollio, who was the first to make works of genius public property by founding a library. Whether this idea began earlier with the kings of Alexandria and Pergamum, who had been great rivals in founding libraries, I could not readily say.

The history of painting

15. The origin of painting is uncertain, and is not part of the intended scope of this work. The Egyptians claim that they invented painting 6,000 years ago, before it crossed to Greece; this is clearly an empty assertion. Some Greeks claim it was discovered in Sicyon,

others in Corinth; but there is universal agreement that it began by the outlining of a man's shadow. Such were the first pictures. The second stage, when a more sophisticated method had been invented, was painting in a single colour, a method, still in use, known as monochrome.

16. Line-drawing was invented by the Egyptian Philocles or the Corinthian Cleanthes. But the first Greeks to practise it were Aridices from Corinth and Telephanes from Sicyon. They used no colour but added extra lines within the outline, and these artists used to write the names of the subjects on their pictures. Ecphantus from Corinth is said to have been the first to daub these pictures with colour made from powdered earthenware.

17. The art of painting had by this time already been perfected in Italy. At any rate, today in the Temple of Ardea paintings survive that antedate the foundation of Rome. And although they are open to the sky, they have, amazingly, survived this long period as if newly painted. Similarly, at Lanuvium there are an *Atlanta* and a *Helen* close together, represented in the nude and painted by the same artist; the former is depicted as a virgin. Both are of outstanding beauty and have not been damaged by the collapse of the temple.

18. The Emperor Caligula, burning with desire for them, tried to remove the pictures but the nature of the plaster would not allow this. Pictures surviving at Caere are even older. Careful assessment of these pictures will prove that none of the arts reached full perfection so quickly, for it is clear that painting was non-existent in Trojan times.

Early Roman painting

19. Roman painting achieved distinction at an early date, and is the origin of the famous Fabii family's surname, 'Pictor'.[1] The first Fabius Pictor decorated the Temple of Health with paintings.[2] This work survived until our own day when the temple was burnt down during the principate of Claudius. Second in order of fame was the decoration of the poet Pacuvius[3] in the Temple of Hercules in the Cattle Market. Pacuvius was a nephew of Ennius[4] and enhanced the

---

1. 'Painter'.
2. *c.* 304 BC.
3. Roman tragic poet (*c.* 220–*c.* 130 BC).
4. Roman epic and dramatic poet (239–169 BC).

standing of painting at Rome because of his fame as a dramatic poet.

20. After Pacuvius, painting was not considered a worthy profession for persons of rank – unless one considers, in our own day, Turpilius, a Roman knight from Venetia, for his beautiful works that survive at Verona. Turpilius painted with his left hand, which was unique among artists.

22. In my view Manius Valerius Maximus Messala made a particularly significant contribution to painting's reputation when in AUC 490[1] he was the first to exhibit a picture in public on a side wall of the Curia Hostilia; this was of the battle in Sicily in which he defeated the Carthaginians and Hiero. Lucius Scipio[2] had a similar impact when he put a picture of his victory in Asia Minor in the Capitol. People say that his brother Africanus was annoyed by this – not without good reason, since his son had been captured in that battle.

23. Lucius Hostilius Mancinus,[3] who had been the first man over the wall at the storming of Carthage, similarly angered Aemilianus by exhibiting in the Forum a diagram of the plan of the city and of the assaults on it: he stood beside it and personally described the details of the siege to the general public who looked on. This sociability gained him the consulship at the next election. The scenery put up for the shows presented by Claudius Pulcher was also greatly admired, as crows were seen trying to alight on the rooftiles, deceived by the realism of the painting.

24. Lucius Mummius, who gained the surname Achaicus from his victory over the Greeks,[4] was the first to bring about official recognition for foreign paintings at Rome. At the sale of booty, King Attalus[5] paid 600,000 denarii for a picture of Bacchus painted by Aristides. Mummius was surprised at the price this fetched and suspected that there must be something special in the picture that had escaped him. So he withdrew it from sale, in spite of strong protests by Attalus, and put it on display in the shrine of Ceres; this, I think, is the first example of a foreign picture passing into state ownership at Rome. After this foreign pictures were commonly hung even in the Forum.

1. 264 BC.
2. Lucius Scipio defeated Antiochus III in 190 BC.
3. Commander of the Roman fleet in the Third Punic War (149–146 BC).
4. In 146 BC Mummius sacked Corinth.
5. King of Pergamum (reigned 159–138 BC).

26. But it was the dictator Caesar who ensured outstanding official importance for pictures, when he dedicated paintings of Ajax and Medea in front of the Temple of Venus Genetrix. Marcus Agrippa, one who was more a rustic at heart than a man of refined tastes, followed Caesar's example: a speech of his, lofty in style and worthy of the greatest of the citizens, survives on the subject of nationalizing all paintings and statues – this would have been preferable to their being spirited away to country houses. And that same severe character bought two pictures, an *Ajax* and an *Aphrodite*, from the city of Cyzicus for 1,200,000 sesterces. He also had small pictures set in the wall of the hot room of his Baths, but these were removed a little while ago when the Baths were being refurbished.

27. The late Emperor Augustus surpassed all others when he put two pictures in the busiest part of the Forum, one representing *War and Triumph*, the other *Castor and Pollux together with Victory*. Similarly, in a wall of the Senate House, which he was dedicating in the Comitium, he set a picture of Nemea seated on a lion, holding a palm-branch in her hand; next to her stood an old man leaning on a stick; and above his head was a two-horse chariot. On the picture Nicias added an inscription to the effect that this was an 'encaustic' painting.

The painter's palette

29. I have already mentioned the colours that the first artists used, when, under the heading of metals, I was writing about pigments called monochromes, so named after the type of painting in which they were employed.

30. The bright colours that the client supplies to the artist are bright red, rich blue, vermilion, green, indigo and bright purple. The rest are dark. Of the whole palette some pigments are natural, some artificial. Natural colours are brown, red ochre, ruddle, white chalk, white marl, Melian white and bright yellow. The rest are artificial. Among the commoner kinds are yellow ochre, burnt white lead, realgar, vermilion, Syrian red and black.

White pigments

36. White chalk, *paraetonium*, derives its name from Paraetonium in Egypt. People say it is sea-foam hardened with mud, which explains why tiny shells are found in it. It is found on the island of Crete and at Cyrene.

37. The best marl, *Melinum*, occurs on the island of Melos. Another kind of white pigment is white lead, the nature of which I have described in my discussion of lead ores.

38. Burnt white lead was discovered by accident when some was burnt in jars in a fire at Piraeus; it is the basis of shadows in paintings.

Black pigments

41. Black pigment, *atramentum*, is also included among artificial colours, although it can come from earth in two ways: it either exudes from the earth like brine, or is actual earth of a sulphur colour. Painters have been known to dig up charred materials from graves to obtain this pigment. Black pigment can be produced in several ways from the soot obtained by burning resin or pitch, and this has led to the construction of factories which do not discharge their smoke into the atmosphere. The most highly rated black pigment is derived from resinous pine-wood.

43. The final stage in the manufacture of all forms of black pigment is to expose it to sunlight. Black for ink is mixed with gum; black for painting walls, with glue.

50. The most renowned painters – Apelles, Aetion, Melanthius and Nicomachus – used only four colours to paint their immortal works:[1] the white pigment they used was marl; the yellow, Attic; the red, brown-red ochre from Sinope; and the black, *atramentum*. Even so each picture sold for the price of a whole town. Now, when purple finds its way on to the walls of rooms and when India furnishes the mud of her rivers and the gore of her snakes and elephants, there is no first-rate painting. Everything was better when resources were more limited. The reason for this change, as I have explained before, is because people nowadays value materials above genius.

51. I shall mention here an example of our generation's folly with regard to painting. The Emperor Nero had commissioned his portrait: it was to be painted on a colossal scale on linen more than 120 feet high, something unprecedented. The finished picture was struck by lightning and destroyed by fires in the Gardens of Maius, along with the larger part of the gardens.

52. When a freedman of Nero was putting on a gladiatorial show

---

1. cf. Cicero, *Brutus*, 70; Cicero mentions the names of other painters who used only four colours.

at Antium, paintings containing life-like portraits of all the gladiators and their assistants decorated the public porticoes. Portraits of gladiators have commanded the greatest interest in art for many generations. It was, however, Gaius Terentius Lucanus who began commissioning pictures of gladiatorial shows and having them publicly exhibited.

Eminent artists

53. I shall now catalogue as briefly as possible, the names of those celebrated in the art of painting.

54. The Greek authorities are inconsistent in this field, putting their painters many Olympiads after sculptors in bronze and metal-engravers: they assign the earliest painter to the 90th Olympiad,[1] although it is said that even Phidias was initially a painter and that there was a shield painted by him at Athens.

55. Candaules[2] paid its weight in gold for a picture by Bularchus of the battle with the Magnetes. In such esteem was painting held. This must have happened at about the time of Romulus, because Candaules died in the 18th Olympiad,[3] or, as some authorities state, in the very same year as Romulus; which shows, unless I am mistaken, that painting was highly regarded and had reached perfection even by that date.

56. Eumarus of Athens was the first artist to distinguish the male from the female sex in painting, and he turned his hand to all kinds of figures. Cimon of Cleonae developed the discoveries of Eumarus and first introduced oblique, that is, three-quarter, views. He also varied the features so that a subject might seem to look backwards, up or down; and he painted the detail of the joints and veins, as well as wrinkles and folds in the clothes.

57. Phidias' brother Panaenus actually painted the Battle of Marathon between the Athenians and Persians. Indeed the use of colour had become so widespread and art so accomplished that Panaenus is said to have included portraits of the leaders who took part in the battle, namely Miltiades, Callimachus and Cynaegirus for the Athenians, and Datis and Artaphernes on the barbarian side.

58. After these artists but before the 90th Olympiad there were

---

1. 420–417 BC.
2. King of Lydia in the eighth century BC.
3. *c.* 708–705 BC.

other famous painters like Polygnotus of Thasos who first depicted women in see-through clothing and with multi-coloured bands around their heads. He made many other improvements to technique: he painted the lips parted so as to reveal the teeth, which brought expression to the face in place of the rigidity of former times.

59. Polygnotus painted the Temple of Apollo at Delphi and the colonnade at Athens called the 'Painted Portico', without a fee, although part of the work was executed by Micon who was paid. Indeed Polygnotus was more highly esteemed, as the Amphictyons, a General Council of Greece, voted him entertainment at the public expense.

Apollodorus, Zeuxis and Parrhasius

60. Apollodorus[1] was the first artist to express realism and to confer fame on the paintbrush in its own right.

61. But it was Zeuxis of Heraclea, in the fourth year of the 95th Olympiad[2], who entered the portals of art thrown open by Apollodorus.

65. In a contest between Zeuxis and Parrhasius, Zeuxis produced so successful a representation of grapes that birds flew up to the stage-buildings where it was hung. Then Parrhasius produced such a successful *trompe-l'oeil* of a curtain that Zeuxis, puffed up with pride at the judgement of the birds, asked that the curtain be drawn aside and the picture revealed. When he realized his mistake, with an unaffected modesty he conceded the prize, saying that whereas he had deceived birds, Parrhasius had deceived him, an artist.

67–8. Parrhasius, who came from Ephesus, also made a substantial contribution to painting. He was the first to introduce proportion, the first to impart liveliness to the expression, elegance to the hair and beauty to the mouth. Artists concede that he was unsurpassed in the drawing of outlines. This is the highest mark of refinement in painting. To paint body and surface within the outlines, although doubtless a great achievement, is one wherein many have acquired fame; but the contours of the figures and the boundaries of the colouring are rarely satisfactorily achieved in painting.

1. For example, Apollodorus invented shading (*skiagraphos*).
2. 398 BC.

Apelles and some interesting stories about his work

79. Apelles of Cos, who flourished around the 112th Olympiad,[1] surpassed all previous painters and those as yet unborn. As an individual he can almost be said to have contributed more to painting than all other artists combined, and even published books on the principles of painting. His art was particularly distinguished by its gracefulness. Other very great painters were his contemporaries. He admired their works and greatly praised all of them, but they themselves used to say that they lacked his elegance – which the Greeks call *charis* – adding that although they possessed all the other qualities, yet none was his equal in this one respect.

80. Apelles also claimed another distinction when he admired the work of Protogenes, which involved immense labour and an inordinate attention to detail. He observed that in all respects his works were the equal of those of Protogenes, or Protogenes' works were superior, but in one thing he, Apelles, stood out, namely, knowing when he had put enough work into a painting. A salutary warning: too much effort can be counterproductive. Apelles' frankness was no less than his skill. However, he took second place to Melanthius in composition and to Asclepiodorus in perspective, that is, the art of determining the distance between objects.

81. An amusing exchange took place between Protogenes and Apelles. The former lived in Rhodes and Apelles sailed there, eager to acquaint himself with Protogenes' works – for he was only known to him by reputation. He made at once for his fellow artist's studio, but the latter was not at home. An old woman was looking after a large board resting on an easel. She announced that Protogenes was not in and asked who was looking for him. Apelles said, 'This person,' as he took up a brush and painted an extremely fine line in colour across the board.

82. On Protogenes' return the old woman showed him what Apelles had done. The story goes that, after a close inspection of the line, Protogenes said that the visitor had been Apelles, since such a line could not be the work of anyone else. Protogenes, using another colour, superimposed an even finer line on the first one. As he left his studio he told the old woman to show this to Apelles if he returned, and to add that he was the person for whom Apelles was searching. So it happened. For Apelles came back and,

1. *c.* 332–329 BC.

red-faced at being beaten, divided the lines with a third colour, leaving no room for any finer line.

83. So Protogenes admitted defeat and rushed down to the harbour in search of his visitor. He decided that the board should be preserved for posterity to be wondered at by all, but particularly by artists. I am told that it was destroyed by the first fire in Augustus' palace on the Palatine.[1] This work had previously been the object of wide admiration, containing, as it did, nothing other than barely visible lines. Hanging among the outstanding masterpieces by many artists it looked blank. For this reason it attracted notice and was more celebrated than any other work on display.

84. It was Apelles' habit not to allow a day to be so full that he had no time to practise his art by drawing a line. This became proverbial. He exhibited his finished pictures in a gallery for passers-by to see. He used to hide behind the pictures and listen to what faults people found, reckoning that the general public were more perceptive critics than he was himself.

85. The story is told of a shoemaker who criticized him because in drawing sandals he had omitted a loop on the edge. Next day the critic was so proud that Apelles had corrected the mistake to which he had previously drawn attention that he found fault with the subject's leg. Apelles was indignant and, looking out from behind the picture, said to him: 'A shoemaker should stick to his last!' This exchange has also become proverbial. Apelles was a courteous man, and this made him all the more acceptable to Alexander. Alexander was a frequent visitor to his studio. For he had issued an edict forbidding any other artist to paint his portrait. When in Apelles' studio, Alexander talked often about painting although he had no specialist knowledge, and Apelles used to advise him politely to keep quiet, saying that the lads who ground the colours were laughing at him.

86–7. Such was Apelles' power over a king who was otherwise bad-tempered. In spite of this Alexander accorded him an unmistakable gesture of honour. The king was particularly fond of Pancaste, one of his mistresses, and, out of admiration for her physical beauty, he commissioned Apelles to paint her in the nude. When Alexander realized that Apelles had fallen in love with his subject, he gave Pancaste to the artist. Alexander's magnanimity in this instance was all the more impressive for his mastery of himself. His action in parting with Pancaste was a feat as great as any of his victories, in

1. AD 4.

that he overcame his own feelings, by giving not only his mistress but his affection. Nor was he influenced by regard for the feelings of his loved one because, having been a king's mistress, she was now a painter's. Some authorities think that she was the model for *Aphrodite Rising from the Foam*. Apelles was generous even to his rivals and was the first to establish Protogenes' reputation at Rhodes.

88. Protogenes was not highly thought of by his own countrymen, as is very often the case with a prophet in his own land. When Apelles inquired the price of some finished works, Protogenes suggested a low figure, but Apelles offered him 50 talents and spread the rumour that he was buying them to sell as his own works. This trick stirred the Rhodians into recognizing the artist, and Apelles let the pictures go only to people who made an increased offer for them. Apelles also painted portraits so completely lifelike that, incredible though it may seem, according to the grammarian Apio, one of the 'physiognomists'[1] – as Greeks call those who foretell a person's future by his face – could tell from the portrait either when the subject would die or how many years he had already lived.

89. Apelles had been disliked by Ptolemy who was in Alexander's entourage. When Ptolemy was king of Egypt, Apelles was once driven by a violent storm into Alexandria. His rivals, out of malice, bribed one of Ptolemy's stewards to give him an invitation to dinner. When Apelles turned up, Ptolemy was furious. He lined up his stewards so that Apelles might identify the one who had issued the invitation. Apelles picked up a piece of spent charcoal from the hearth and drew a likeness on the wall. Ptolemy recognized the features of his steward as soon as he started drawing.

90. Apelles also painted a portrait of King Antigonus who had lost one eye. He was the first artist to devise a way of hiding the defect: he drew the portrait in three-quarters view so that the missing eye would not appear in the picture, and he showed only that part of the king's face that he could present intact. Among his works there are pictures of people on their deathbed. It is, however, difficult to say which paintings are the most celebrated.

95. There is, or at least there once existed, a picture of a horse by Apelles. It was painted for a competition in which he sought judgement not from men but from dumb animals. For, seeing that his rivals were getting the upper hand by devious means, he showed

1. *metoposcopi*: literally, 'men who examine foreheads'.

the pictures individually to some horses he had brought in, and they neighed only at Apelles' picture. As this frequently happened on subsequent occasions it proved to be a good test of the artist's skill.

97. Other painters have profited from his discoveries, but there was one which nobody could imitate. On the completion of his pictures he used to cover them with a layer of black varnish, but so thin that, although its reflection enhanced the brilliance of the colours and protected them from dust and dirt, it could be seen only on minute inspection.

Aristides, Protogenes and minor artists

98. A contemporary of Apelles was Aristides, of Thebes, the first painter to portray the mind and express the personality of a human being – what the Greeks call *ethos* – and also the emotions.

101. Protogenes also flourished at the same time, as has been observed. He was born at Caunus, of a people subject to Rhodes. At the beginning of his career he was very poor and put a great deal of effort into his paintings, and, therefore, was not very productive. There is no certainty about the identity of his teacher. Some authorities claim that he painted ships until he was fifty years old. For when he was painting the porch of the Temple of Athena on a very famous site at Athens (where he painted his well-known *Paralus* and *Hammonias*,[1] which some people call the *Nausicaa*), he added some small representations of warships in what painters designate as 'secondary subjects'.[2]

102. Protogenes' portrait of Ialysus[3] is his best painting and this has been dedicated in the Temple of Peace at Rome.

104. To avoid burning this picture, King Demetrius[4] did not set fire to Rhodes when the city could be taken only from the side where the masterpiece was housed. So Demetrius let slip a chance of victory for the sake of a picture![5] At that time Protogenes was in his tiny garden on the outskirts of Rhodes town – that is, in the middle

1. The 'patron' heroes of the sacred triremes used at Athens in the service of the state for official business. The *Hammonias* replaced the older ship known as the *Salamnia*.

2. *parerga*: details other than the main subject of the painting.

3. The mythical founder of the town Ialysus in Rhodes.

4. Of Macedonia, surnamed Poliorcetes.

5. Demetrius' inconclusive siege of Rhodes took place in 305 BC. From the proceeds of the sale of the siege-engines the Rhodians financed the Colossus of Rhodes.

of Demetrius' camp – and he would not have been disturbed by the raging battle or have broken off from the works he had begun, had Demetrius not summoned him and asked what gave him confidence to remain outside the walls. Protogenes replied that he knew that the king was waging war against the Rhodians, not the arts. Demetrius, pleased to be able to protect the hands he had spared, set up guardposts to ensure his security. So as not to call him too often away from his work, although an enemy, Demetrius visited Protogenes. He abandoned his hopes of victory, and in the midst of the fighting and the battering of the walls, he observed the artist at work.

Italian painters

115. I ought also to mention the painter of the temple at Ardea, especially since he was made a citizen of the town and honoured with the following inscription on the picture:

Marcus Plautius, a worthy man, decorated the shrine of Queen Juno, consort of Jupiter most high, with paintings worthy of this place. Reputed to have been born in wide Asia, now and evermore Ardea praises him for his artistic skill.

116. These lines are written in archaic Latin. Nor must Spurius Tadius go unmentioned, who belongs to the age of the late Emperor Augustus. He was the first to introduce the extremely attractive feature of painting room walls with representations of country houses, porticoes, landscaped gardens, groves, woods, hills, fishponds, canals, rivers, coasts, and anything else one could wish for. He also painted people walking, sailing, or on land; he painted them visiting country houses, riding on asses or in carriages, as well as fishing, fowling, hunting or even gathering the grape harvest.

117. Some of his works show splendid country houses approached by roads across marshes, with men sinking to their knees as they carry women on their shoulders for a wager. Several of these pictures are skilful and very witty. Tadius also began to paint pictures of seaside cities for terraces; these provided an extremely agreeable effect at very little cost.

118. Fame then was achieved only by those who painted movable pictures, and this wisdom of an earlier age is worthy of respect – namely that they did not care to decorate walls for the exclusive enjoyment of the owners of houses, creating works that would remain in one place and could not be saved if there was a fire.

Protogenes was happy with a cottage in a small garden. There were no frescos in Apelles' house; people had no wish then to paint entire walls. Each artist painted with the interests of his city at heart, and the artist himself was the common property of the world.

Women painters

147. There have also been women painters: Timarete, the daughter of Micon, who painted the picture *Artemis at Ephesus* in the old-fashioned style; Irene, daughter and pupil of the painter Cratinus, who painted *Maiden at Eleusis, Calypso, Old Man, Theodorus the Juggler*, and *Alcisthenes, the Dancer*; Aristarete, the daughter and pupil of Nearchus, who painted an *Asclepius*. When Marcus Varro was a young man, Iaia from Cyzicus, who never married, painted with a brush at Rome, and also drew on ivory with an engraving-tool: she did mainly female portraits, including a large picture on wood, *Old Woman at Neapolis*, and a self-portrait done with the aid of a mirror.

148. No one produced a picture faster than she did, and her artistic skill was so great that in the prices her pictures fetched she far exceeded the most famous portrait-painters of the same period, namely, Sopolis and Dionysius, whose pictures fill the art galleries.

Modelling and moulding in clay and wax

151. Quite enough has been said here about the art of painting. It is now time to say something about modelling. By taking advantage of the earth itself, Butades, a potter from Sicyon, was the first to introduce the modelling of portraits in clay at Corinth. This was thanks to his daughter. She was in love with a young man, and when he was going abroad she drew a silhouette on the wall round the shadow of his face cast by the lamp. Her father pressed clay on this to make a relief and fired it with the rest of his pottery. This is said to have been preserved in the Shrine of the Nymphs until Mummius sacked Corinth.

153. The first person to make a plaster likeness of a human being from the actual face and, having poured wax on to this plaster mould, to make final corrections to the wax cast, was Lysistratus of Sicyon, the brother of the Lysippus I have mentioned. He it was who began the practice of rendering exact likenesses; before his time artists were more concerned with flattering the sitter. Lysistratus also invented the taking of casts from statues, and this practice became so widespread that no statues or figures were made without a clay model. So clearly the knowledge of modelling in clay was older than that of bronze-casting.

156. Marcus Varro praises Pasiteles,[1] who said that modelling was the mother of metal-engraving, of bronze statues and of sculpture. Although Pasiteles was a leading light in all these arts, he never executed anything before fashioning a model. Varro also asserts that this art had achieved perfection in Italy, especially in Etruria.

The use of clay for statues, pottery and other items

158–9. Terracotta statues still survive in many places. Indeed there are numerous temple pediments in Rome and municipal towns remarkable for their modelling, elegance and durability; they are worthy of more respect than gold and are certainly more blameless. Today in our sacred rites, even despite our rich resources, the first libation is poured not from fluorspar or crystal vessels, but from small earthenware dishes that are the gift of Mother Earth. Her kindness is beyond description, even if we consider her gifts one by one and leave out the benefits she bestows in the form of different kinds of corn, wine, fruit, herbs, shrubs, drugs and metals. Even for clay, increasing demand outruns supply; it is used for wine jars, pipes for water, drains for baths, tiles for roofs, fired bricks for house walls and foundations, or the vessels turned on a wheel that caused King Numa to establish a seventh guild of potters.

160. Many people have preferred to be buried in terracotta coffins – Marcus Varro, for example, who in the Pythagorean style was buried in leaves of myrtle, olive and black poplar. Most of humanity use earthenware containers. Samian pottery is still esteemed for its tableware.

161. In a temple at Erythrae even today two amphoras are exhibited that were dedicated because of their delicacy – the result of a competition between a master and his apprentice as to which could make the thinner earthenware. The pottery of Cos is the most famous in this respect.

163. One dish cost the tragic actor Aesop 100,000 sesterces. I have no doubt that those who read this will be indignant, but – heavens above! – Vitellius, when emperor, commissioned a dish that cost a

1. A Greek sculptor from south Italy who worked in the first century BC. He became a Roman citizen in 90/89 BC. Pasiteles invented exact copying from casts by means of pointing. He made an ivory statue of Jupiter.

million sesterces. To fire it, a special kiln had to be built in the fields. Such is the growth of luxury that even earthenware costs more than vessels of fluorspar!

164. It was because of this dish that Mucianus, in his second consulship,[1] protested and attacked the memory of Vitellius for dishes as big as marshes, although that dish was no more infamous than the poisoned dish with which Asprenas caused the death of 130 guests. This was the basis of Cassius Severus' prosecution.

165. The towns of Rhegium and Cumae are famous for their pottery. The priests of Cybele, known as Galli, mutilate themselves – if we believe Marcus Caelinus – with a fragment of Samian pottery, and only by using this do they avoid any dangerous complications.

Different types of earth and their use in walling and brick-making

169. In Africa and Spain there are earthen walls described as 'compacted' because they are made by packing earth down between two sets of shuttering, so that the material is stuffed in rather than raised up. These walls are durable, resistant to rain, wind and fire, and stronger than quarried stone. Even now Hannibal's watchtowers can still be seen in Spain, and likewise earthen towers on mountain-ridges. The same source provides substantial turves suitable for fortifying camps and building embankments against the force of flooding rivers. And everyone knows that dividing walls can be made with daub and wattle as if constructed with unfired bricks.

170. Bricks should not be made with sandy, gravelly or stony soil, but with clay and white soil, or with red earth, or even with sand if it is coarse. The best time for brick-making is spring since in midsummer the bricks are liable to crack. For buildings, only two-year-old bricks are recommended. When the material for the bricks has been pounded it should be well soaked before moulding.

171. Three types of bricks are made: the *didoron*, which we use, 18 inches long by 12 inches wide; secondly the *tetradoron*; and finally the *pentadoron*. The term *doron* is an old Greek word meaning 'the palm of the hand'. In Greece small bricks are used for private buildings, large bricks for public. At Pitana, in Asia Minor – as also in the city-states of Maxilua and Callet in Further Spain – bricks were produced that when dried would float in water; these were

1. AD 70.

made of a pumice-like material which is extremely useful when it can be worked.

172. Except where a 'concrete' structure is possible, Greeks prefer brick walls since these last for ever if built vertically true.

Sulphur

174. Among the other kinds of minerals, the one with the most remarkable properties is sulphur,[1] which breaks down many other substances. Sulphur occurs in the Aeolian Islands between Sicily and Italy, around Neapolis and in Campania on the hills called Leucogaei. There it is dug out of galleries in mines and purified with fire.

175. There are four kinds of sulphur. First there is rhombic sulphur; the Greek name for this – *apyron* – means 'untouched by fire'; this kind alone forms a solid mass, while all the other kinds are liquid and are treated by boiling in oil. Rhombic sulphur is obtained by mining, and is a translucent green colour; it is the only kind that the doctors use. Then there is 'clod' sulphur; this is used only in fullers' shops. The third kind is *egula*, employed only for bleaching woollens which are held over it; *egula* imparts whiteness and softness. The fourth kind is used especially for making lamp-wicks.

Further uses of sulphur

117. Sulphur has a place in religious ceremonies – namely the purification of houses by fumigation. Its potency is also seen in hot springs, and no other substance is more easily ignited, which clearly shows that it contains a great fiery force. Thunderbolts also give off a smell of sulphur and their light is of a sulphurous nature.

Bitumen

178. Bitumen has similar properties to sulphur. In some places it resembles mud, in others a mineral. It issues with the consistency of mud from the Dead Sea, and is found as a mineral in Syria round about the coastal town of Sidon. Both varieties solidify, thickening to a dense consistency. There is, however, a liquid form of bitumen, such as that from Zacynthus and the kind imported from Babylon – there is also a white variety in Babylon. The bitumen from

---

1. This mineral was, surprisingly, not included by Theophrastus in his treatise *On Stones.*

Apollonia is similarly a liquid, and all these liquid kinds are known to the Greeks as *pissasphalton* because of their similarity to a vegetable pitch.

179. There is also a bitumen of the consistency of oil; it is found in a spring at Agrigentum in Sicily and pollutes the spring's waters. The inhabitants collect it on bunches of reeds to which it quickly adheres, and use it instead of lamp-oil and also as a cure for scab in beasts of burden. Some authorities include naphtha among the types of bitumen.

180. Bitumen can be recognized by its brilliant hue, its heaviness and its overpowering smell.

Alum

183. No less important are the uses of alum. This term refers to a salt exudation from the earth. There are several kinds. In Cyprus there occurs a white alum and a variety of a somewhat darker colour. Although the difference in colour is only marginal, the use to which each is put is very different: the white, liquid kind of alum is useful for dyeing woollens a bright colour, whereas the black is best for dark or dull colours. Black alum is also used in purifying gold.

184. All alum is produced from water and slime, which is a substance exuded by the earth. This collects naturally in hollows during the winter, and then the summer sunshine causes it to crystallize.[1] The first part to precipitate is white in colour. This occurs in Spain, Egypt, Armenia, Macedonia, Pontus, Africa and the islands of Sardinia, Melos, Lipara and Strongyle. The most highly prized alum is Egyptian; the next best is Melian, of which there are two types – namely liquid and dense.

185. Liquid alum[2] has astringent, hardening and corrosive properties. It cures mouth ulcers and is an anti-perspirant.

186. One kind of solid alum, which the Greeks call *schistos*[3] breaks up into a kind of whitish filament, and because of this some people have called it 'hairy alum' (*trichitis*). This is produced from the same ore-mineral as the copper known as *chalcitis*; it is like a sort of sweat coagulated into foam.

1. Pliny is clearly aware of the principles of precipitation and crystallization.
2. Potash alum.
3. 'splittable'.

Kaolin and chalk

194. Kaolin from Chios is also found among medicines; it has the same effect as Samian earth. It is used especially as a face powder for women.

199. There is another chalk-like mineral called silversmith's powder, which is used to polish silver. The lowest grade of chalk is that which our ancestors used to mark the finishing-line in Circus races and the feet of slaves for sale when they had been imported from overseas. Examples of such slaves are Publilius[1] of Antioch, the founder of Roman farce; his cousin Manilius Antiochus,[2] the founder of Roman astronomy; and Staberius Eros, the first Roman grammarian. All of these our ancestors saw brought over in the same ship.

200. But let us leave aside these men, commended to our attention by their literary achievement. Other examples seen on the auctioneer's platform include Chrysogonus, a freedman of Sulla; Amphion, a freedman of Quintus Catulus; and a whole catalogue of persons – although this is not the appropriate time to reproduce it – who have gained wealth at the expense of Roman citizens' blood in the licence that resulted from the proscriptions.

201. This is the mark on these herds of slaves put up for sale, and the disgrace of overweening fortune. We have seen them rise to such power that the honour of the praetorship was decreed them by the Senate at the command of the Emperor Claudius' wife Agrippina. And then we saw them sent back with the fasces wreathed in laurels to places from which they came to Rome with their feet marked with chalk.

1. Publilius Syrus (*fl. c.* 45 BC).
2. Probably father of Gaius Manilius who wrote the *Astronomica* (in the time of Augustus and Tiberius).

BOOK XXXVI

STONES, MINERALS AND MONUMENTS

Marble

1. The nature of stones remains to be considered – a particular source of irresponsible behaviour in men, even leaving aside any reference to gems, amber and vessels of rock-crystal[1] or fluorspar. Everything I have treated so far may be thought of as created for mankind's sake. Nature, however, made mountains for her own benefit, as a kind of structure for holding down the inner parts of the earth, and, at the same time as a means of checking river torrents and wave erosion – that is, to restrain the wildest elements by the hardest material of which she is composed. We quarry mountains and drag them off solely for our pleasure – mountains that we once considered it a remarkable feat even to have crossed.

2. Our ancestors thought the conquest of the Alps by Hannibal, and later by the Cimbri, little short of miraculous. Now these same Alps are quarried for a thousand types of marble. Promontories are opened up to the sea, Nature is levelled. We carry off materials which were meant as barriers between nations; ships are built to transport the marble. Thus mountain-ranges are carried here and there over the waves, Nature's wildest domain. Nevertheless we have a better excuse for this than we have for climbing to the clouds in search of vessels in which to keep drinks cold and for hollowing out rocks that approach heaven so that we may drink from ice.

3. When we hear the prices fetched for these drinking vessels and we see the volume of marble transported by sea or road, let each of us reflect how many people's lives would be happier without these! Indeed, mortals do these things, or rather endure them, for no purpose or pleasure other than that of reclining among variegated marbles – as if we were not deprived of this pleasure by night's darkness, which accounts for half our life.

4. Such thoughts make us blush even for the men of former

---

1. The primary meaning of the Greek word *krystallos* is 'ice'; it came to be applied to rock-crystal because of that mineral's transparent appearance.

times. The sumptuary laws passed by Claudius as censor[1] still exist; they forbade the serving at dinner of dormice and other things too trivial to mention. But no law was passed that prohibited the import of marble, or the crossing of seas to obtain this commodity.

5. Perhaps someone may interject, 'But no marble was being imported!' That is not the case. Some 360 columns were taken to the stage of a temporary theatre intended to be in use for scarcely a month; this happened while Scaurus was aedile. The authorities said nothing. Indeed, they made concessions in the interests of pleasing the populace! But why this excuse? How do vices infiltrate other than through official channels? How else have ivory, gold and precious stones come to be used by private persons? Have we left anything wholly to the gods?

6. Well, so be it. Let them indulge the populace. Were not the authorities also silent when the largest of those columns – nearly 40 feet high and of Lucullan marble – were erected in the hall of Scaurus' house? And this was not carried out surreptitiously. A sewerage contractor compelled Scaurus to indemnify him against liability for damage to the drains while they were being transported to the Palatine. Would it not have been better to give some protection to our morals rather than allow so bad a precedent? But the authorities still kept quiet, when these great masses of marble were dragged to a private house past temples with terracotta pediments! Nor can we think that Scaurus, with a simple lesson in vice, surprised an inexperienced state unable to foresee the outcome of this immoral deed. For it was at that time that Marcus Brutus, in some quarrel, nicknamed the famous orator Crassus the 'Palatine Venus': Crassus had been the first to erect columns of foreign marble on that very same Palatine – although these columns were only of Hymettan marble, not more than six in number and no longer than 12 feet.

8. The moral considerations were naturally disregarded because standards had already collapsed; people saw that attempts to stop what had been forbidden were to no purpose, and preferred to have no laws at all rather than ones that were ineffective. Comparison with subsequent history will demonstrate that we have improved. For who now has a hall with such massive columns? However, before I discuss marble, I think that the good qualities of sculptors working in this medium should be set out. So to begin with I shall review the artists.

1. 169 BC.

Marble statues and sculptors

9. The first to achieve fame as sculptors in marble were Dipoenus and Scyllis, who were born on Crete while Media was still powerful and before Cyrus had come to the throne of Persia, that is, in the 50th Olympiad.[1] They went to Sicyon which was long the home of all sculptors' studios. The Sicyonians had negotiated a state contract with them for making statues of the gods, but before these were finished the sculptors complained of mistreatment and left for Aetolia.

10. Forthwith a famine struck Sicyon, bringing desolation and terrible suffering in its train. The people sought a remedy from Apollo at Delphi and his reply was that Dipoenus and Scyllis should finish the images of the gods. The Sicyonians' request was then granted on payment of hefty fees and after a display of obsequiousness. The statues were of Apollo, Diana, Hercules and Minerva; the *Minerva* was later struck by lightning.

11. Before these artists there had been a sculptor on the island of Chios called Melas, who was followed by his son Micciades and his grandson Archermus. The sons of Archermus – Bupalus and Athenis – had the highest reputation as sculptors at the time of the poet Hipponax,[2] who, it is established, flourished around the 60th Olympiad.[3] If one traces their genealogy back to the time of their great-grandfather, one will find that the beginnings of sculpture coincide with the 1st Olympiad.[4]

12. Hipponax was well known for his ugliness. So, with jokes in poor taste, people exhibited his portrait to groups of friends for ridicule. This infuriated Hipponax, who attacked Dipoenus and Scyllis so savagely in his bitter poems that some believe they were driven to hang themselves. This is untrue, for they later carved many statues on neighbouring islands, for example on Delos.

13. In Chios itself there is said to be a head of Diana credited to them; it is placed high up, and on entering the building people think her expression is sad, but as they leave it strikes them as cheerful. In Rome there are statues by them in the angles of the pediment of the Temple of Apollo on the Palatine and on almost all the buildings commissioned by the late Emperor Augustus.

1. *c.* 580–577 BC.
2. Poet from Ephesus, Hipponax invented the scazon, or choliambus, also known as the 'halting iambic' – an iambic trimeter, that is, ending with a spondee.
3. *c.* 540–537 BC.
4. 776 BC.

14. Ambracia, Argos and Cleonae were covered with statues by Dipoenus. All these sculptors used only white marble from the island of Paros, a stone that eventually became known as *lychnites*, because, according to Varro, it was quarried in galleries by the light of oil-lamps.[1] Subsequently many other varieties of white marble have been found, some indeed only recently in the quarries at Luna. There is a bizarre story that in the quarries on Paros, on one occasion when a single block was split with wedges, it was found to contain an image of Silenus.

Phidias

15. I should add that sculpture in stone is much older than painting or bronze sculpture, both of which began with Phidias in the 83rd Olympiad[2] – that is, about 332 years later. Phidias himself is said to have worked in marble and to have carved the outstandingly beautiful *Venus* in the buildings dedicated to Octavia in Rome. It is established also that Phidias taught Alcamenes, an Athenian and a famous sculptor of the first rank; several of his statues are to be seen in temples at Athens, while outside the walls is the famous statue of Venus known in Greek as *Aphrodite in the Gardens*. Phidias himself is said to have put the finishing touches to this.

17. Another of Phidias' pupils was Agoracritus of Paros; he pleased him because his youthful good looks so delighted his master that the latter is said to have allowed him to take credit for several of his own statues. The two pupils entered a competition for making a statue of Venus; Alcamenes won, not for the quality of the work, but because the judges favoured a fellow-citizen against a foreigner. And so Agoracritus is said to have sold his statue with the stipulation that it should not stay in Athens; he called it *Nemesis*. It was set up at Rhamnus, a town in Attica. Marcus Varro preferred this statue to any other. In the same town there is also a statue by Agoracritus in the Shrine of the Great Mother.

18. Among all those who have heard of his *Olympian Jupiter*, Phidias is undoubtedly the most celebrated of all sculptors. However, so that those who have not seen his statues may realize that he is justly praised, I shall cite evidence which is admittedly not substantial, but sufficient to prove his talent. To this end I shall not have recourse to the beauty of his *Olympian Jupiter*, or to the size of his

1. *lychnos*: 'oil-lamp'.
2. *c.* 448–445 BC.

Minerva at Athens, even though, made of ivory and gold, it is about 39 feet high. Instead I shall cite the shield of this *Minerva*. On its convex side Phidias engraved a battle of the Amazons; and on the concave side, battles between gods and giants. And I shall also cite her sandals, on which he depicted battles between Lapiths and Centaurs. Thus no detail was considered unworthy of his artistic concern.

Praxiteles

20. I mentioned Praxiteles[1] among the makers of bronze statues, but he excelled above all as a sculptor in marble. There are statues by him at Athens in the Ceramicus.[2] Superior to any other statue, not only to others made by Praxiteles himself, but throughout the world, is the *Venus* which many people have sailed to Cnidus to see. He had made two statues and was offering them for sale at the same time. One was clothed, and for this reason was preferred by the people of Cos who had an option to buy, although Praxiteles offered it at the same price as the other – this was thought the only decent and proper response. So the people of Cnidus bought the *Venus* when the Coans refused, and its reputation became greatly enhanced.

21. Subsequently King Nicomedes[3] wanted to buy it from them, promising to cancel all the state's debts, which were vast. The Cnidians, however, preferred to endure anything rather than sell the statue. Nor without just cause, for with it Praxiteles made Cnidus famous. The shrine that houses it is completely open so that the statue of the goddess can be seen from all sides, and it was made in this way, so it is believed, with the goddess's approval. It is admirable from every angle. There is a story that a man who had fallen in love with the statue hid in the temple at night and embraced it intimately; a stain bears witness to his lust.

22. There are also in Cnidus other marble statues by famous artists: a *Bacchus* by Bryaxis, and a *Bacchus* and a *Minerva* by Scopas. There is, however, no greater proof of the merit of Praxiteles' *Venus* than that among all these works it alone is mentioned.

1. *fl.* 364–361 BC.

2. A cemetery for heroes. According to Pausanias (I, 3, 1) it took its name from the hero Keramos, a supposed son of Dionysus and Ariadne.

3. Nicomedes IV Philopator of Bithynia (reigned *c.* 94–75/4 BC) caused the Mithridatic War (88 BC) by attacking Pontus at Rome's behest. He bequeathed his kingdom to Rome.

Scopas and the Mausoleum

25. Scopas matches the foregoing in fame.

30. His rivals and contemporaries were Bryaxis, Timotheus and Leochares, whom I must discuss alongside him because they all worked on the relief sculptures of the Mausoleum. This is the tomb that Artemisia commissioned for her husband Mausolus, satrap of Caria, who died in the second year of the 107th Olympiad.[1] These sculptors in particular made the Mausoleum one of the Seven Wonders of the World.

31. Scopas carved the east side, Bryaxis the north, Timotheus the south, and Leochares the west. Before they completed the commission, the queen died. However, they did not leave the building unfinished, since they believed that it would be a monument to their own glory and their art. There was also a fifth sculptor. For above the colonnade rises a pyramid as high again as its substructure, tapering in twenty-four stages to its apex. Here there stands a four-horse chariot in marble made by Pythius, the addition of which brings the overall height to about 150 feet.[2]

The 'Laocoon' group

37. There are not a great many more famous sculptors. The reputation of some, although their work is distinguished, has been obscured by the number of other artists associated with them on a single commission, because the credit is shared, nor is it possible to give equal credit to members of a group. This is the situation with regard to the *Laocoon* in the Emperor Titus' palace, a work superior to any painting or bronze. Laocoon, his sons and the wonderful coils of the snakes were carved from a single block, following an agreed plan, by three artists of the highest calibre – namely Hagesander, Polydorus and Athenodorus, all of whom came from Rhodes.[3]

---

1. 352–349 BC.

2. Pythius was also responsible for the overall design of the Mausoleum.

3. The *Laocoon* group, which is now in the Vatican Museum, in fact comprises five blocks. Laocoon, a priest of Apollo, protested against the Trojans taking in the wooden horse. While he was making a sacrifice to Neptune, two huge serpents came out of the sea and attacked his two sons. When he tried to fight them off, the serpents squeezed him to death with their coils. According to one account the serpents were sent by Apollo to punish Laocoon for having married in spite of being a priest.

Veined marble and marble veneers

46–7. Marble marked with different colours first appeared, I believe, in the quarries of Chios when they were building their walls. The art of cutting marble into thin slabs was possibly invented in Caria. The earliest example I can find of its use was in the palace of Mausolus at Halicarnassus, where the brick walls were decorated with marble veneer from the island of Proconnesus.

48. Cornelius records that the first man in Rome to cover whole walls with marble veneer, in his house on the Caelian Hill, was Mamurra, a Roman knight from Formiae who was Julius Caesar's chief engineer in Gaul. That this invention was promoted by such a man makes it an object of contempt, for this is the Mamurra censured in the poems of Catullus of Verona. His house, as a matter of fact, announces more clearly than Catullus that 'he has all that Transalpine Gaul had'.[1] Nepos adds that Mamurra was the first to have a complete set of marble columns throughout his house, and that they were all made of solid marble from Carystus or Luna.

50. I believe that Marcus Scaurus' stage was the first structure to have marble walls, although I could not readily say whether these were of marble veneer or of polished solid blocks – as is the wall of the Temple of Jupiter the Thunderer on the Capitol today. For I find no evidence of marble veneer in Italy at such an early date.

Marble dressing

51. But whoever was the first to discover how to cut marble and divide self-indulgence was a man of ill-directed ingenuity. The cutting appears to be done by iron, but really it is achieved by sand: the saw exerts pressure on the sand along a very thin line and it is the sand's movement to and fro that actually does the cutting.[2] The most highly approved sand is Ethiopian – material for cutting marble is sought from that far away! Indeed men even go to India for sand, although at one time it was considered an offence against our strict morals to seek even pearls from that country.

52. Indian sand is second in quality. Ethiopian is finer and cuts without leaving any raggedness, while the Indian does not impart such a smooth surface to the marble. Those involved in polishing

---

1. This is a paraphrase of Catullus, *Poems*, XXIX, 3–4.
2. This technique is still used today.

marbles are recommended to rub the surface with Indian sand when it has been oxidized by heating.

53. Nowadays fraudulent craftsmen dare to cut marble with any kind of sand from any river, causing waste, which few people realize. For the coarser the sand, the less accurate the sections it produces, the more marble it wears away and, because of the rough surfaces produced, the more work it leaves for the polishers. This results in thinner veneers.

Granite

55. There is no point in discussing the varieties and colours of marble since they are well known, and it is not easy to list them because of their very large number. For most places (that have marble at all) have a characteristic variety. Not all types occur in quarries; many are mined. Some are extremely valuable like the green Lacedaemonian, which is brighter than any other marble.

63. Red granite mottled with gold spots is found in that part of Africa designated as Egypt; it is naturally suited to being made into whetstones used for the grinding of eye-salves. Another red granite is found near Syene in the district of Thebes and was formerly called *pyrrhopoecilus*.[1]

Egyptian obelisks

64. The kings of Egypt, out of a kind of rivalry, made monoliths of red granite which they called obelisks and dedicated to the sun-god. An obelisk symbolizes the sun's rays, and this is the meaning of the Egyptian word for monolith.[2] Mesphres[3] who ruled at Heliopolis, was the first king to put up an obelisk, having been ordered to do so in a dream. This fact is actually inscribed on the stone; for the carvings and hieroglyphs we see are Egyptian letters.

65. Afterwards other kings also cut monoliths. Sesothis erected four, which were about 80 feet high, at Heliopolis. Rameses II, however, in whose reign Troy was captured, set up one which was 230 feet high. He also set up another at the exit from the precinct where the palace of Mnevis once was; this is 200 feet in height and

1. Mottled red.
2. *Tekhen*: 'sunbeam', 'obelisk'.
3. Tuthmosis III

is extraordinarily thick, each side being 17 feet in width. Some 120,000 men are said to have carried out this work.

66. When the obelisk was about to be set upright, the king himself was afraid that the lifting equipment would not be strong enough to take the weight. To make his workmen pay attention to the dangers, he tied his son to the top so that in ensuring the child's safety the work-force would treat the monolith with necessary care. This monument commanded such admiration that when Cambyses was storming that city and flames had reached the base of the obelisk, he ordered the fire to be put out, thus showing that he had respect for this mighty mass even though he had none for the city.

67. There are also two other obelisks, one set up by Zmarres, the other by Phius; these do not have any inscriptions and are about 70 feet high. Ptolemy Philadelphus erected an obelisk almost 120 feet high at Alexandria.[1] Necthebis had quarried this monolith and left it uninscribed. Transporting it down river and setting it up proved greater achievements than quarrying it. Some authorities say that Satyrus the architect floated it on a raft. Callixenus[2] says it was carried down by Phoenix, who dug a canal and brought the Nile right up to where the obelisk lay. Two wide barges were loaded with cubes of granite – each one with a side of 1 foot – until the weight of the blocks was double that of the obelisk, their total volume being also twice as great. The ships were thus able to float beneath the obelisk which was suspended by its ends from both banks of the canal. Then the blocks of granite were unloaded and the lightened barges took the weight of the obelisk. The monument was set up on six granite supports cut from the same quarries. The consultant responsible for this scheme was paid 50 talents. This obelisk was placed by the king in the Arsinoëum as a tribute to his wife and sister Arsinoë.

69. There are two further obelisks[3] at Alexandria in the enclosure of Caesar's temple, near the harbour. These were quarried for King Mesphres and are 63 feet high.

Obelisks at Rome

70. Transporting obelisks to Rome by sea was a more difficult task

---

1. Ptolemy II Philadelphus (308–246 BC) also built the Pharos, the Museum and the Library as well as a canal from the Nile to the Red Sea.
2. A Rhodian Greek (*fl. c.* 155 BC) who wrote about Alexandria.
3. Both survive: one is Cleopatra's Needle in London; the other is in New York.

by far. The ships attracted much interest. The late Emperor Augustus dedicated the ship that carried the first obelisk and preserved it in a permanent dock at Puteoli to mark this marvellous feat. It was later destroyed by fire. For several years the Emperor Claudius preserved the ship used by Gaius Caesar for bringing the third obelisk because it was the most amazing vessel that had ever been seen afloat. Cement caissons were installed on board at Puteoli; the vessel was then towed to Ostia, and scuttled to help construct the port. There is another problem with obelisks – namely, finding ships that can carry them up the Tiber. Experiment has shown that the River Tiber is just as deep as the Nile.

71. The obelisk set up by the late Emperor Augustus in the Circus Maximus was quarried for King Psemetnepserphreus[1] who was king at the time that Pythagoras was in Egypt; it is 85 feet high, excluding its base which is an integral part of the obelisk. The monolith in the Campus Martius is about 10 feet shorter; it was quarried for Sesothis. According to learned Egyptians, both carry hieroglyphs that give an account of natural science.

An obelisk used as a large-scale gnomon

72. Augustus used the obelisk in the Campus Martius in a remarkable way – namely to cast a shadow and thus mark the length of days and nights. A paved area was laid out commensurate with the height of the monolith in such a way that the shadow at noon on the shortest day might extend to the edge of the paving. As the shadow gradually grew shorter and longer again, it was measured by bronze rods fixed in the paving. This device deserves study; it was the result of a brainwave of Facundus Novius. Novius placed a gilded ball on the apex of the monolith so that the shadow would be concentrated at its tip; otherwise the shadow cast would have been very indistinct. He got this idea, so it is said, from seeing the shadow cast by a man's head.

73. These measurements, however, have not agreed with the calendar for some thirty years; either the orbit of the sun itself is out of phase or has been altered by some change in the behaviour of the heavens, or the whole earth has moved slightly off-centre. I hear this phenomenon has also been observed in other places.

---

1. This name is probably a corruption of the names Psamtik and Neferibre.

The pyramids and the Sphinx

75. The pyramids in Egypt deserve mention, even if only incidentally. They are a pointless and absurd display of royal wealth, since the general view is that they were built either to deny money to the kings' successors and the rivals who plotted against them, or else to keep the people employed. These men showed much vanity in this enterprise. The remains of several unfinished pyramids survive. One is in the name of Arsinoë, and two are in the name of Memphis, not far from the labyrinth described below. Two are in what was formerly Lake Moeris, a gigantic excavation recorded by the Egyptians among their extraordinary and remarkable feats. The apex of each pyramid is said to stand out of the water. The other three pyramids, famous throughout the whole world, can be seen from all points of the compass by all travellers on the Nile. They stand on a rocky hill in the desert. Close by is a village called Busiris where there are people who regularly climb the pyramids.

77. In front of the pyramids is the Sphinx,[1] which is even more noteworthy. The local inhabitants worship it as a deity, yet say nothing about it. They believe that King Harmais is buried inside it and like to think it was transported there. It is, however, carved from local rock. The face of this legendary creature is coloured red – a mark of reverence. The circumference of its head, at its forehead, is 102 feet; its length, 243 feet; and it measures about 62 feet from its stomach to the top of the asp on its head.

80. The largest pyramid is built on an area of almost 5 acres. Each of the four sides measures 783 feet from corner to corner, and the height from ground-level to the top is 725 feet.

81. No trace of the method of building these pyramids remains. The most significant problem is how the blocks were raised to such a great height: some think that as the pyramid rose, soda and salt were heaped up to form ramps, and that when the structure was finished these were flooded and dissolved by river water. Others, however, think that causeways were built of mud-bricks, which, on completion of the work, were distributed for building private houses; for they reckon that it would be impossible for the Nile, which flows at a lower level, to have flooded the area.

1. The Sphinx represents King Cephren – builder of one of the three great pyramids at Giza – in the guise of the sun-god, and stands as guardian to his tomb. It was disinterred from the shifting sands and restored by Thutmose IV (1425–1408 BC).

The tower at Pharos

83. Another tower built by a king also received praise. This is on the island of Pharos, and commands the harbour of Alexandria; it is said to have cost 800 talents.[1] King Ptolemy, in a spirit of generosity, allowed the name of the architect, Sostratus of Cnidus, to be inscribed on the building itself.[2] Its purpose is to provide a beacon for ships sailing by night, to warn them of shallows and to mark the entrance to the harbour. Similar beacons now burn brightly in several places, for example at Ostia and Ravenna. The danger from a continuously burning beacon is that it may be thought to be a star since the appearance of the fire from a distance is very similar. The same architect Sostratus is said to have been the very first to build a promenade supported by piers, at Cnidus.

Labyrinths

84. I should like also to mention labyrinths, which are outstanding among the works to which man has devoted his resources. These are not mythical, as one might suppose. A labyrinth still exists in Egypt in the nome of Heracleopolis. This was the first ever built – 3,600 years ago. Various explanations have been suggested for its construction. Whatever the truth, there can be no doubt that Daedalus[3] took it as the model for the labyrinth he built in Crete;[4] but he copied only a hundredth part of its passages, which wind, advance and retreat in the most intricate way. Not only is a narrow area bafflingly packed with many miles of 'walks' or 'rides', such as we see in mosaic floors or in the game played by boys in the Campus Martius,[5] but also at frequent intervals doors are let into the walls to deceive people into retracing the very paths they have already taken in their wanderings.

86. The labyrinth in Crete was built after the Egyptian. There was a third labyrinth on Lemnos and a fourth in Italy; all were covered by a vaulted roof of carefully dressed stone. The Egyptian

1. The tower, called the Pharos, was a lighthouse about 400 feet tall. A substantial part of it survived until the thirteenth century.

2. Sostratus (*fl.* first half of the third century BC) was also the first to build a pier into the sea.

3. The name is derived from the Greek *daidala*: 'cunning works'.

4. i.e., the labyrinth in which King Minos concealed the Minotaur.

5. Pliny refers to the game known as the *Lusus Troiae*. Virgil also draws a comparison between this game and the Cretan labyrinth (*Aeneid*, V, 588 ff.).

labyrinth has an entrance and columns of Parian marble, which I, at any rate, find surprising. The rest of the building is of granite from Syene, and the great blocks have been so arranged that they are capable of resisting the passage of centuries. In addition the inhabitants of Heracleopolis have helped by treating this work with remarkable respect, despite their hatred of it.

87–8. It is impossible to describe the overall plan or separate parts of the labyrinth, since it represents the administrative regions which the Egyptians call 'nomes'. There are twenty-one nomes, each symbolized by a vast hall. The labyrinth also contains temples within its forty precints of all the Egyptian gods as well as Nemesis, and several pyramids each about 60 feet high and with a base area of 4 acres. When one is tired from walking, one arrives at the 'maze' part of the labyrinth. There are lofty upper rooms reached by ramps and porticoes with ninety steps to ground-level. Inside are columns of porphyry, images of gods, statues of kings and representations of monsters. Some of the halls are so constructed that when the doors are opened there is a spine-chilling rumble of thunder within. Most of the way through the labyrinth proper lies in darkness. There are other massive buildings outside the labyrinth enclosure, forming what the Greeks call a 'wing'. There are also underground chambers which were constructed by tunnelling galleries.

The Temple of Diana

95. A genuine and remarkable example of the Greek notion of grandeur survives in the Temple of Diana, whose construction took 120 years and involved the whole of Asia Minor. It was built on marshy ground so that it would not be affected by earthquakes or subsidence. But to avoid laying the foundations of such a large building on unstable ground, a compacted raft of charcoal and then a layer of unshorn fleeces was placed underneath them. The temple is 425 feet in length and about 225 feet in breadth. There are 127 columns, some 60 feet high, each built by a different king. Of these, thirty-six were decorated with relief sculpture – one by Scopas. The consultant architect was Chersiphron.

96. Chersiphron's greatest feat was his success in raising the architraves of this huge building into position. He achieved this by filling reed bags with sand and building a gently inclined ramp up to and above the level of the capitals. Then he gradually emptied the sacks at the bottom so that the architrave gently settled into

place. The greatest difficulty occurred as he tried to position the lintel above the doors. This was the largest stone and refused to settle on its bed. Chersiphron was distraught and wondered whether his last decision should be to commit suicide.

97. The story goes that he became exhausted thinking about the problem, but that in the night while he slept the goddess Diana appeared to him and urged him to continue living since she had laid the stone. Next day, when it was light this was seen to be true; the position of the stone appeared to have been corrected by its weight alone. The other decorative features of the building are enough to fill a number of books but they are in no way related to natural forms.

The buildings of Rome

101. Now is the time to proceed to the wonders of Rome, to examine what we have learned over 800 years and to show that we have conquered the world with our buildings too.

Sewers

104. I regard the fact that Milo had debts of 70 million sesterces as one of the strangest quirks of human character. In his day leading citizens still marvelled at the network of sewers, generally acknowledged as the greatest achievement of all. Hills were tunnelled into in the course of its construction, and Rome was a 'city on stilts' beneath which men sailed when Marcus Agrippa was aedile.[1]

105. Seven rivers join together and rush headlong through Rome, and, like torrents, they necessarily sweep away everything in their path. With raging force, owing to the additional amount of rainwater, they shake the bottom and sides of the sewers. Sometimes water from the Tiber flows backwards and makes its way up the sewers. Then the powerful flood-waters clash head-on in the confined space, but the unyielding structure holds firm.

106. Huge blocks of stone are dragged across the surface above the tunnels; buildings collapse of their own accord or come crashing down because of fire; earth tremors shake the ground – but still, for 700 years from the time of Tarquinius Priscus, the sewers have survived almost completely intact.

Now I come to a memorable event that should not be omitted

1. 33 BC.

and is all the more worthy of inclusion here in that it has been passed over by the best-known historians.

107. When Tarquinius put the project in hand with the assistance of the people as labourers – at a time when no one knew that the network of sewers would be noted for its size or the time it took to construct – there was a mass outbreak of suicides as citizens tried to escape from their exhaustion. As a result the king invented a strange countermeasure never before thought of or indeed used since. He crucified the bodies of suicides, so that they might be gazed on by their fellow citizens and torn to pieces by wild animals and birds of prey.

108. In consequence, the sense of shame that is characteristic of the Roman people, and has so often retrieved a desperate situation in battle, helped them also on that occasion.

Houses

109. The most carefully researched authorities are agreed that, in the consulship of Marcus Lepidus and Quintus Catulus,[1] there was no finer house in Rome than that owned by Lepidus himself; but, thirty-five years later, believe me, the same house did not even make the top hundred.

110. If anyone doubts this assessment, let him consider the great volume of marble, the countless paintings and the royal fortunes that went into these hundred houses – houses which were later to be surpassed time and time again by others right down to the present day. Fires, to be sure, punish extravagance, and yet human nature cannot be made to understand that there are things more perishable than man himself.

Nero's House of Gold

111. However, two houses stand out: on two occasions we have seen imperial palaces dwarf the whole of Rome – namely, those of Gaius and Nero. Nero's palace, so as not to be inadequate in any respect, was a 'House of Gold'.

112. Indeed one cannot help thinking how small in comparison with these palaces were the parcels of land that the state formerly granted to undefeated generals for their private houses.

113. Yet even the madness of Gaius and Nero was surpassed by

1. 78 BC.

the resources of a private person: Marcus Scaurus, who, as aedile, arguably did more than anyone to undermine morality, and whose elevation to a position of influence may have been a greater misdeed on the part of his stepfather Sulla than the killing of so many thousands of people in his proscriptions.

Scaurus' theatre

114. As aedile, Scaurus built the greatest building ever constructed by man. It was greater not only than any temporary works but also than those planned to last for ever. This was a theatre that had a three-tiered stage with 360 columns – in a city that previously had not tolerated 6 columns of Hymettan marble without criticizing a distinguished citizen. The lowest tier of the stage was of marble, the middle one of glass – an unheard-of example of extravagance even in later times – and the top one of gilded planks. The columns of the lowest tier were each almost 38 feet high.

115. The bronze statues in the spaces between the columns were 3,000 in number. The auditorium held 80,000 – yet Pompey's Theatre seating 40,000 is sufficiently large now, even though the city is many times larger and the population much more numerous. The rest of the theatre's equipment – including costumes interwoven with gold thread, scenery and other properties – was on such a scale that when surplus items of everyday use were taken to Scaurus' villa at Tusculum and the house itself was set ablaze and burnt down by angry servants, he lost 30 million sesterces.

116. Scaurus gained one advantage from that fire: no one could emulate his madness in the future.

Curio's double theatre

117. Curio, therefore, had to exercise his ingenuity and devise something else. It is worth taking the trouble to learn what he conceived and to be glad of our modern moral code and, reversing the usual terminology, to call ourselves 'older and better'.[1] Curio built two vast wooden theatres side by side, each balanced on a revolving pivot. Before midday, a performance of a play was staged in both; the theatres faced in opposite directions so that the actors should not

1. A play on the word *maior*. Compared with Scaurus or Curio, Pliny lives like Cato and other worthy men: to this extent he is *maior*. He is, however, younger (*minor*) in age.

drown each other's lines. Then suddenly the theatres revolved (it is agreed that after the first few days this happened while some of the audience actually remained in their seats), and their corners came together to form an amphitheatre.[1] Here Curio staged fights between gladiators – although the Roman people found themselves in even greater danger than the gladiators, as Curio spun them round.

118. It is difficult to know what should amaze us more, the inventor or the invention, or the sheer audacity of the conception. Most amazing of all is the madness of a people rash enough to sit in such treacherous and unstable seats!

119. What contempt for human life this shows! How can we justify our complaints about Cannae! What a disaster this could have been! Here the whole Roman people, as if put on board two ships, were supported by a pair of pivots and watched themselves fighting for their lives and likely to perish at any moment should the mechanism be put out of gear!

120. And the aim of all this was merely to win favour for the speeches Curio intended to make as tribune: he wanted to be able to continue to sway the undecided voters. On the speaker's platform he would stop at nothing in addressing those whom he had persuaded to participate in this dangerous activity.

Aqueducts and water-supply

121. Let us now move on to achievements which are unsurpassed because of their real value.

122. The most recent and costly project, begun by the Emperor Gaius and completed by Claudius, has surpassed all previous aqueducts: the springs called Curtius and Caeruleus, as well as the Anio Novus, were made to flow into Rome from the fortieth milestone at such a high level as to provide water for all the hills of the city; the work cost 350 million sesterces.

123. If we ponder the abundant water-supply in public buildings, baths, open channels, private houses, gardens and country estates; if we consider the distances travelled by the water, the building of the arches, the tunnelling through mountains and the construction of level routes across valleys; we can only conclude that this is a supreme wonder of the world.

124. One of the most remarkable feats of Claudius, neglected by his successor, who hated him, is, in my opinion at least, the channel

1. Such a state could not have revolved in the manner that Pliny describes. See further J. F. Healy, *Pliny on Science and Technology* (Oxford, 1999), 165 f. and Fig. 12.2.

that he dug through a mountain to drain the Fucine Lake.[1] This involved incalculable expense and the employment of an army of labourers over a period of many years, because where the inside of the mountain was earthy, the spoil had to be raised to the surface by hoists, and elsewhere bedrock had to be hewn away. These immense operations had to be carried out in darkness – operations imaginable only by those who saw them and which are beyond the power of words to describe.

The discussion of minerals continued

125. Amongst the many marvels of Italy itself is one attested by Papirius Fabianus, an outstanding natural scientist. He asserts that marble actually *grows* in quarries; quarrymen also say that gashes in the sides of mountains fill up spontaneously. If this is true, there are grounds for hoping that there will always be sufficient marble to satisfy extravagant life-styles.

Magnetite

127. Iron is attracted by the magnet: the substance that overcomes all else rushes into a kind of vacuum and, as it approaches the magnet, leaps towards it and is held fast and embraced by it. Some Greeks call the magnet by another name, 'ironstone'; and some call it the 'stone of Hercules'. According to Nicander it was known as *magnes* after its discoverer, Magnes, who found the mineral on Mount Ida. It is also found in many other places, including Spain. The discovery is said to have been made when the nails of Magnes' sandals and the tip of his staff stuck to the stone as he was grazing his herds.

128. Sotacus describes five kinds of magnetite. The highest grade is the Ethiopian kind, which is worth its weight in silver.

129. One can tell an Ethiopian magnet by its ability to attract another magnet to itself.

1. This channel was intended to carry the water of the Fucine Lake into the River Liris. The project employed 30,000 men and took eleven years to complete. The emperor Claudius opened the channel with great ceremony (AD 52). Later modifications to the channel were only partially successful.

The Sarcophagus stone

131. At Assos, in the Troad, is found the Sarcophagus stone which splits along a line of cleavage.[1] It is established that corpses buried in it are, with the exception of the teeth, consumed within a period of forty days. Mucianus[2] is the authority who asserts that mirrors, strigils, clothes and shoes buried with the corpses are also turned to stone.[3] There are rocks of the same type in Lydia and in the East, which, when attached even to living persons, eat away their bodies.

Asbestos

139. Asbestos looks like alum and is completely fire-proof; it also resists all magic potions, especially those concocted by the Magi.

Haematite

146. Sotacus, among the oldest authorities, records five kinds of haematite apart from the magnet. He gives first place to the Ethiopian variety.

Selenite

160. Selenite, although it is called a rock, is, however, far more tractable and can be split into plates of any desired thinness. At one time it was produced only in Hither Spain – not throughout the whole province, but only within a radius of 100 miles from the city of Segobriga. Today it also comes from Cyprus, Cappadocia, Sicily and – a recent discovery – Africa. All these types, however, are inferior to the Spanish kind. Cappadocia produces the largest pieces, but they are opaque.

161. In Spain selenite is mined to a great depth by sinking shafts; it is also found just below the surface, embedded in the prevailing rock of the locality, from which it is torn or cut away. However, for the most part it is in a form that can be dug up since it occurs free-standing, like rough blocks in quarries; no block to date has

---

1. Pliny may be referring to limestone.
2. Governor of Syria in AD 69. He compiled a book of *mirabilia*, many of which he became acquainted with during his period of office in the East.
3. Water seepage might have deposited calcium carbonate, thus hardening the objects inside. A deposit of slaked lime could also have produced a hard coating.

been found more than 5 feet in length. It is evident that selenite comes from a liquid that, like rock-crystal, has been frozen and hardened into stone by an exhalation within the earth; for when wild animals fall down these shafts, the marrow in their bones assumes the form of selenite after one winter.

162. Sometimes a black variety of selenite is found; this is bright and, although noted for its softness, has the remarkable property of withstanding heat and cold; it does not deteriorate, provided it is not damaged, although this applies also to blocks of many other kinds of stone. Men have found yet another purpose for selenite: it is used spread as flakes during the Games to give the surface of the Circus Maximus a much-praised bright appearance.

Onyx marble (phengites)

163. In Nero's principate a stone came to light in Cappadocia that had the hardness of marble, was white, and translucent even where marked by dark yellow veins. It was called *phengites*. Nero used it to rebuild the Temple of Fortune; known as the Shrine of Sejanus, it was originally dedicated by King Servius Tullius, and became incorporated in Nero's 'House of Gold'. Because of the stone's translucent nature, during the day it was light in the temple even when the doors were closed. But, unlike with windows of selenite, the light was, as it were, trapped inside the stone and not allowed to pass through.

Mosaics and glass

189. Mosaics came into use in Sulla's time. At any rate there still exists a mosaic floor, made from very small tesserae, which Sulla commissioned for the Temple of Fortune at Praeneste. Subsequently mosaics were transferred from ground-level to vaulted ceilings, and these are now made of glass. This is a technical innovation. At all events, Agrippa used encaustic to paint the terracotta in the hot rooms of his baths in Rome, and had the rest whitewashed. I must now describe the nature of glass.

190. That part of Syria adjoining Judaea and called Phoenicia includes a swamp known as Candebia, bounded by the lower slopes of Mount Carmel. This is believed to be the source of the River Belus, which, after a distance of 5 miles flows into the sea next to the colony of Ptolemais. Its course is sluggish and its water not fit for drinking, although considered holy for religious rites. The river

is muddy but deep, only revealing its sands when the tide retreats. The sand does not glisten until it has been tossed about by the waves and had its impurities removed.

191. Then, and only then, when the sand is thought to have been cleansed by the scouring action of the sea, is it ready for use. The beach extends for not more than half a mile, but for many years this area was the sole producer of glass. A ship belonging to traders in soda once called here, so the story goes, and they spread out along the shore to make a meal. There were no stones to support their cooking-pots, so they placed lumps of soda from their ship under them. When these became hot and fused with the sand on the beach, streams of an unknown translucent liquid flowed, and this was the origin of glass.[1]

192. Soon, in keeping with man's inventive genius, he was no longer content to mix in only soda. He began to add compounds of magnesium. Similarly, in many places lustrous stones[2] were added to the melt, and then again shells[3] and quarried sand. Authorities claim that in India glass is made from broken rock-crystal and for this reason no glass can be compared with Indian.

193. Copper and soda, especially Egyptian, are added to the mixture, which is melted by a fire of light, dry wood. Glass, like copper, is processed in a series of furnaces and forms dark lumps.[4] The molten glass is so sharp that before one can feel any pain, it cuts through to the bone any part of the body it splashes. The lumps of glass are fused again in the workshop and coloured. Some glass is shaped by blowing,[5] some fashioned by a lathe, and some engraved like silver. Sidon was once renowned for its glass factories; glass mirrors, among other things, were invented there.

194. This was the old method of glass manufacture. But now in Italy a white sand, which occurs in the River Volturnus, is found along 6 miles of the seashore between Cumae and Liternum; the

1. Glass is the product of the fusion of sand (silica), soda (an alkali) and an alkaline earth such as lime. Impurities in the sand may have provided the lime which Pliny omits from his account. Without this ingredient water-glass would have been produced, that is, a solution of sodium silicate. On the production of glass and this passage, see R. C. A. Rotländer, 'Glasherstellung bei Plinius dem Alteren', *Glastechnische Berichte*, 52 (1979), pp. 265–70.

2. Probably quartz.

3. These might have been used as a source of lime.

4. The stages involved in this process were probably (1) firing at a low temperature, (2) melting, and (3) slow cooling. A different furnace could have been used for each stage.

5. Glass blowing was not introduced until the middle of the first century BC.

softest of this sand is ground in a mortar or mill. Then it is mixed with three-quarters of a part of soda, by either weight or volume, and, after fusion, it is taken in its molten state to other furnaces. There it becomes a lump called 'sand-soda'; this is remelted and becomes pure glass – a lump of clear glass. Now sand is blended in a similar way in the provinces of Gaul and Spain.

195. When Tiberius was emperor, it is said, a method of blending glass was invented to make it flexible, but the craftsman's workshop was completely destroyed for fear that this might detract from the value of metals such as copper, silver and gold. This story went the rounds for a long time thanks to frequent repetition rather than because there was any truth in it. However that may be, in Nero's principate a technique of glass-making was invented that produced two small cups called 'stoneware', which were sold for 6,000 sesterces.

198-9. The most highly prized glass is transparent, and its appearance is as close as possible to that of rock-crystal. But although glass has replaced gold and silver for drinking-vessels, it cannot bear heat unless cold water is first poured in. Glass globes containing water become so hot when they face the sun that they can burn clothes. When heated to a moderate temperature, pieces of broken glass can be stuck together. Even so, they can never again be completely fused but melt into separate globules, as happens in the manufacture of glass pebbles called 'eyeballs'; in some instances these glass pebbles are multi-coloured and multi-patterned. When glass is heated with sulphur it coalesces into stone.

The power of fire

200. Having described everything that exists as a result of man's talent for making art copy Nature, I marvel to think that scarcely anything is brought to a finished state without the involvement of fire. It takes sand and melts it, as occasion offers, into glass, silver, cinnabar, lead, pigments and drugs. Ore-minerals are smelted to produce copper. Fire produces iron and tempers it, purifies gold, and burns limestone to make mortar that binds blocks together in buildings.

201. Other substances benefit from being subjected to fire more than once, the same substance taking on different forms after the first, second or third firings. Charcoal starts to acquire special powers when burnt and quenched; when it is apparently dead its potency increases. Fire is a vast, violent element of Nature, and it is a moot point as to whether its essence is more destructive or creative.

202–3. Fire has applications in the medical domain. It is well established that plagues brought on by an eclipse of the sun can be relieved in many ways by lighting fires. Empedocles and Hippocrates have shown this in different places. 'For pains in the stomach or bruises,' says Marcus Varro – I quote his very words – 'let your hearth be your medicine-box.' A draught of lye made from its ashes will set you right, and one can see how gladiators benefit from drinking this after the end of a contest. Moreover, anthrax, a disease that recently caused the deaths of two ex-consuls, is cured by a mixture of oak-charcoal and honey. So even in substances that are spent and cast aside, there are some beneficial properties – as in the case of charcoal and ash.

204. I shall not pass over an example of a hearth celebrated in Roman literature. The story goes that during the reign of Tarquinius Priscus a penis suddenly emerged from the ashes on the hearth and that a captive girl who was sitting there – Ocresia, a servant of Queen Tanaquil – rose up pregnant from in front of the fire. This was how Servius Tullius, who succeeded to the throne, came to be born. It is said that later, in Tullius' house, flames licked round the child's head as he slept and that he was consequently believed to be the son of a god who protected the household.

BOOK XXXVII

PRECIOUS STONES

1. The only topic that remains of those I set out to cover is that of precious stones. Therein Nature reaches its utmost concentration and in no department does she arouse more wonder. So much store do men set by the variety, colours, texture and elegance of gems that they consider it criminal to tamper with precious stones by engraving them as signets. Some they consider beyond price on any human scale of valuation. Consequently, for very many people a single precious stone can provide a matchless and perfect view of Nature.

2. Myths, based on a misinterpretation of Prometheus' bonds, state that the wearing of rings had its origin on the rocks of the Caucasus. It was a fragment of this rock that was first mounted in an iron setting and placed on a finger; this, we are informed, constituted the first ring and gemstone.

3. This was the beginning of the vogue for precious stones, which became such an obsession that Polycrates of Samos, the tyrant of the islands and coasts, regarded the sacrifice of a single gem as sufficient expiation for his prosperity; even he, happy as he was, admitted that his good fortune was excessive. Clearly Polycrates thought that he ought to settle his account with capricious Fortune, and would be amply indemnified against her ill-will, should he suffer this one calamity – weary as he was of constant happiness. So he sailed out to sea and threw the ring into deep water.[1]

4. The ring, however, was snapped up by a huge fish fit for a king, which returned it – an evil omen – to its owner, in his own kitchen, thanks to the hand of treacherous Fortune. The gem is generally agreed to have been a sardonyx, and it is on show – if we believe that this is the original gemstone – in Rome in the Temple of Concord, set in a gold horn. The precious stone was given by Livia, but it is virtually lost in a collection of more highly valued gems.

1. See Herodotus, III, 40 ff.

The fashion for pearls and other forms of extravagance

12. Pompey's victory over Mithridates turned fashion in the direction of pearls and precious stones. The victories of Lucius Scipio and Gnaeus Manilius had achieved the same for engraved silver, garments of golden cloth and dining-couches inlaid with bronze; and Mummius' victory, for Corinthian bronzes and paintings. To bring this fully home, I shall add statements from the official records of Pompey's triumphs. Pompey's third triumph was held on his birthday, 29 September,[1] to celebrate his defeat of the pirates, Asia, Pontus and all the peoples previously mentioned.[2] In this triumph he carried a gaming-board complete with pieces; the board, which was 4 feet long and a foot wide, was made of precious stones. On that board there was a gold moonstone thirty pounds in weight – which proves that our natural resources have become exhausted, since nowadays no precious stones approach these in size.

14–15. There was also Pompey's portrait done in pearls. Never, in my opinion, would his surname 'Great' have survived among the men of that age had he celebrated his first triumph in this fashion.

16–17. In other respects he acted most worthily at his triumph. He gave 200 million sesterces to the state, 100 million to the commanders and quaestors who had guarded the coasts, and 6,000 to each soldier. However, he made it easier for us to tolerate the behaviour of the Emperior Gaius, when, in addition to other articles of women's clothing, he wore slippers decorated with pearls – likewise, the behaviour of the Emperor Nero, when he had sceptres, actors' masks and travelling couches decorated with pearls. Indeed we seem to have forfeited the right to be critical even of cups and other household utensils inlaid with gems – or even of translucent rings. For surely any extravagance must be considered less pernicious than Pompey's.

Myrrhine ware – fluorspar

18. That same victory first brought myrrhine ware to Rome. Pompey was the first to dedicate fluorspar bowls and cups from his triumph to Capitoline Jupiter. Vessels of fluorspar immediately passed into everyday use, and even display stands and tableware were eagerly sought. This kind of extravagance increases daily. An

1. 61 BC.
2. VII, 98 (not included in this selection).

ex-consul drank from a fluorspar cup for which he had paid 70,000 sesterces, although it held only 3 pints. He was so enamoured of it that he used to chew the rim. Yet this damage increased its value, and no item of fluorspar today bears a higher price-tag on it.

19. The amount of money he dissipated in acquiring his other pieces can be gauged from their large number; there were so many that when they were confiscated from that man's children by Nero and put on show, they filled the private theatre in Nero's gardens across the Tiber! This structure was large enough even for Nero to sing in before an audience while he was rehearsing for Pompey's Theatre.

20. When the ex-consul Titus Petronius was at the point of death, he broke a fluorspar ladle for which he had paid 300,000 sesterces, thus depriving the emperor's dining-room table of this legacy.[1] Nero, however, as was fitting for an emperor, outdid everyone else by paying a million sesterces for a single bowl. That a commander-in-chief and Father of his Country paid so much to drink is a matter worthy of record.

21. The East exports fluorspar vessels. There the mineral is found in many otherwise unremarkable places, especially in the kingdom of Parthia. The best specimens of fluorspar, however, occur in Carmania. The actual mineral is thought to be a liquid that is solidified underground by heat. Pieces of fluorspar are never larger than a small display stand, and usually seldom even the size of the drinking vessels to which I have alluded. They shine, but not intensely – indeed, they can more accurately be said to glisten. Their value lies in their variegated colours. As the veins swirl round they vary repeatedly from purple to white to a mixture of the two, the purple becoming fiery, and the milk-white, red, as though the new colour were passing through the vein.

22. Some people reserve special admiration for pieces whose edges reflect colours as we see them in the inner part of a rainbow. The smell of fluorspar is also one of its attractions.

Rock-crystal

23. An opposite cause produces rock-crystal; for it is hardened by intense cold. At any rate, this mineral is found only where the winter snows freeze solid and it is undoubtedly a kind of ice, which is the reason for the Greek term *krystallos*.[2] The East exports

---

1. See Tacitus, *Annals*, XVI, 19.
2. Ice.

rock-crystal; the kind that occurs in India is preferred to any other. Rock-crystal is also found in Asia Minor, and a very poor variety occurs round about the adjacent areas of Alabanda and Orthosia, and similarly in Cyprus. In Europe a high-quality rock-crystal is found in the Alps.

24. Juba is the authority for the statement that rock-crystal is found also in the Red Sea on an island called Necron which faces Arabia, as well as on a neighbouring island which produces the mineral peridot. There, he claims, one of Ptolemy's officers, named Pythagoras, dug up a piece measuring 20 inches. Cornelius Bocchus mentions that rock-crystal of surprising weight was found in Lusitania in the Ammaiensian Mountains when wells were being sunk to water-level.

25. Xenocrates of Ephesus makes the surprising observation that in Asia Minor and Cyprus rock-crystal is turned up by the plough, for previously it had not been thought to occur in earth except among rocks. Sudines claims that it occurs only in locations facing south. What is generally agreed is that it is not found in places where there is a lot of water, however cold, even if the rivers freeze solid.

26. Rock-crystal must therefore be derived from moisture from the sky in the form of pure snow. For this reason it cannot stand heat, and is used only for cold drinks. Why rock-crystal is formed with hexagonal faces is hard to explain, and the difficulty is complicated by the fact that its terminal points are not symmetrical, while its faces are so perfectly smooth that not even the most skilful lapidary could achieve such a finish.

27. The largest piece of rock-crystal seen by us to date is the one that Augustus' wife Livia dedicated in the Capitol; it weighs about 150 pounds. Xenocrates records that he saw a container which could hold 7 gallons and some authorities refer to one from India holding 4 pints. I can confirm unequivocally that rock-crystal is formed in the Alps, in such inaccessible places that it has to be extracted by men hanging from ropes. The signs of rock-crystal's presence are well known to prospectors.

29. Rock-crystal is responsible for yet another mad obsession. Not many years ago a married woman of modest means paid 150,000 sesterces for a single ladle. Nero, hearing the news that all was lost, in a final outburst of anger dashed two crystal cups to the ground and broke them; this was an act of vengeance on the part of one who wished to punish his own generation by preventing anyone else from drinking out of these cups. There is no way of repairing

rock-crystal when shattered. Glass has now come to resemble rock-crystal and has increased its own value, while not lessening that of its rival.

Greek accounts of the origin of amber are incorrect; it is a resin from pine-trees

30. Next among items of luxury comes amber, although as yet it is exclusively an adornment for women. Not even luxury has been able to invent a reason for its use.

31. Amber provides an opportunity for exposing the false accounts of the Greeks. My readers should bear with me patiently, since it is important to realize that not everything handed down by the Greeks merits admiration.

42. It is now known that amber occurs on islands in the Northern Ocean, and is called by the Germans *glaesum*. Accordingly one of these islands was named Glaesaria by our men when Germanicus Caesar was involved in naval operations there: its native name was Austeravia. Amber is formed from a liquid that exudes from the inside of a type of pine-tree – just as the gum in a cherry-tree, or the resin in a pine, breaks out when there is an excess of liquid. The distillation is hardened by the cold or by moderate heat, or by the sea after a spring tide has carried off the resin from the islands. At any rate, the amber is washed up on the shores, whirling along in such a way that it seems to be suspended and not to settle on the sea-bed.

43. Even our ancestors believed that amber was a distillation[1] from a tree, and so named it *sucinum*. The fact that it smells of pine when rubbed and burns like a pine-torch with its smoky smell are indications that the tree belongs to a species of pine.

45. There is a Roman knight still living who was sent to get amber by Julianus when he was in charge of a gladiatorial show given by the Emperor Nero. He travelled the trade-routes and the coasts, and brought back such a large amount that the nets deployed to keep the wild beasts off the parapet of the amphitheatre were knotted with pieces of amber. Indeed the arms, stretchers and all the equipment used on one day – the daily display was varied – had amber trimmings.

46. The heaviest piece brought to Rome by Julianus weighed about 13 pounds. That amber is first exuded as a liquid is proved by

1. *sucus.*

certain objects, such as ants, gnats and lizards, being visible inside it. These undoubtedly stuck to the fresh sap and remained trapped inside when it hardened.

48. When the hot exhalation is released by rubbing amber with the fingers, it attracts straw, dry leaves and bark from the linden-tree, just as a magnet attracts iron.

49. Amber is so greatly valued among luxuries that even a statuette of a man, however small, costs more than a number of living, healthy slaves. With Corinthian bronzes it is the alloy of copper with silver and gold that delights. I have mentioned the attractive qualities of fluorspar and rock-crystal vessels. Pearls are worn on the head, gems on the finger. In short all other things for which we have a weakness please us because we can show them off or use them; with amber, however, there is only the inner satisfaction of knowing that it is a luxury.

51. Amber is found to have some use in medicine, but this is not the reason it pleases women.

Gemstones

54. I shall now discuss stones acknowledged as gems, beginning with the most highly prized. In addition, for the greater benefit of mankind I shall, incidentally, refute the unspeakable lies of the Magi, who have made several assertions about gems, from the seductive pretence of their efficacy as alternative medicine to their supernatural qualities.

Diamonds

55. The diamond, known for a long time only to kings and then to very few of them, has greater value than any other human possession, and not merely than any other gemstone. Adamas[1] was also the name given to the 'knot of gold' found, very occasionally, in gold mines and apparently formed only in gold. Ancient authorities thought it occurred only in the mines of Ethiopia between the Temple of Mercury and the island of Meroë; they claimed that examples were no larger than cucumber seeds and not unlike them in colour.

1. Some hard substance – possibly the metal platinum, although Pliny's comparison of its colour to cucumber seed does not accord with this interpretation. See also Theophrastus, *On Stones*, 32.

56. Now six varieties of diamond are known. There is the Indian octahedral diamond, which is not formed in gold and has a certain affinity with rock-crystal, which it resembles in its transparency and in its smooth faces meeting at six corners. It tapers to a point in two opposite directions, and is all the more remarkable because it is like two spinning-tops joined at their bases. It can even be as large as a hazel-nut. Similar to the Indian, but smaller, is the Arabian diamond, which is formed under similar conditions. The other kinds have the pale appearance of silver and are formed only in gold of the finest quality.

57. All these gems can be tested on an anvil: they so repel blows that the hammer-head may split in two and the anvil become displaced. The hardness of the diamond defies description, and similarly that property by which it resists fire yet never becomes hot. This is the reason for its name, because according to the meaning of the Greek term,[1] it is an 'unconquerable force'. The Greeks call one kind of diamond 'millet seed'[2] because it resembles one in size. A second kind is known as 'Macedonian' and is found in the gold mines of Philippi; this is equal in size to a cucumber seed.

58. Next comes the type known as 'Cypriot', which is found in Cyprus and has a colour verging on that of copper. After this there is ironstone, which has the lustre of iron; it is heavier than the others and differs in other respects.

'Industrial' diamonds

60. When a diamond is successfully broken, it disintegrates into splinters so small as to be scarcely visible. These fragments are greatly sought after by engravers and are inserted by them in their iron tools because they cut into the hardest surfaces with little effort.

Emeralds

62. Next in value after diamonds come the pearls of India and Arabia and in third place, for several reasons, the emerald. Certainly

1. The term *adamas* has a variety of meanings and may refer to any hard substance. Diamonds were already known from India. Cf. Manilius, *Astronomicon*, IV, 926. Contrary to Pliny's statement, diamonds can be broken by a hammer striking them. They are, however, affected only by temperatures above about 700° C.
2. *cenchros.*

no colour is more delightful in appearance. For although we enjoy looking at plants and leaves, we regard emeralds with all the more pleasure because compared with them there is absolutely nothing that is more intensely green.

63. Moreover, emeralds alone of precious stones satisfy our gaze without causing a surfeit of colour. And after straining our eyes by looking at another object, we can restore our vision to normal by gazing at an emerald. Lapidaries find this the most agreeable means of refreshing their eyes. So soothing is the mellow green of the gem for tired eyes.

64. Emeralds are generally concave in shape, so that they concentrate one's vision. For this reason it is universally accepted that emeralds should be preserved in their natural state and should not be engraved. In any case, emeralds from Scythia and Egypt are so hard that they are not able to be marked. When emeralds are tabular, they reflect objects just like mirrors. The Emperor Nero used to watch fights between gladiators which were reflected by the surface of an emerald.

66. The story is told that in Cyprus there was a statue of a lion on the burial mound of a prince named Hermias, near the tunny-fisheries; the lion had inset eyes which were made of emerald and they could be seen shining so brightly, even below the water, that the tunny-fish fled in terror. For a long time the fishermen were baffled by this strange behaviour until they changed the gemstones used for the eyes.

Beryls

76. Many consider that beryls are similar to, if not the same as, emeralds. Beryls occur in India, but are rarely found elsewhere. All are cut to a hexagonal shape by skilled craftsmen since their colour, which is dull when the surface is unbroken, is enlivened by the reflection from the facets. If they are cut in any other way they lack brilliance. The most esteemed beryls are those that imitate the pure green of the sea.

78. The Indians are passionately fond of elongated beryls. They prefer to make beryls into long prisms rather than gemstones, because they equate excellence with length.

79. Some people think that beryls are natural prisms. The Indians have discovered a means of counterfeiting gemstones, especially beryls, by colouring rock-crystal.

Tourmaline

103. Tourmaline is found around Orthosia and throughout Caria and the neighbouring regions, but the best quality occurs in India. I find that there are other varieties, such as violet-red and rose-red tourmaline. When these are heated in the sun or when they are rubbed between the fingers, they are said to attract straws and papyrus fibres.

104. It is also claimed that the ruby, although far less valuable, has the same power of attraction.

Amethysts

121. Indian amethysts hold first place, although amethysts are found also in that part of Arabia called Petra, bordering on Syria, as well as in Lesser Armenia, Egypt and Galatia. The poorest and least valuable varieties occur in Thasos and Cyprus. Some explain the name 'amethyst' by the fact that its bright colour approaches that of wine, but is finally a violet shade somewhat short of the red colour of wine. However, all amethysts are transparent, have a beautiful violet colour and are easily engraved.

124. The Magi falsely claim that amethysts prevent drunkenness, and that they are so named because of this.[1]

Sapphires, topaz and the rainbow stone

126. Ethiopia sends us blue sapphires and also the topaz, a bright, golden, transparent stone.

136. Next is the gem known as the 'rainbow' stone. It is dug up on an island in the Red Sea, 60 miles from the city of Berenice. In all but one respect it is rock-crystal; and some authorities have called it 'root-crystal'. It is called 'rainbow' stone from the fact that when it catches the sunlight in a room, it refracts the light and throws the colours of a rainbow on the nearby walls; it continually changes its colours, and this kaleidoscopic effect provokes ever-increasing astonishment.

137. It is generally agreed that the rainbow stone has hexagonal faces like rock-crystal, but there are some authorities who claim that its faces are rough and that the angles are unequal. They say that in full sunlight this stone refracts the beams that shine on it, while

1. The Greek word *amethystos*, means 'not drunk'.

simultaneously lighting adjacent objects by throwing out a kind of bright light in front of it. As I have said, however, it only produces colours in a darkened place. The best rainbow stone is that which produces the largest rainbow effect, most like the natural phenomenon.

Methods of classification

186. There is yet another classification for gemstones that I would like to adopt, having varied the presentation of my subjects from time to time. For some stones take their names from parts of the body: thus hepatite is named after the liver,[1] and a number of varieties of steatite after the fat[2] found in animals. Again, *triophthalmos*[3] is a type of onyx that has the appearance of three human eyes grouped together.

187. Some stones derive their names from animals: *carcinias* is so-called because it has the colour of a crab,[4] and *echitis* because it has the colour of a viper.[5] *Scorpitis* has the colour or appearance of a scorpion.

188. A resemblance to inanimate objects occurs in the case of *ammochrysus* which looks like gold[6] mixed with sand.[7] *Cenchrites* gives the appearance of being sprinkled with grains of millet;[8] *dryites* looks like the trunk of an oak-tree.[9] *Cissitis* is a transparent, colourless stone in which leaves of ivy[10] can be seen, and these cover the whole stone.

New discoveries

193. New gemstones without names can occur unexpectedly, like the one which, according to Theophrastus,[11] was once discovered in

1. *hepar.*
2. *stear.*
3. Three-eyed.
4. *carcinos.*
5. *echis.*
6. *chrysos.*
7. *ammos.*
8. *cenchros.*
9. *drys.*
10. *cissos.*
11. *De Lapidibus,* 32. Theophrastus does not actually mention Alexander.

the gold mines near Lampsacus and sent to Alexander on account of its beauty.

Some general observations concerning gems: tests to determine their genuineness

196. I shall now make some general observations of relevance to our study of all precious stones, following the views of our authorities on these matters. Concave and convex gems are thought to be less valuable than those with a flat surface. An elongated shape is most prized; then the lenticular, as it is called, and next a flat, round shape. Gems with sharp angles are least favoured.

197. Distinguishing between genuine and false gems is a difficult problem, because people have discovered how to make genuine stones of one variety into false stones of another. Thus a sardonyx can be made by sticking three gems together in such a way that the deception is undetectable: a black stone is taken from one category of gemstones, a white from another, and a red from yet another – all of them of high quality in their own group. What is more, there are papers by authors, whom I shall not name, which outline how emeralds and other transparent gems can be produced by colouring rock-crystal; or how a sardonyx can be made from a sard, and similarly how other gems can be made from other precious stones. Indeed there is no more lucrative fraud against society.

198. However, I shall demonstrate ways of detecting false gems, since it is proper that even luxury should be protected against fraud. Apart, therefore, from the observations that I made when describing the stones of highest quality in each category, our authorities are of the opinion that transparent stones should generally be tested early in the day, or, if this is impossible, well before midday, certainly not later.

199. There are many different types of test; first by weight, since genuine stones are heavier; then by coldness, as genuine stones feel colder in the mouth; after this, by physical characteristics: counterfeit stones have bubbles deep inside, a rough surface with fine streaks, inconsistent lustre and a brightness that peters out before it reaches the eyes.

200. The most effective test is to break off a fragment of the gem so that it can be heated on an iron plate, but gem-dealers understandably refuse to allow this, and, likewise testing with a file. Flakes of obsidian do not scratch genuine gemstones, whereas on counterfeit gems every scratch appears white. Gems differ so much

that some cannot be engraved with an iron tool, and others only with a blunt iron tool; but all can be engraved with a diamond. Heat from drilling has the greatest effect on precious stones. The rivers on which gems occur are the Acesines and Ganges, and of all lands, India is the chief source of precious stones.

In praise of Italy

201. Now that I have finished my account of Nature's works, it is time to make some critical assessment of her products and of the lands that furnish them. In the whole world, wherever the vault of heaven turns, Italy is the most beautiful of all lands, endowed with all that wins Nature's crown. Italy is the ruler and second mother of the world – with her men and women, her generals and soldiers, her slaves, her outstanding position in arts and crafts, her abundance of brilliant talent, her geographical location and healthy, temperate climate, her easy accessibility for all other peoples, and her shores with their many harbours and the kindly winds that blow towards her. All these advantages ensue from Italy's situation – for the land-mass juts out in the most advantageous direction, midway between East and West – and from her abundant supply of water, flourishing forests, passable mountains, harmless wild creatures, fertile soil and rich pastures.

202. No land is more distinguished with regard to what man may reasonably expect to enjoy – chiefly crops, wine, olive-oil, fleeces, flax, clothes and young cattle. Italian horses are preferred to others on the training-ground; in ores of gold, silver, copper and iron, Italy was second to none while it was lawful to exploit them. Now she keeps them within her like a pregnant woman, and, for her dowry, bestows on us many different juices and crops and fruits with varying tastes.

203. After Italy, leaving aside the marvels of India, I would put Spain, or at least the coastal regions. Although Spain is partly rough country, yet where it is productive, it is rich in crops, oil, wine, horses and every kind of ore-mineral. Thus far Gaul is the equal of Spain, but Spain has the edge because of her deserts with their esparto grass and selenite, as well as luxury in the form of pigments. Here there is a stimulus to work, a place where slaves can be trained, where men's bodies are tough and their hearts eager.

The world's most expensive products

204. Among the products of Nature, the most expensive to be derived from the sea is the pearl; on the earth's surface, it is rock-crystal; from the earth's interior, diamonds, emeralds and gem-stones as well as vessels of fluorspar; of things that come from the soil, the cochineal insect and silphium, together with nard and silk. From leaves the most expensive product is cassia; from shrubs, amomum; from trees or shrubs, amber, balsam, myrrh and frank-incense; from roots, aromatic costus. From those animals that breathe, the most expensive produce found on land is ivory; in the sea, the turtle's shell. In the case of animal hides and coats, the most expensive are the skins dyed by the Chinese and the Arabian she-goat's beard that we call ladanum. From amphibious creatures the most expensive products are scarlet and purple dyes made from shellfish. Birds make no outstanding contribution, except plumes for use in war, and grease from the geese of Commagene. I must not pass over the fact that gold, with which all mankind is madly obsessed, is scarcely tenth in the list of valuable commodities, while silver, with which gold is bought, is almost twentieth.

Valediction to Nature

205. Greetings, Nature, mother of all creation, show me your favour in that I alone of Rome's citizens have praised you in all your aspects.

KEY TO PLACE-NAMES

ANCIENT NAME	MODERN NAME OR REGION
Acesines, R.,	Chenab, R.,
Acherusian Marsh	Lago di Fusaro, Italy
Acmodae	Shetlands
Agrigentum (Akragas)	Agrigento, Sicily
Ambracia	Arta, north-west Greece
Ameria	Amelia, Italy
Anio, R.,	Teverone, Italy
Antium	Anzio, Italy
Apamea	Dinar, Turkey
Apollonia	Uluborlu, Turkey
Appian Way	Rome–Benevento–Brindisi
Aquitania	Province between Pyrenees and Garonne River
Ardea	Town south of Rome in Latium, Italy
Ariminium	Rimini, Italy
Arretium	Arrezzo, Italy
Aspendus	Near Belkis, 25 miles east of Antalya, Turkey
Assos	Behramkale, north-west Turkey
Assyria	South-east Turkey/north-east Iraq
Astura	Town in Latium, Italy
Avernus, L.,	8 miles west of Naples, Italy
Baetica	Andalucia, Spain
Baetis, R.,	Guadalquivir, R., Spain
Berytus	Beirut, Lebanon
Bilbilis	Calatayud, north-east Spain
Bithynia	North-west Anatolia, Turkey
Bononia	Bologna, Italy
Borysthenes, R.,	Dnieper, R., USSR
Brundisium	Brindisi, Italy
Bruttium	Calabria, Italy

Byzantium	Istanbul, Turkey
Caere	Cerveteri, Italy
Caesarea	Between Tel Aviv and Haifa, Israel
Caledonia	Scotland
Calenum	Calvi, Italy
Camulodunum	Colchester, England
Cannae	Town in Apulia, south Italy
Cantabria	North-west Spain
Cantabrian Sea	Bay of Biscay
Canusium	Canossa, Italy
Cappadocia	East Anatolia, Turkey
Caralis	Cagliari, Sicily
Carmania	South Iran
Carthage	North-east of Tunis, on Mediterranean coast
Carystus	Karystos, Greece
Casinum	Cassino, Italy
Caspian Gates	Narrow pass on west shore of Caspian Sea, near Derbent, USSR
Cassiterides	The Tin Islands (location unknown)
Catina	Catania, Sicily
Ceos	Kea, Greece
Cerasus	Giresun, Turkey
Chalcedon	Kadiköy, Turkey
Cilicia	South Anatolia, Turkey
Cimmerian Strait	Kerch Strait, USSR
Circeii	Town in Latium, between Rome and Naples, Italy
Clazomenae	Urla Iskelesi, 20 miles west of Smyrna, Turkey
Cnidus	Cape Krio, south-west Turkey
Cnossus	4 miles south-east of Iraklion, Crete
Colchis	Town in west Georgia on Black Sea
Coliacus	Cape Comorin, India
Colossae	4 miles east of Denizli, west Turkey
Commagene	East Anatolia, Turkey
Compsa	Conza, Italy
Comum	Como, Italy
Coptos	Qift, Egypt
Corfinum	Near Popoli, south Italy
Croton	Crotone, Italy

Cumae	Cuma, Italy
Cyllene	Killíni, Greece
Cyrenaica	East Libya
Cyrene	Shahhat, Libya
Cyzicus	Kapidagi, north-west Turkey
Drepanum	Trapani, Sicily
Ecbatana	Hamadan, west Iran
Engeda	En Gedi, Israel
Ephesus	Selçuk, west Turkey
Erythrae	Town in Turkey, 40 miles west of Smyrna
Etruria	Tuscany, Italy
Euboea	Evvia, Greece
Falerum	See Massicus, Mt,
Fidenae	Castel Giubileo, Italy
Fucine Lake	Lago Fucino, Italy
Gades	Cadiz, Spain
Gaetulia	South Algeria
Galatia	Central Anatolia, Turkey
Gallaecia	Galicia, Spain
Gallia Comata	Transalpine Gaul
Gallia Narbonensis	Narbonne, France
Gauli	Gozo, one of the Maltese Islands
Gemelli Mountains	Monte di Mele, Italy
Gerra	Oqair, Saudi Arabia
Gesoriacum	Boulogne, France
Glanis, R.,	Chiana, R., Italy
Gortyna	Town in Crete, south-west of Iraklion
Halys, R.,	Kisil-Irmak, R., Turkey
Hebrus, R.,	Maritsa, R., north Greece
Hebudes	Hebrides
Hellespont	Dardanelles, Turkey
Heraclea	Aylvalik, Turkey
Hermus, R.,	Gediz, R., Turkey
Hibernia	Ireland
Hierosolyma	Jerusalem, Israel
Hippo Diarryhtus	Bizerte, north Tunisia

Hybla	Paterno, Sicily
Hymettus	Mountain range in Attica, Greece
Idumaea	South Israel
Ilium (Troy)	Hisarlik, Turkey
Joppa	Jaffa, Israel
Judaea	Israel
Kenchreae	Port on the Saronic Gulf, Greece
Laconia	South-east Peloponnesus
Leontini	Lentini, Sicily
Liburnia	Croatia, Yugoslavia
Lilybaeum	Marsala, Sicily
Lixus, R.,	Al-Araish, R., north Morocco
Locri	Town in Magna Graecia, south Italy
Lucania	South Campania, Italy
Lucrine Lake	Lago Lucrino, Italy
Luna	West of Carrara, Italy
Lusitania	Portugal and west Spain
Lycabettus	Mountain in Athens, Greece
Machaerus	Mukawir, Jordan
Maeotic Lake	Sea of Azov, USSR
Mareotis, L.,	Mariout, L., south of Alexandria, Greece
Marona	Maronia, north Greece
Masada	West shore of Dead Sea, Israel
Massicus, Mt,	Campania, Italy
Massilia	Marseille, France
Mauretania	Morocco
Media	Ustan Yakum and Ustan Duwum, Iran
Melita	Malta
Memphis	14 miles south of Cairo, Egypt
Mendes	Nile Delta, north Egypt
Meninx	Jerba, Tunisia
Meroe	Near Kabushiyah, Sudan
Mesopotamia	North Iraq
Messana	Messina, Sicily
Metapontum	North-west shore of Gulf of Tarentum, Italy
Mictis	St Michael's Mount, Cornwall

Miletus	Yeniköy, Turkey
Misenum	Miseno, Italy
Moeris, L.,	Birket Qarun, L., Upper Egypt
Mona	Anglesey
Monapia	Isle of Man
Mutina	Modena, Italy
Myra	Town in south Asia Minor
Myrina	5 miles west of Gryneion, north-west Asia Minor
Nar	Nera, central Italy
Neapolis	Naples, Italy
Nemausus	Nîmes, France
Nomentanum	Mentana, Italy
Noricum	Province south of the Danube
Nuceria	Nocera, Italy
Numantia	Castillo-Peña Redonda, Spain
Olisipo	Lisbon, Portugal
Orcades	Orkneys
Orchomenus	Near Levádhia, central Greece
Pachynum	Cape Passaro, Sicily
Padus, R.,	Po, R., north Italy
Palmyra	Town in Syria, north-east of Damascus
Pandateria	Island off Pozzuoli, Campania, south Italy
Paphlagonia	North Anatolia/Black Sea
Paphos	Near Ktima, Cyprus
Paraetonium	Matruh, north-west Egypt
Parium	Town on the south-west shore of the Sea of Marmara, Turkey
Paropanisus	Hindu Kush, central Asia
Parthia	North-east Iran
Pelorum	Capo di Faro, Sicily
Pergamum	Bergama, west Turkey
Perusia	Perugia, Italy
Phaleron	East of Piraeus, Greece
Pharos	Peninsula in Lower Egypt
Pharsalus	Farsala in Thessaly, north-east Greece
Phaselis	Town in south Turkey
Pherae	Town in north-east Greece, south-east of Larissa

Phlegraean Fields	Solfatara, Italy
Phoenicia	South Lebanon
Picenum	Pizzino, Italy
Pillars of Hercules	Straits of Gibraltar
Placentia	Piacenza, Italy
Pontus	North Anatolia, Turkey
Populonia	Piombino, Italy
Praeneste	Palestrina, Italy
Proconnesus	Island in the Sea of Marmara, Turkey
Provincia	Provence, France
Puteoli	Pozzuoli, Italy
Rhamnus	Town north-east of Athens
Rhegium	Reggio Calabria, Italy
Rhyndacus, R.,	Atranos, R., Turkey
Saguntum	Sagunto, Spain
Samaria	Central Israel
Samosata	Samsat, south-east Turkey, on the west bank of the Euphrates
Sardis	50 miles east of Izmir, Turkey
Sarnus, R.,	Sarno, R., Italy
Scamander, R.,	Cayster, R., Asia Minor
Scythia	North shore of the Black Sea
Sebennys	Town on Nile Delta
Segobriga	Priego, Spain
Setia	Sezze, Italy
Sicyon	Town near Kiáto on the south shore of the Gulf of Corinth
Sinope	Sinop, north Turkey
Sisapo	Almaden, Spain
Smyrna	Izmir, Turkey
Soli	South-west of Tarsus, Turkey
Soracte	Mountain north-east of Rome, Italy
Stoechades Islands	Iles d'Hyères, France
Strongyle	Stromboli, Italy
Suessa Pometia	In region of Latium, Italy
Sunium, Cape	Cape on the southern tip of Attica, Greece
Surrentum	Sorrento, Italy

Susa	Shush, south-west Iran
Sybaris	Near Terranova di Sibari, Italy
Syene	Assuan, Egypt
Tanagra	Near Thebes, Greece
Tanais, R.,	Don, R., USSR
Taprobane	Sri Lanka
Tarentum	Taranto, Italy
Tarraco	Tarragona, north-east Spain
Tauromenium	Taormina, Sicily
Thapsus	Sousse, Tunisia
Thebes	Karnak-Luxor
Thebes	Thívai, central Greece
Therasia	Aegean island in the Cyclades
Thracian Strait	Straits of Constantinople
Tiberis, R.,	Tiber, R., Italy
Tibur	Tivoli, Italy
Ticinus	Ticino, Italy
Tingi	Tangier, Morocco
Tmolus, Mt,	Boz Dag, Turkey
Torone	Town in Chalcidice, north-east Greece
Tralles	Aydin, Turkey
Trasumenus, L.,	Lago di Trasimeno, Italy
Trebia	Trebbia, Italy
Tripolis	Tripoli, Libya
Tusculum	Near Frascati, Italy
Tyros (Tylos)	Bahrain island
Utica	15 miles north-west of Tunis, Tunisia
Velitrae	Velletri, Italy
Venafrum	Venafro, Italy
Vercellae	Vercelli, Italy
Volsinii	Bolsena, Italy
Volturnus, R.,	Volturno, R., Italy
Zacynthus	Zákinthos, Greece

INDEX

PENGUIN CLASSICS

www.penguinclassics.com

- *Details about every Penguin Classic*

- *Advanced information about forthcoming titles*

- *Hundreds of author biographies*

- *FREE resources including critical essays on the books and their historical background, reader's and teacher's guides.*

- *Links to other web resources for the Classics*

- *Discussion area*

- *Online review copy ordering for academics*

- *Competitions with prizes, and challenging Classics trivia quizzes*

PENGUIN CLASSICS ONLINE